Progress in Nonlinear Differential Equations and Their Applications
Volume 22

Partial Differential Equations and Functional Analysis

In Memory of Pierre Grisvard

Jean Cea
Denise Chenais
Giuseppe Geymonat
Jacques Louis Lions
Editors

Springer Science+Business Media, LLC

Jean Cea
Denise Chenais
University of Nice
Laboratoire de Mathématiques
F-06108 Nice Cedex
France

Giuseppe Geymonat
Laboratoire de Mathématiques
Ecole Normale Supérieure
94235 Cachan Cedex, France

Jacques Louis Lions
Collège de France
3, rue d'Ulm
75005 Paris, France

Library of Congress Cataloging-in-Publication Data

Partial differential equations and functional analysis : in memory of
Pierre Grisvard / J. Cea ... [et al.], editors.
p. cm. -- (Progress in nonlinear differential equations and
their applications ; v. 22)
ISBN 978-1-4612-7536-7 ISBN 978-1-4612-2436-5 (eBook)
DOI 10.1007/978-1-4612-2436-5
1. Functional analysis. 2. Differential equations, Partial.
I. Cea, Jean, 1932- . II. Series.
QA321.P37 1996 96-1384
515'.353--dc20 CIP

Printed on acid-free paper

Birkhäuser
Originally published by Birkhäuser Boston in 1996
Softcover reprint of the hardcover 1st edition 1996

Typeset by TeXniques, Boston, MA

9 8 7 6 5 4 3 2 1

Pierre Grisvard

De 1990 jusqu'à sa disparition, le 22 Avril 1994, Pierre Grisvard a été chargé de la rénovation de l'Institut Henri Poincaré à Paris. Les membres du personnel de cet Institut qui ont eu l'occasion de collaborer avec lui, ont tenu à être présents dans cet ouvrage :

Cela aura été un moment important, privilégié, dans nos vies professionnelles d'avoir eu à travailler pour, et avec, Pierre Grisvard, un honneur pour nous qu'il nous appelle "ses collaborateurs". Notre grade lui importait peu.

Sa personnalité chaleureuse, sa droiture, son humour et sa modestie nous faisaient naturellement accepter son autorité. Ce n'est pas tant l'expérience professionnelle acquise pendant tout ce temps passé ensemble qui nous importe que le sentiment d'avoir connu un "Grand Monsieur". Nous souhaitons du fond du coeur que cette maison ne l'oublie pas.

"Son équipe" de l'Institut Henri Poincaré

Contents

Preface

Pierre Grisvard, one of the most distinguished French mathematicians, died on April 22, 1994. A Conference was held in November 1994 out of which grew the invited articles contained in this volume. All of the papers are related to functional analysis applied to partial differential equations, which was Grisvard's specialty. Indeed his knowledge of this area was extremely broad.

He began his career as one of the very first students of Jacques Louis Lions, and in 1965, he presented his "State Thesis" on interpolation spaces, using in particular, spectral theory for linear operators in Banach spaces. After 1970, he became a specialist in the study of optimal regularity for partial differential equations with boundary conditions. He studied singularities coming from coefficients, boundary conditions, and mainly non-smooth domains, and left a legacy of precise results which have been published in journals and books.

Pierre Grisvard spent most of his career as a full professor at the University of Nice, where he started in 1967. For shorter or longer periods, he visited several foreign countries, and collaborated with some of the most famous mathematicians in his field. He was also an excellent organizer and directed a large number of Ph.D. students.

Finally, this volume contains a bibliography of Grisvard's works as well as one paper which he wrote and which has not been published before.

Denise Chenais

Avant Propos

Pierre Grisvard avait suivi à Nancy en 1960 mon cours de Méthodes Mathématiques de la Physique. Je ne disposais, avant la fin d'année, d'aucune indication sur sa valeur. D'où ma très agréable surprise à la lecture de sa copie d'examen, copie parfaite dans le fond et dans la forme. Aucun doute pour moi: cet étudiant timide, discret et d'aspect fragile, pouvait devenir un vrai mathématicien. Diagnostic rapidement confirmé: en quelques mois, il résolvait parfaitement le problème que je lui avais soumis comme thèse de troisième cycle (une question sur les semi-groupes, thème dans lequel il allait devenir un maître).

En fait, la timidité masquait une imagination puissante. Le chercheur discret pouvait exposer avec force et lumineuse clarté les questions les plus complexes. L'étudiant d'apparence fragile était persévérant, courageux et avait une très forte volonté.

C'est ainsi que'avec son calme modeste et tranquille, il apporta rapidement des contributions décisives à des sujets délicats: problème d'évolution dans des espaces L^p, $p \neq 2$; interpolation entre espaces et sous espaces; problèmes aux limites dans des domaines avec singularités, questions auxquelles son nom restera attaché.

On a retrouvé dans les archives de Pierre Grisvard un travail qui est ici publié pour la première fois. Ce travail, hélas le dernier, et fort intéressant, témoigne, une fois encore, du talent et du courage de notre ami prématurément disparu.

Jacques-Louis Lions

Introduction

Le 22 Avril 1994 Pierre GRISVARD est mort à Paris à l'age de 54 ans. Il était professeur de Mathématiques à l'Université de Nice (France) depuis 1967 et Directeur de l'Institut Henri Poincaré à Paris, à la rénovation duquel il travaillait avec enthousiasme depuis 1990.

Après des études secondaires au Lycée Henri Poincaré de Nancy, il obtient une Licence de Physique. Il doit pour ceci suivre le certificat "Méthodes Mathématiques de la Physique", enseigné par Jacques-Louis Lions. La fascination de J.L. Lions et des autres professeurs de mathématiques de l'Université de Nancy (notamment F. Bruhat, M. Hervé, J. Delsarte) le poussent finalement à s'orienter vers les mathématiques, tout en restant très attaché à la modélisation mathématique des phénomènes physiques. Il est donc l'un des premiers élèves de J.L. Lions à qui il reste très attaché.

Ses premiers travaux et sa Thèse d'Etat soutenue à Paris en 1965, portent sur des propriétés difficiles des espaces d'interpolation. Déjà il utilise de façon très originale et profonde les résultats et les techniques de la théorie spectrale des opérateurs linéaires dans les espaces de Banach, qu'il développera plus tard dans l'étude des équations opérationnelles abstraites (en particulier dans une série de travaux en collaboration avec G. Da Prato).

Tout en s'intéressant à de très nombreux domaines de l'analyse fonctionnelle, P. Grisvard devient l'un des principaux spécialistes mondiaux de l'études des singularités de solutions d'équations aux dérivées partielles. Celles-ci peuvent être dues aux fait que les équations sont posées sur des domaines à frontière non régulière (frontières présentant des coins, polygones, polyèdres), ou bien au fait que les données sont singulières (masses concentrées en des points), ou encore à des changements brusques de conditions aux limites (problèmes mêlés) ou de coefficients (problèmes de transmission). Dans ce type de situation, il est vain de chercher une solution très dérivable. Pour illustrer le type de difficultés rencontrées dans ces problèmes, on peut considérer dans le plan le cas particulier de l'ouvert $\Omega = \{(r\ \cos\theta, r\ \sin\theta);\ 0 < r < 1, 0 < \theta < \omega\}$ avec $\omega > \pi$. Si $\phi(r)$ est une fonction de troncature régulière égale à 1 pour r petit et à 0 au voisinage de 1, la fonction $u_s = \phi(r)\ r^\alpha \sin{(\alpha\theta)}$ est dans $H_0^1(\Omega)$ si $\alpha = \pi/\omega$, et est solution variationnelle de $\Delta u_s = f$. Cependant, même si f est de classe $\mathcal{C}^\infty$, elle n'appartient pas à l'espace $H^2(\Omega)$, mais seulement à l'espace $H^{\alpha+1-\epsilon}(\Omega)$ pour tout $\epsilon > 0$. De plus, la fonction $u_s^* = (r^{-\alpha} - r^\alpha)\ \sin{(\alpha\theta)} \in L^2(\Omega)$, qui est harmonique dans Ω et nulle au bord, fournit un exemple insolite de non unicité de la solution d'un problème de Dirichlet. Enfin on a $\int u_s^*\ \Delta u_s\ dx = \pi$ alors qu'une application hâtive de la formule de Green donnerait la valeur 0. Dans ses travaux, P.Grisvard détermine les conditions de compatibilité sur f nécessaires et suffisantes pour que $u \in H^1(\Omega)$ solution de $\Delta\ u = f \in \mathcal{C}^\infty\ (\overline{\Omega})$ soit dans $H^m(\Omega)$ pour $m \geq 2$. Toutefois ces conditions non locales s'avèrent d'un faible intérêt pratique. P. Grisvard développe alors une technique consistant à décomposer la solution d'un problème singulier en une partie régulière (permettant tous les traitements usuels de solutions d'équations régulières), et d'une partie singulière associée au problème spécifique étudié, qu'on calcule de la manière la plus explicite possible. Ce point de vue a été à sa suite très développé, en particulier par ses nombreux élèves et a donné des résultats très fructueux, riches en applications. Il a montré son efficacité dans divers contextes parmi lesquels on peut citer le calcul numérique, la théorie de la rupture fragile, le contrôle frontière hyperbolique.

Pierre Grisvard a ainsi ouvert plusieurs voies de recherche qui sont devenues classiques, et qui continueront d'être exploitées après sa disparition, qui paraît tout à fait irréelle aux nombreux scientifiques qui avaient l'habitude de dialoguer avec lui.

Giuseppe Geymonat et Denise Chenais

Pierre Grisvard
Liste de publications

1. Espaces intermédiaires entre espaces de Sobolev avec poids, *Annali della Scuola Normale Superiore di Pisa* III **17** (1963), 255–296.
2. Identités entre espaces de traces, *Mathematica Scandinavica* **13** (1963), 70–74.
3. Une remarque sur les espaces intermédiaires entre des espaces de Sobolev avec poids, *C.R.A.S.* **256** (1963), 2745–2748.
4. Théorèmes de traces, *C.R.A.S.* **256** (1963), 2990–2992.
5. Théorème de traces et applications, *C.R.A.S.* **256** (1963), 3226–3228.
6. Espaces de traces à plusieurs variables, *C.R.A.S.* **257** (1963), 349–352.
7. (avec Geymonat) Problèmes aux limites elliptiques dans L^p, *Polycopié d'Orsay* (1964), 1–167.
8. Commutativité des procédés d'interpolation "réel" et "complexe," *C.R.A.S.* **258** (1964), 4900–4902.
9. Semi-groupes faiblement continus et interpolation, *C.R.A.S.* **259** (1964), 27–29.
10. Les intégrales spectrales dans l'étude des problèmes aux limites, *Convegno sulle equazioni a derivate parziali*, Edizioni Cremonese (1965), 66–68.
11. Sur l'utilisation du calcul opérationnel dans l'étude des problèmes aux limites, *Séminaire de Mathématiques Supérieures*, Université de Montréal (1965), 87–99.
12. Espaces d'interpolation et équations opérationnelles, *C.R.A.S.* **260** (1965), 1536–1538.
13. Commutativité de deux foncteurs d'interpolation et applications, (thèse), *Journal de Mathématiques Pures et Appliquées* **45** (1966), 143–290.
14. Caractérisation de quelques espaces d'interpolation, *Archive for Rational Mechanics and Analysis*, (1) **35** (1967), 40–63.
15. (avec Geymonat), Problemi ai limiti ellittici negli spazi di Sobolev con peso, *Le Matematiche* (2) **22** (1967), 1–38.
16. (avec Geymonat) Alcuni risultati di teoria spettrale per i problemi ai limiti lineari ellittici, *Rendiconti del Seminario Matematico della Università di Padova* **38** (1967), 121–173.
17. Equations opérationnelles abstraites dans les espaces de Banach et problèmes aux limites dans les ouverts cylindriques, *Annali della Scuola Normale Superiore di Pisa* **21** (1967), 307–347.

18. (avec Baouendi) Sur une équation d'évolution changeant de type, *Journal of Functional Analysis* (3) **2** (1968), 352–395.
19. Equations différentielles abstraites, *Annales Scientifiques de l'Ecole Normale Supérieure* (1969), 311–395.
20. (avec Goulaouic) Existence de traces pour les éléments d'espaces de distributions définis comme domaines d'opérateurs différentiels maximaux, *Inventiones Mathematicae* **9** (1970), 308–317.
21. Equations opérationnelles abstraites et problèmes aux limites dans des domaines non réguliers, *Actes du Congrès International des Mathématiciens, Nice* **2** (1970), 731–736.
22. Problème de Dirichlet dans un cone, *Ricerche di Matematica* **20** (1971), 175–192.
23. (avec Boutet de Monvel) The asymptotic behaviour of the eigenvalues of an operator related to width of balls, *Alta Matematica, Symposia Mathematica* **7** (1971), 559–576.
24. Spazi di tracce ed applicazioni, *Rendiconti di Matematica* (4) **5** (1972), 657–729.
25. (avec Souffrin et Zerner) Note on the growth rate of convective modes in a self gravitating gas sphere, *Astronomie et Astrophysique* **17** (1972), 309.
26. Interpolation non commutative, *Accademia dei Lincei, VIII* **52** (1972), 11–15.
27. Alternative de Fredholm relative au problème de Dirichlet dans un polygone ou un polyèdre I, *Bollettino delle Unione Matematica Italiana* (4) **5** (1972), 132–164.
28. Théorèmes de trace relatifs à un polyèdre, *C.R.A.S.* **278** (1974), 1581–1583.
29. Problème de Dirichlet dans un domaine non régulier, *C.R.A.S.* **278** (1974), 1615–1617.
30. Alternative de Fredholm relative au problème de Dirichlet dans un polyèdre II, *Annali della Scuola Normale Superiore di Pisa* (1975), 359–388.
31. (avec Da Prato) Sommes d'opérateurs linéaires et équations différentielles opérationnelles, *Journal de Mathématiques Pures et Appliquées* **54** (1975), 305–387.
32. (avec Da Prato) Equations d'évolution abstraites non linéaires de type parabolique, *C.R.A.S.* **283** (1976), 709–711.
33. The semi-group theory and the integration of evolution equations, *Mathematical and Numerical Methods in Fluid Dynamics ICTP* (1976), 137–166.
34. Régularité de la solution d'un problème aux limites unilatéral dans un domaine convexe, Séminaire Goulaouic–Schwartz, 9 Mars 1976.
35. Smoothness of the solution of a monotonic boundary value problem for a

second order elliptic equation in a general convex domain, *Lecture Notes in Mathematics* **564**, Springer–Verlag (1976), 135–151.

36. Problèmes aux limites pour l'opérateur de Laplace dans un polygone plan, *Séminaire d'Analyse Numérique de Lyon-Saint-Etienne* (1976), 1–45.
37. Behavior of the solution of an elliptic boundary value problem in a polygonal or polyhedral domain, *Numerical Solutions of Partial Differential Equations III*, Academic Press (1976), 207–274.
38. (avec Iooss) Problèmes aux limites unilatéraux dans des domaines non réguliers, *Publications des Séminaires de Mathématiques de l'Université de Rennes* (1) **IX** (1976), 1–26.
39. Equazioni d'evoluzione astatte di tipo parabolico secondo G. Da Prato e P. Grisvard, *Rendiconti del Seminario Matematico e fisico di Milano* **47** (1977), 221–231.
40. (avec Da Prato) On the abstract evolution equation, *Communications in Partial Differential Equations* **3** (1978), 1077–1082.
41. (avec Da Prato) Equations d'évolution abstraites non linéaires de type parabolique, *Annali di Matematica Pura ed Applicata* **120** (1979), 329–396.
42. Boundary value problems in non smooth domains, *Lecture Notes #19, University of Maryland* (1979).
43. Singularités des problèmes aux limites dans les polyèdres, *Séminaire Goulaouic–Meyer–Schwartz*, 16 Mars 1982.
44. Singular solutions of elliptic boundary value problems in polyhedra, *Portugaliae Mathematica* **41** (1982), 367–382.
45. (avec Da Prato) On extrapolation spaces, *Accademia dei Lincei VIII* **72** (1982), 330–332.
46. Comportement de la solution d'un problème aux limites elliptique dans un domaine plan avec fissure, *Problèmes non linéaires appliqués CEA/INRIA/EDF* (1982), 184–192.
47. (avec Geymonat) Diagonalisation d'opérateurs non autoadjoints et séparation des variables, *C.R.A.S.* **296** (1983), 809–812.
48. (avec Da Prato) Maximal regularity for evolution equations by interpolation and extrapolation, *Journal of Functional Analysis* (2) **58** (1984), 107–124.
49. (avec Geymonat) Eigenfunctions expansions associated to some nonselfadjoint operators and separation of variables, *Lecture Notes in Mathematics 1121, Singularities and Constructive Methods for their Treatment* (1985), 123–136.
50. Elliptic problems in non-smooth domains, *Monographs and Studies in Mathematics* **24**, Pitman (1985).

51. Résolvante du Laplacien dans un polygone et singularités des équations elliptiques et paraboliques, *C.R.A.S.* **301** (1985), 181–183.

52. Problèmes aux limites dans les polygones, mode d'emploi, *Bulletin de la Direction des études et recherches de l'EDF, Série C: Mathématiques et Informatique* **1** (1985), 21–59.

53. Le problème de Dirichlet dans l'espace $W^{1,p}$, *Portugaliae Mathematica* **43**, Fasc. 4 (1985/86).

54. An approach to the singular solutions of elliptic problems via the theory of differential equations in Banach spaces, *Lecture Notes in Mathematics* **1223**, Differential Equations in Banach Spaces, Proceedings Bologna (1985), 131–155.

55. Le problème de Dirichlet pour les équations de Lamé, *C. R. A. S.* **304** (1987), 71–73.

56. Contrôlabilité exacte dans les polygones et polyèdres, *C. R. A. S.* **304** (1987), 367–370.

57. Contrôlabilité exacte avec conditions mêlées, *C. R. A. S.* **305** (1987), 363–366.

58. Edge behavior of the solution of an elliptic problem, *Mathematische Nachrichten* **132** (1987), 381–299.

59. Singularities in Elasticity Theory, in *Applications of Multiple Scaling in Mechanics*, édité par P. G. Ciarlet et E. Sanchez–Palencia, Collection *Recherches en Mathématiques Appliquées* **4**, Masson, Paris (1987), 134–150.

60. (avec C. Bardos et J. Céa) Error estimates related to singular perturbations of first-order equations and systems, *Comput. Math. Applic.* (9–11) **13** (1987), 801–829.

61. Solutions singulières du système de Lamé, *Les Annales de l'ENIT* (2) **2**, Tunis (1988), 25–34.

62. Contrôlabilité exacte des solutions de l'équation des ondes en présence de singularités, *Journal de Mathématiques Pures et Appliquées* **68**, Paris (1989), 215–259.

63. Singularités en élasticité, *Archive for Rational, Mechanics and Analysis* (2) **107** (1989), 157–180.

64. Majorations en norme du maximum de la résolvante du Laplacian dans un polygone, *Nonlinear Partial Differential Equations and their Applications, College de France Seminar* **XII**, édité par Brezis et Lions, Longman Scientific and Technical (1991).

65. Some known results that fail on non convex domains, *Nonlinear Partial Differential Equations and their Applications, College de France Seminar* **X**, édité par Brezis et Lions, Longman Scientific and Technical (1991).

66. Régularité maximale pour le problème de Neumann dans les espaces de Sobolev fractionnaires, préprint #218, Nice.
67. Singular solutions of elliptic problems and application to the exact controllability of hyperbolic problems, function spaces, differential operators, and non linear analysis, *Pitman Research Notes in Mathematics Series* **211**, édité par L. Päivärinta, Longman Scientific and Technical (1989), 169–191.
68. Singularities in Elliptic Boundary Value Problems, *Delft Progress Report* **13** (3) (1989), 349–373.
69. Exact controllability of the wave equation in presence of corners and cracks, *Semi-group Theory and Evolution Equations*, édité par Clément, Mitidieri, et De Pagter, Marcel Dekker (1991), 213–226.
70. Stiff eigenvalue problems in solid mechanics, in *Progress in Partial Differential Equations: The Metz Surveys*, *Pitman Research Notes in Mathematics Series* **249**, édité par Chipot et Saint Jean Paulin (1991), 88–104.
71. (avec Geymonat) Expansions on generalized eigenvectors of operators arising in the theory of elasticity, *Differential and Integral Equations* 4 (1991), 459–481.
72. Boundary control of cracked domains, *Evolutions Equations, Control Theory and Biomathematics*, édité par Clément et Lumer, *Lecture Notes in Pure and Applied Mathematics* **155**, Marcel Dekker (1993), 235–240.
73. Understanding differential systems through observation, Cours CIMPA, 1992.
74. Singularities in Boundary Value Problems, *RMA* #22, Masson (1992), 214 pp.
75. (avec Clément) Sommes d'opérateurs et régularité L^p dans les problèmes aux limites, *C.R.A.S.* **314**, série I (1992), 821–824.
76. Singularités des problèmes aux limites et contrôlabilité exacte des systèmes hyperboliques, texte de l'exposé au Colloque Jean Céa d'Avril 1992.
77. Singularities in boundary value problems and exact controllability of hyperbolic systems, dans *Optimization, Optimal Control and Partial Differential Equations* **107**, édité par Barbu, Bonnans et Tiba, Birkhäuser Verlag (1992), 77–84.
78. Singular behavior of elliptic problems in non hilbertian Sobolev spaces, prépublication #321, Nice, accepté au JMPA.
79. Problèmes aux limites dans des domaines avec points de rebroussement, soumis.

Allocution prononcée au nom des élèves de Pierre Grisvard

Monsieur le Président, Mesdames, Messieurs, mes chers collègues,

Au nom des élèves de Pierre Grisvard, je voudrais tout d'abord remercier tous les organismes publics ou privés, sociétés savantes et universités qui ont contribué au financement de ce colloque ainsi que tous les collègues qui ont donné beaucoup de leur temps et de leur énergie pour son organisation. Je voudrais remercier tous les conférenciers qui ont, sans hésitation, répondu à notre invitation. Ils contribuent sans aucun doute par leur présence et à travers leurs exposés à la solennité et à la qualité scientifique que nous désirions donner à ces journées dédiées à la mémoire de celui qui fût notre maître mais aussi un ami pour la plupart d'entre nous. Je dis merci également à toutes les personnes présentes ici, mathématiciens ou non, collègues et amis de Pierre, qui ont tenu à se joindre à nous et partager ce moment de souvenir et de grande émotion. Avant de dire quelques mots au nom de ses élèves je voudrais d'abord les citer et parler brièvement de leur domaine de recherche. J'ai pensé que cela était nécessaire à deux titres au moins. En premier lieu cela permettra d'apprécier un peu mieux encore à travers, ne serait ce que les thèses soutenues par ses élèves, une part importante de sa contribution tant au plan scientifique qu'au plan de la formation par la recherche. En second lieu l'école de Pierre Grisvard est une vraie diaspora souvent méconnue tout comme l'est son apport sur le plan mathématique. Il est cependant vrai qu'il avait contribué lui même à cet état de fait en s'imposant une règle qu'il n'a jamais transgressée : en effet, ceux qui l'ont bien connu le savent bien, Pierre s'était interdit, quelle que soit sa propre contribution à une recherche, de publier un papier avec l'un quelconque de ses élèves que ce fût en cours de thèse ou après ou plus tard. C'etait une des caractéristiques de sa personnalité sur laquelle je reviendrai plus loin. Dans l'ordre chronologique ses élèves ont été, du moins à ma connaissance et je demande, par avance, pardon à ceux que par mégarde j'aurais oubliés :

1970
GOUDJO, C. : (BENIN) 3ème cycle : problèmes aux limites dans les espaces avec poids.

1971
ARANDA, (ARGENTINE) 3ème cycle: Injections de classe Lp des espaces de Sobolev relatifs à des ouverts non bornés.
CATTANEO, (ARGENTINE) 3ème cycle: Injections de Hilbert-Schmidt dans les espaces de Sobolev en ouverts non bornés et applications.

1973
MOUSSAOUI, M. (ALGERIE) 3ème cycle: Singularités du problème de Neumann

1974

MERIGOT, M.(FRANCE) Thèse d'état: Solutions en norme L^p des problèmes elliptiques dans des polygones plans.

1976

LEMRABET, K.(ALGERIE) 3ème cycle: Etude globale d'un problème de transmission dans un polygone ou un polyèdre.

MOULAY, M.S. (ALGERIE) 3ème cycle: Régularité de la solution d'un problème quasi-elliptique.

SADALLAH, K. (ALGERIE) 3ème cycle: Problème mixte pour l'équation de la chaleur dans un polygone en espace temps.

1977

ZOLESIO, J.L. (FRANCE) 3ème cycle: Interpolation d'espaces de Sobolev avec des conditions aux limites de type mêlé.

TRAORE, D. (MALI) 3ème cycle: Problèmes aux limites dans des domaines non bornés.

MOUSSAOUI, M.(ALGERIE)Thèse d'état: Régularité hölderienne de problèmes aux limites et problèmes à dérivée oblique dans un polygone.

1978

KHELIF, A (ALGERIE) 3ème cycle: problèmes aux limites pour le laplacien dans un domaine à points cuspides.

NAJMI, M. (MAROC) 3è cycle: Résolution d'un problème de Dirichlet fortement non linéaire dans un domaine non régulier.

BAILET, J. (FRANCE) 3ème cycle: Problèmes aux limites non linéaires avec conditions aux limites de type mêlé.

LABBAS, R. (ALGERIE) 3ème cycle: Problèmes à deux points pour une équation différentielle abstraite.

1979

BRAHIMI, M. (ALGERIE) 3ème cycle: Problèmes aux limites obliques et non linéaires pour l'équation de Laplace.

1983

BOUHAFA, H (TUNISIE) 3ème cycle: Etude de la partie singulière de la résolvante du problème de Dirichlet dans un polygone.

1985

GOUDJO, C.(BENIN)Thèse d'état: Contribution à l'étude des solutions singulières des problèmes aux limites.

AIBECHE, A (ALGERIE) 3ème cycle: Quelques problèmes non linéaires dans des domaines polygonaux; comportement singulier des solutions.

1986

DING, H (CHINE) Thèse nouveau régime: Etude qualitative du comportement d'un corps elastique sous l'effet de charges concentrées.

1987

LABBAS, R (ALGERIE) Thèse d'état: Problèmes aux limites pour une équation différentielle opérationnelle du second ordre.

LEMRABET, K (ALGERIE)Thèse d'état: Justification de problèmes aux limites de Ventcel en élasticité linéaire.

1989
TCHA-KONDOR, O (TOGO) thèse nouveau régime : Eléments finis et calculs analytiques.

1990
NIANE, M.T. (SENEGAL) thèse nouveau régime : Régularité et contrôlabilité exacte de l'équation des plaques vibrantes.
GOUDJO, A. (BENIN) thèse nouveau régime: Singularités d'arête en thermique et résolution de quelques problèmes hyperboliques.
NIANE, M.T. (SENEGAL) Thèse d'état Dakar: Régularité, contrôlabilité exacte et contrôlabilité spectrale de l'équation des ondes et de l'équation des plaques vibrantes
MEROUANI, B. (ALGERIE)Thèse d'état Alger : Etude analytique et numérique de l'équation transcendante associée au système de Lamé.

1992
NAJMI, M (MAROC) Thèse d'état : Problèmes de régularité-singularité dans les espaces höldériens.

THESES EN COURS
CHAIRA, A (MAROC) travaux en contrôlabilité en vue d'une thèse d'état.
TCHA-KONDOR, O (NICE) travaux en théorie spectrale en vue d'une habilitation.
BELAHDJI, K (ALGERIE) : Régularité Lp pour le laplacien dans des domaines à points cuspides en vue d'une nouvelle thèse.

Sur ces vingt sept élèves, vingt cinq sont enseignants dont vingt et un au moins exercent dans l'enseignement supérieur et parmi eux huit sont professeurs.

Je voudrais à présent dire quelques mots sur ce que Pierre a été et représente toujours pour nous. Bien entendu cela ne peut être qu'à travers les relations de travail et d'amitié que nous avons entretenues durant un peu plus de vingt cinq ans.

Je ne saurais donc prétendre ici à une objectivité absolue. Je suis cependant tout à fait convaincu que la grande majorité de ses élèves partage bien mon point de vue. P. Grisvard était un homme attachant à plus d'un titre et tout d'abord par son immense talent d'enseignant. La clarté et la concision, l'élégance dans la présentation et son sens inné de la pédagogie ont marqué profondément ses élèves, ses étudiants, bref beaucoup de tous ceux qui ont eu le privilège et le plaisir de suivre un de ses cours ou une de ses conférences. Il avait le don de rendre simples et accessibles à ses auditeurs les idées essentielles qui sous tendaient ses cours et ses conférences. Tout semblait alors si facile et faisait naître une envie de travailler sous sa direction. Au plan de la recherche il avait une grande force de conviction et il savait plus que tout autre communiquer cette passion et cette tenacité qui sont la marque des grands scientifiques, tout comme il était remarquable par son souci constant de la minutie et de la rigueur. Bien sûr cela pouvait paraitre pour ses élèves, du moins au début de leur thèse, comme une véritable manie mais il la compensait bien largement par une écoute et une disponibilité bien rares de nos jours. Les sujets de thèses qu'il proposait étaient toujours bien précis bien délimités, mais il n'avait de cesse d'inciter ses élèves à s'interesser aux domaines voisins comme à enrichir leurs connaissances et completer leur formation. Il avait en fait un souci permanent de former des gens qui se devaient d'être autonomes, capables dès la fin de leur thèse de travailler et de reflechir par eux-mêmes, de s'ouvrir sur les autres et de partager leur savoir. Mais s'il y avait un aspect qui était le plus attachant chez lui, c'est bien son côté humaniste et la fidélité à ses engagements. P. Grisvard étaient de ceux qui, comme de nombreux collègues présents ici, s'étaient engagés au cours des années soixante à apporter leur aide aux pays en développement. Ils étaient guidés par l'idée que la meilleure contribution qu'ils pouvaient apporter était de participer à la formation de mathématiciens dans leurs propres départements ou institutions universitaires. Je crois que Pierre est resté fidèle à cet engagement jusqu'à ses derniers jours. Avec ses élèves venus se former en France, il y a eu toujours un contrat moral, plus ou moins explicite qu'une fois formés on se devait de servir dans son propre pays. Il hésitait rarement par ailleurs à se déplacer d'un pays à l'autre de l'Afrique, du Maroc à la Tunisie en passant par l'Algérie, du Sénégal au Mali en passant par la Côte d'Ivoire, le Benin ou le Nigeria, apporter ses encouragemens et son optimisme. Il se rendait compte sur place et comprenait combien les conditions de travail étaient difficiles dans ces petits départements isolés et sans resources. Il était toujours là pour appuyer, soutenir et dynamiser toutes les initiatives. Ceci explique en fait pourquoi il s'attachait tant à exiger de chacun d'entre nous un travail sérieux de formation et un effort particulier pour l'acquisition de la plus large autonomie possible. Sur les vingt deux élèves, originaires de pays en développement, et malgré les problèmes vécus ces dernières années par les algériens, dix-sept exercent dans leur pays d'origine. Il a atteint là un des objectifs de sa vie et je crois qu'il en tirait et à juste titre un de ses plus grands motifs de satisfaction. Je ne saurais finir sans parler de Pierre simplement en tant qu'homme. J'ai toujours été impressionné par son ouverture d'esprit, son attachement à la liberté de chacun et à la solidarité entre les personnes. A côté d'une très vaste connaissance de sa propre culture, il avait une immense curiosité pour toutes les autres qu'il a eu l'occasion, à travers ses nombreux déplacements, de cotoyer.

Tout en étant critique, il a toujours été extrêmement respectueux de la spécificité et de la richesse de chacune. Il était aidé en cela par sa capacité à apprendre les langues. Je n'oublierai pas qu'en 1972-73, après un séjour d'à peine un an à Alger paraissait un des tous premiers polycopiés en arabe. C'était un cours de Calcul Différentiel dont les premiers chapitres étaient écrits de sa main. Pierre a su, comme je l'ai constaté au cours des années, transmettre à ses enfants ses qualités d'ouverture et d'amour de la connaissance des autres. La solidarité de Pierre ne s'exprimait pas uniquement à travers son activité professionnelle, comme tout ce temps et toute cette énergie consacrés à la bonne marche du CIMPA dont il a été le premier directeur scientifique. Elle s'exprimait aussi par son militantisme dans diverses associations et par diverses autres actions comme par exemple cette initiative d'aider à ouvrir une école primaire et collège dans un quartier pauvre de la banlieue de Dakar, initiative à laquelle ont participé nombre de collègues ici présents. Pierre Grisvard a beaucoup contribué à travers ses actions en France comme ailleurs au renom de l'Université française en général et à celui de l'Université de Nice en particulier. Pour ses enfants c'est là un motif légitime de fierté. Sa disparition est sûrement une grande perte pour la communauté mathématique française mais elle l'est beaucoup plus encore pour ces départements de mathématiques disséminés en Afrique pour l'émergence desquels il a tant donné. Encore une fois, c'est avec une grande émotion que je vous remercie tous d'être venus ici, avec nous, ses élèves, lui rendre un hommage et saluer sa mémoire.

M. Moussaoui
Professeur, Ecole Centrale de Lyon

Problèmes aux limites dans des domaines avec points de rebroussement

P. Grisvard

Introduction

Les problèmes aux limites elliptiques dans des domaines polygonaux ou polyédraux ont été abondamment étudiés. Les résultats maintenant classiques d'Agmon–Douglis–Nirenberg [1] et Lions–Magenes [7], établis pour des domaines réguliers, ont été adaptés. Ceci répondait à des nécessités pratiques. En effet dans les applications industrielles, les problèmes aux limites qui interviennent sont souvent posés dans des domaines polyédraux.

Cependant on n'a pas ainsi épuisé tous les types de domaines rencontrés en pratique. Dans les problèmes de lubrification on rencontre en particulier des domaines avec points de rebroussement. On peut citer par exemple une situation typique dans les roulements à billes, ou plutôt à rouleaux: un cylindre métallique roule sur un plan ou sur l'intérieur d'un autre cylindre. Le tout baigne dans un lubrifiant. L'écoulement du lubrifiant est gouverné par les équations de Navier–Stokes dans un domaine présentant un segment de points de rebroussement constitué par la ligne de contact entre les deux cylindres.

Ce travail est une contribution dans cette direction. On y étudiera le problème de Dirichlet pour l'équation de Laplace et pour le bilaplacien dans un domaine plan modèle présentant un point de rebroussement:

$$\Omega = \left\{(x,y) \in \mathbb{R}^2 \ ; \quad 0 < x < a \ , \quad 0 < y < \varphi(x)\right\}$$

où $\varphi(x) = x^\alpha$ avec $\alpha > 1$. On cherchera la solution dans les espaces de Sobolev construits sur $L^p(\Omega)$, $1 < p < +\infty$.

On démontrera que $u \in H_0^1(\Omega)$ solution de $\Delta u = f$ avec f donné dans $L^p(\Omega)$, appartient à l'espace $W^{2,p}(\Omega)$. Ce résultat n'est pas nouveau lorsque $p = 2$. Il a été démontré par Ibuki [5] puis généralisé à une fonction φ arbitraire de classe C^2 telle que $\varphi(0) = 0$, $\varphi(x) > 0$ pour $x \in]0, a]$, par Khelif [6]. Ce résultat sera ensuite étendu à un domaine tridimensionnel modèle $Q = \Omega \times]0, \pi[$. Ce domaine présente une "arête de rebroussement."

La technique de démonstration repose au départ sur le même changement de variables déjà utilisé par Ibuki [5], qui réduit le problème à une perturbation du problème de Dirichlet pour l'équation de Laplace dans une demi-bande infinie

$$\Omega_0 = \left\{(x,y) \in \mathbb{R}^2 \ ; \ x > 0 \ , \ 0 < y < 1\right\} \ .$$

Ce problème est analysé dans les espaces de Sobolev relatifs à L^p par des techniques de multiplicateurs de Fourier. L'extension à un domaine tridimensionnel repose sur les techniques de Dore-Venni [3].

Dans le même esprit on démontrera que $u \in H_0^1(\Omega)$ solution de $\Delta^2 u = f$ avec f donné dans $W^{-1,p}(\Omega)$, appartient à l'espace $W^{3,p}$ au voisinage du point de rebroussement. Appliqué à la fonction courant ce résultat implique la régularité $W^{2,p}$ pour les équations de Navier–Stokes linéarisées près du point de rebroussement (en dimension deux).

Plan

I. Le problème de Dirichlet pour l'équation de Laplace dans le domaine modèle

1.1. Le problème de référence

On démontre ici le

Théorème 1.1. *Pour $f \in L^p(\Omega_0)$ donné, il existe $u \in W^{2,p}(\Omega_0) \cap W_0^{1,p}(\Omega_0)$ unique, solution de $\Delta u = f$, dans Ω_0.*

On utilise une réflexion impaire dans la direction de x. On pose donc

$$U(x,y) = u(x,y) \text{ si } x > 0 \text{ et } -u(-x,y) \text{ si } x < 0$$

et

$$F(x,y) = f(x,y) \text{ si } x > 0 \text{ et } -f(-x,y) \text{ si } x < 0 \,.$$

On vérifie aisément l'équivalence des propriétés suivantes

(a) $u \in W^{2,p}(\Omega_0) \cap W_0^{1,p}(\Omega_0)$ et $\Delta u = f$,

(b) $U \in W^{2,p}(]0,1[\times\mathbb{R}) \cap W_0^{1,p}(]0,1[\times\mathbb{R})$ et $\Delta U = F$.

Par ailleurs il est clair que $f \in L^p(\Omega_0)$ si et seulement si $F \in L^p(]0,1[\times\mathbb{R})$.

L'existence et l'unicité de U vérifiant (b) est bien connue (c.f. entre autres Grisvard [4]). L'affirmation du Théorème 1.1 en résulte immédiatement.

1.2. Le changement de variables

Suivant Ibuki [5], on pose

$$\xi = \frac{1}{\alpha-1} x^{-\alpha+1}, \quad \eta = yx^{-\alpha}.$$

On étudie l'effet de ce changement de variables sur l'équation $\Delta u = f$ posée dans Ω. L'image de Ω par ce changement de variables est l'ouvert

$$\Omega_a = \left\{(\xi,\eta) \in \mathbb{R}^2 \,;\, \xi > \frac{1}{\alpha-1} a^{1-\alpha}, \quad 0 < \eta < 1\right\}.$$

On pose naturellement $v(\xi,\eta) = u(x,y)$ et $g(\xi,\eta) = f(x,y)$. On a donc

$$u(x,y) = v\left(\frac{1}{\alpha-1} x^{-\alpha+1}, yx^{-\alpha}\right).$$

Il vient

$$D_y u = x^{-\alpha} D_\eta v, D_\eta^2 u = x^{-2\alpha} D_\eta^2 v,$$

donc

$$D_y u = c\xi^\beta D_\eta v, D_y^2 u = c^2 \xi^{2\beta} D_\eta^2 v \text{ où } c = (\alpha-1)^\beta \text{ et } \beta = \frac{\alpha}{\alpha-1}.$$

Il vient également

$$\begin{aligned} D_x u = &- x^{-\alpha} D_\xi v\left(\frac{1}{\alpha-1} x^{-\alpha+1}, yx^{-\alpha}\right) \\ &- \alpha y x^{-\alpha-1} D_\eta v\left(\frac{1}{\alpha-1} x^{-\alpha+1}, yx^{-\alpha}\right) \end{aligned}$$

et

$$\begin{aligned}
D_x^2 u =& \alpha x^{-\alpha-1} D_\xi v + x^{-2\alpha} D_\xi^2 v + 2\alpha y x^{-2\alpha-1} D_\xi D_\eta v \\
&+ \alpha(\alpha+1) y x^{-\alpha-2} D_\eta v + \alpha^2 y^2 x^{-2\alpha-2} D_\eta^2 v \\
=& x^{-2\alpha}\Big\{ D_\xi^2 v + 2\alpha y x^{-1} D_\xi D_\eta v \\
&+ \alpha^2 y^2 x^{-2} D_\eta^2 v + \alpha x^{\alpha-1} D_\xi v + \alpha(\alpha+1) y x^{\alpha-2} D_\eta v \Big\} \\
=& x^{-2\alpha}\Big\{ D_\xi^2 v + 2\alpha\eta x^{\alpha-1} D_\xi D_\eta v + \alpha^2\eta^2 x^{2\alpha-2} D_\eta^2 v \\
&+ \alpha x^{\alpha-1} D_\xi v + \alpha(\alpha+1)\eta x^{2\alpha-2} D_\eta v \Big\} \\
=& c^2 \xi^{2\beta}\Big\{ D_\xi^2 v + 2\alpha c^{-\frac{1}{\beta}} \frac{\eta}{\xi} D_\xi D_\eta v + \alpha^2 c^{-\frac{2}{\beta}} \frac{\eta^2}{\xi^2} D_\eta^2 v \\
&+ \alpha c^{-\frac{1}{\beta}} \frac{1}{\xi} D_\xi v + \alpha(\alpha+1) c^{-\frac{2}{\beta}} \frac{\eta}{\xi^2} D_\eta v \Big\}.
\end{aligned}$$

Au total l'équation $\Delta u = f$ devient

$$\begin{aligned}
c^2\xi^{2\beta}\Big\{ \Delta v + 2\alpha c^{-\frac{1}{\beta}} \frac{\eta}{\xi} D_\xi D_\eta v + \alpha^2 c^{-\frac{2}{\beta}} \frac{\eta^2}{\xi^2} D_\eta^2 v + \alpha c^{-\frac{1}{\beta}} \frac{1}{\xi} D_\xi v \\
+ \alpha(\alpha+1) c^{-\frac{2}{\beta}} \frac{\eta}{\xi^2} D_\eta v \Big\} = g.
\end{aligned}$$

Il convient également d'étudier l'effet du même changement de variables sur les espaces fonctionnels. Le résultat suivant est évident.

Lemme 2.1. *On a $f \in L^p(\Omega)$ si et seulement si $\xi^{-\frac{2\beta}{p}} g \in L^p(\Omega_a)$.*

En d'autres termes une fonction est de puissance p sommable en x, y si et seulement si elle est de puissance p sommable en ξ, η après multiplication par $\xi^{-\frac{2\beta}{p}}$

On devra donc étudier l'équation en v ci-dessus en supposant que $\xi^{-\frac{2\beta}{p}} g \in L^p(\Omega_a)$. Pour éviter de manipuler des espaces avec poids, il sera plus commode de considérer une équation dont le second membre est proportionnel à $\xi^{-\frac{2\beta}{p}} g$. Pour cela on pose

$$w = \xi^\gamma v \,, \gamma = \frac{2\beta}{p'}$$

et on cherche l'équation de w. Il vient

$$\begin{aligned}
&D_\eta v = \xi^{-\gamma} D_\eta w \,, \quad D_\eta^2 v = \xi^{-\gamma} D_\eta^2 w, \\
&D_\xi v = \xi^{-\gamma} D_\xi w - \gamma^{\xi^{-\gamma-1}} w \,, \quad D_\xi D_\eta v = \xi^{-\gamma} D_\xi D_\eta w - \gamma^{\xi^{-\gamma-1}} D_\eta w \,, \\
&D_\xi^2 v = \xi^{-\gamma} D_\xi^2 w - 2\gamma \xi^{-\gamma-1} D_\xi w + \gamma(\gamma+1) \xi^{-\gamma-2} w
\end{aligned}$$

d'où

$$c^2\xi^{-\gamma}\left\{\Delta w - \frac{2\gamma}{\xi}D_\xi w + \frac{\gamma(\gamma+1)}{\xi^2}w + 2\alpha c^{-\frac{1}{\beta}}\left[\frac{\eta}{\xi}D_\xi D_\eta w - \gamma\cdot\frac{\eta}{\xi^2}D_\eta w\right]\right.$$
$$\left. +\alpha^2 c^{-\frac{2}{\beta}}\frac{\eta^2}{\xi^2}D_\eta^2 w + \alpha c^{-\frac{1}{\beta}}\left[\frac{1}{\xi}D_\xi w - \frac{\gamma}{\xi^2}w\right] + \alpha(\alpha+1)c^{-\frac{2}{\beta}}\frac{\eta}{\xi^2}D_\eta w\right\} = \xi^{-2\beta}g\,.$$

Autrement dit on a montré l'existence d'un opérateur différentiel linéaire du second ordre L à coefficients bornés (pour $\xi > 1$ pour fixer les idées) tel que

$$c^2\left\{\Delta w + \frac{1}{\xi}Lw\right\} = \xi^{\gamma-2\beta}g = \xi^{-\frac{2\beta}{p}}g\,.$$

Il est naturel à ce point de poser $h = \xi^{-\frac{2\beta}{p}}g$.

En résumé on a établi la

Proposition 2.2. *Il existe un opérateur différentiel linéaire du second ordre à coefficients bornés L tel que l'équation $\Delta u = f$ dans Ω soit équivalente à l'équation $c^2\left\{\Delta w + \frac{1}{\xi}Lw\right\} = h$ dans Ω_a, où on a posé*

$$h = \xi^{-\frac{2\beta}{p}}f, w = \xi^{\frac{2\beta}{p'}}u, \quad \textit{avec } \beta = \frac{\alpha}{\alpha-1} \quad \textit{et } c = (\alpha-1)^\beta\,.$$

1.3. Résolution du problème transformé

On déduit du théorème 1.1 le résultat suivant.

Théorème 3.1. *Pour a assez petit la propriété suivante est vérifiée: L'opérateur $\Delta + \frac{1}{\xi}L$ est un isomorphisme de $W^{2,p}(\Omega_a)\cap W_0^{1,p}(\Omega_a)$ sur $L^p(\Omega_a)$.*

Démonstration. On sait que Δ est un isomorphisme de $W^{2,p}(\Omega_0)\cap W_0^{1,p}(\Omega_0)$ sur $L^p(\Omega_0)$. Par ailleurs Ω_a est un translaté de Ω_0 (de $\frac{1}{\alpha-1}a^{1-\alpha}$ dans la direction de x). Comme Δ et les conditions aux limites de Dirichlet sont invariants par translation, il en résulte que Δ est un isomorphisme de $W^{2,p}(\Omega_a)\cap W_0^{1,p}(\Omega_a)$ sur $L^p(\Omega_a)$ pour tout a, et la norme de Δ^{-1} est indépendante de a.

Pour a assez petit, donc $\frac{1}{\alpha-1}a^{1-\alpha}$ assez grand, l'opérateur $\frac{1}{\xi}L$ a une norme inférieure à celle de Δ^{-1}. En d'autres termes $\frac{1}{\xi}L\Delta^{-1}$ est une contraction stricte. Il en résulte que $\Delta + \frac{1}{\xi}L$ est un isomorphisme. □

Ainsi, partant de f dans $L^p(\Omega)$ ou, ce qui revient au même, de h dans $L^p(\Omega_a)$, il existe $w \in W^{2,p}(\Omega_a)$ unique solution de

$$c^2\left\{\Delta w + \frac{1}{\xi}Lw\right\} = h \text{ dans } \Omega_a \text{ avec } w = 0 \text{ sur } \partial\Omega_a\,.$$

Il reste à étudier propriétés de dérivabilité de u correspondant à w.

1.4. Effet du changement de variables inverse

On a par définition $w = \xi^{\frac{2\beta}{p'}} u\left(c^{-\frac{1}{\alpha}}\xi^{-\frac{\beta}{\alpha}}, c^{-1}\eta\xi^{-\beta}\right)$ d'où

$$D_\eta w = c^{-1}\xi^{-\beta}\xi^{\frac{2\beta}{p'}} D_y u(x,y), D_\eta^2 w = c^{-2}\xi^{-2\beta}\xi^{\frac{2\beta}{p'}} D_y^2 u(x,y) .$$

De manière équivalente

$$\xi^{-2\beta} w = \xi^{-\frac{2\beta}{p}} u, \xi^{-\beta} D_\eta w = c^{-1}\xi^{-\frac{2\beta}{p}} D_y u, D_\eta^2 w = c^{-2}\xi^{-\frac{2\beta}{p}} D_y^2 u$$

ou encore

$$\begin{aligned} w = c^{-2}\xi^{-\frac{2\beta}{p}} x^{-2\alpha} u, D_\eta w = c^{-2}\xi^{-\frac{2\beta}{p}} x^{-\alpha} D_y u, \\ D_\eta^2 w = c^{-2}\xi^{-\frac{2\beta}{p}} D_y^2 u . \end{aligned}$$

Du fait que $w, D_\eta w$ et $D_\eta^2 w$ sont de puissances p sommables dans Ω_a, le lemme 2.1 implique que $x^{-2\alpha}u, x^{-\alpha}D_y u$ et $D_y^2 u$ sont de puissances p sommables dans Ω. Ecrivant que $D_x^2 u = f - D_y^2 u$, on voit immédiatement que $D_x^2 u$ est aussi de puissance p sommable. Il reste donc à étudier la sommabilité de $D_x u$ et $D_x D_y u$ à la puissance p.

On a

$$\begin{aligned} D_\xi w &= \frac{2\beta}{p'}\xi^{\frac{2\beta}{p'}-1} u - \frac{\beta}{\alpha} c^{-\frac{1}{\alpha}} \xi^{\frac{2\beta}{p'}-\frac{\beta}{\alpha}-1} D_x u - \beta c^{-1}\eta\xi^{-\beta-1}\xi^{\frac{2\beta}{p'}} D_y u \\ &= \frac{2\beta}{p'}\xi^{2\beta-1}\left[\xi^{-\frac{2\beta}{p}} u\right] - \frac{\beta}{\alpha} c^{-\frac{1}{\alpha}}\xi^{2\beta-1-\frac{\beta}{\alpha}}\left[\xi^{-\frac{2\beta}{p}} D_x u\right] \\ &\quad - \beta c^{-1}\eta\xi^{\beta-1}\left[\xi^{-\frac{2\beta}{p}} D_y u\right] \\ &= \frac{2\beta}{p'}\xi^{2\beta-1}\left[\xi^{-\frac{2\beta}{p}} u\right] - \frac{\beta}{\alpha} c^{-\frac{1}{\alpha}}\xi^{\beta}\left[\xi^{-\frac{2\beta}{p}} D_x u\right] \\ &\quad - \beta c^{-1}\eta\xi^{\beta-1}\left[\xi^{-\frac{2\beta}{p}} D_y u\right] . \end{aligned}$$

C'est encore

$$\begin{aligned} D_\xi w = \frac{2\beta}{p'c^2}\xi^{-1}\left[\xi^{-\frac{2\beta}{p}} x^{-2\alpha} u\right] - \frac{\beta}{\alpha c} c^{-\frac{1}{\alpha}}\left[\xi^{-\frac{2\beta}{p}} x^{-\alpha} D_x u\right] \\ - \frac{\beta}{c^2}\eta\xi^{-1}\left[\xi^{-\frac{2\beta}{p}} x^{-\alpha} D_y u\right] . \end{aligned}$$

On sait que $D_\xi w$ est de puissance p sommable. D'après les calculs précédents on sait déjà que

$$\xi^{-\frac{2\beta}{p}} x^{-2\alpha} u \text{ et } \xi^{-\frac{2\beta}{p}} x^{-\alpha} D_y u$$

sont de puissance p sommable en ξ, η. Il en résulte à plus forte raison que $\xi^{-\frac{2\beta}{p}} x^{-\alpha} D_x u$ est de puissance p sommable en ξ et η donc que $x^{-\alpha} D_y u \in L^p(\Omega)$, par application du lemme 2.1.

On a enfin

$$D_\xi D_\eta w = \frac{2\beta}{cp'} \xi^{\beta-1} \left[\xi^{-\frac{2\beta}{p}} D_y u\right] - \frac{\beta}{\alpha c} c^{-\frac{1}{\alpha}} \left[\xi^{-\frac{2\beta}{p}} D_x D_y u\right] - \frac{\beta\eta}{c^2} \eta\xi^{-1} \left[\xi^{-\frac{2\beta}{p}} D_y^2 u\right]$$

d'où encore

$$D_\xi D_\eta w = \frac{2\beta}{c^2 p'} \xi^{-1} \left[\xi^{-\frac{2\beta}{p}} x^{-\alpha} D_y u\right] - \frac{\beta}{\alpha c} c^{\frac{-1}{\alpha}} \left[\xi^{-\frac{2\beta}{p}} D_x D_y u\right] - \frac{\beta\eta}{c^2} \eta\xi^{-1} \left[\xi^{-\frac{2\beta}{p}} D_y^2 u\right].$$

Comme on sait que $D_\xi D_\eta w$, $\xi^{-\frac{2\beta}{p}} x^{-\alpha} D_y u$ et $\xi^{-\frac{2\beta}{p}} D_y^2 u$ sont de puissance p sommable en ξ, η, il en est de même pour $\xi^{-\frac{2\beta}{p}} D_x D_y u$. Ceci établit que $D_x D_y u \in L^p(\Omega)$, par application du lemme 2.1.

Au total on a établi la

Proposition 4.1. *Le fait que* $w \in W^{2,p}(\Omega_a)$ *implique que*

$$x^{-2\alpha} u, x^{-\alpha} D_x u, x^{-\alpha} D_y u, D_x^2 u, D_x D_y u, D_y^2 u$$

appartiennent à $L^p(\Omega)$.

1.5. Bilan

On part de $f \in L^p(\Omega)$. On a donc $f \in H^{-1}(\Omega)$ car l'inclusion duale $H_0^1(\Omega) \subset L^{p'}(\Omega)$ est valable du fait que les fonctions de $H_0^1(\Omega)$ sont prolongeables par 0 hors de Ω (bien que Ω n'ait peut être pas la propriété du prolongement usuelle). On en déduit sans peine l'existence et l'unicité de $u \in H_0^1(\Omega)$ solution de $\Delta u = f$. Son comportement loin de l'origine est donné par les résultats classiques et son comportement près de l'origine a été étudié dans les §§ précédents.

Théorème 5.1. Δ *est un isomorphisme de* $W^{2,p}(\Omega) \cap W_0^{1,p}(\Omega)$ *sur* $L^p(\Omega)$.

Remarque 5.2. La proposition 4.1 donne en outre une propriété de décroissance de u et ∇u près de l'origine.

1.6. Un résultat tridimensionnel

On utilise des résultats récents de Dore–Venni [3] et de Coifman–Weiss [2].

On se place dans un espace de Banach E qui a la propriété U.M.D. Sans entrer dans les détails les espaces L^p avec $1 < p < +\infty$ ont cette propriété. On considère deux opérateurs A et B linéaires fermés non bornés de domaines $D(A)$ et $D(B)$ dans E. On suppose que

(i) $A+tI$ et $B+tI$ sont inversibles pour tout $t \geq 0$ et que $\|(A+tI)^{-1}\|$ et $\|(B+tI)^{-1}\|$ ont un comportement en $0\left(\frac{1}{t}\right)$ lorsque $t \to +\infty$.

(ii) $(A+tI)^{-1}$ et $(B+sI)^{-1}$ commutent pour tout t et tout $s \geq 0$.

(iii) Il existe $\theta(A)$ et $\theta(B)$ tels que $\theta(A) + \theta(B) < \pi$ et $\|A^{is}\| \leq Ce^{|s|\theta(A)}$, $\|B^{is}\| \leq Ce^{|s|\theta(B)}$ pour tout $s \in \mathbb{R}$.

On remarque que l'hypothèse (i) permet de définir les puissances imaginaires impliquées dans (iii).

Sous les hypothèses (i), (ii), et (iii) l'opérateur

$$L : x \mapsto Ax + Bx$$

défini sur $D(L) = D(A) \cap D(B)$ est inversible. C'est la théorème de Dore–Venni [3].

Les hypothèses (i) et (ii) sont faciles à vérifier en pratique. Il n'en va pas de même pour l'hypothèse (iii). Selon un procédé proposé par Clément (communication personnelle) on peut faire appel au résultat suivant de Coifman–Weiss [2]. On suppose que $-\Lambda$ génère un semi-groupe de contractions qui préservent la positivité dans un espace $X = L^p(M)$ (où M est n'importe quel espace mesuré) alors on a

$$\|\Lambda^{is}\| \leq O\left(|s|e^{|s|\frac{\pi}{2}}\right)$$

pour tout $s \in \mathbb{R}$.

Ce cadre abstrait va être utilisé avec

$$E = L^p(Q) = L^p\Big( ]0,\Pi[; L^p(\Omega)\Big)$$

où $Q = \Omega \times]0,\Pi[$ et

$$Au = -\Delta_2 u \text{ pour } u \in D(A) = L^p\left( ]0,\Pi[; W^{2,p}(\Omega) \cap W_0^{1,p}(\Omega)\right)$$

$$Bu = -D_z^2 u \text{ pour } u \in D(B) = W^{2,p}\Big( ]0,\Pi[; L^p(\Omega)\Big) \cap W_0^{1,p}\Big( ]0,\Pi[; L^p(\Omega)\Big).$$

On va établir le

Théorème 6.1. *Δ_3 réalise un isomorphisme de*

$$\left\{u \in W_0^{1,p}(Q); D_x^2 u, D_y^2 u, D_z^2 u, D_x D_y u \in L^p(Q)\right\}$$

sur $L^p(Q)$.

On a désigné par Δ_2 et Δ_3 respectivement l'opérateur de Laplace dans les deux variables x, y et dans les trois variables x, y, z.

Démonstration. Considérons d'abord l'opérateur Λ défini dans $X = L^p(\Omega)$ par

$$D_\Lambda = \left\{u \in H_0^1(\Omega); \Delta_2 u \in L^p(\Omega)\right\}$$
$$\Lambda u = -\Delta_2 u \text{ pour } u \in D_\Lambda .$$

Il est connu que, même sans hypothèse sur Ω, $-\Lambda$ est générateur d'un semi-groupe de contractions qui préservent la positivité dans X. Il en résulte que $\Lambda + tI$ est inversible pour $t > 0$ et que

$$\|(\Lambda + tI)^{-1}\|_{X \to X} \leq \frac{1}{t} .$$

De plus on a

$$\|\Lambda^{is}\|_{X \to X} = O\left(|s| e^{|s| \frac{\pi}{2}}\right) .$$

Le théorème 5.1 ci-dessus permet de préciser que

$$D_\Lambda = W^{2,p}(\Omega) \cap W_0^{1,p}(\Omega) .$$

Par ailleurs dans le cas $p = 2$, Λ est un opérateur autoadjoint positif; Λ^{is} est alors une contraction. Au total on a

$$\|\Lambda^{is}\|_{L^2(\Omega) \to L^2(\Omega)} = O(1)$$
$$\|\Lambda^{is}\|_{L^q(\Omega) \to L^q(\Omega)} = O\left(|s| e^{|s| \frac{\pi}{2}}\right)$$

pour tout $q < +\infty$. En interpolant, il en résulte qu'il existe $\theta_\Lambda < \frac{\pi}{2}$ tel que

$$\|\Lambda^{is}\|_{X \to X} = O\left(e^{|s| \theta_\Lambda}\right) .$$

Ces propriétés de l'opérateur Λ impliquent les propriétés concernant A dans (i) et (ii) avec $\theta_A < \frac{\pi}{2}$.

On raisonne de la même manière sur l'opérateur M défini dans $Y = L^p(]0, \pi[)$ par

$$D_M = W^{2,p}(]0, \pi[) \cap W_0^{1,p}(]0, \pi[)$$
$$Mu = -u'' \text{ pour } u \in D_M .$$

On en déduit que $M + tI$ est inversible pour $t > 0$ et que

$$\|(M + tI)^{-1}\|_{Y \to Y} \leq \frac{1}{t}$$
$$\|M^{is}\|_{Y \to Y} = O\left(e^{|s| \theta_M}\right)$$

avec $\theta_M < \frac{\pi}{2}$. Ces propriétés impliquent celles concernant B dans (i) et (iii) avec $\theta_B < \frac{\pi}{2}$.

Par ailleurs la propriété (ii) est évidente. La conclusion est que

$$L = A + B = -\Delta_2 - D_z^2 = -\Delta_3$$

réalise un isomorphisme de $D_A \cap D_B$ sur E. Une fonction $u \in D_A \cap D_B$ est une fonction telle que

$$u, D_x u, D_y u, D_x D_y u, D_x^2 u, D_y^2 u, D_z u, D_z^2 u \in L^p(Q)$$

et $u = 0$ sur ∂Q. □

Remarque 6.2. Il n'est pas évident que les dérivées croisées $D_x D_z u$ et $D_y D_z u$ sont dans $L^p(Q)$, car on ne dispose pas d'un théorème de prolongement à $\mathbb{R}^3$. Cependant dans le cas $p = 2$ on vérifie facilement par développement en série de Fourier en z que

$$D_z u \in L^2(]0, \pi[; V)$$

où $V = H_0^1(\Omega)$.

1.7. Autres conditions aux limites

Les mêmes techniques peuvent être appliquées dans le cas des conditions de Neumann. Voyons quel est le problème transformé dans les changements de variables et de fonctions du §2.

La condition $D_y u = 0$ sur $y = 0$ devient $D_\eta w = 0$ sur $\eta = 0$.

La condition $\frac{\partial u}{\partial \nu} = 0$ sur la courbe $y = \varphi(x) = x^\alpha$ équivaut à

$$\alpha x^{\alpha-1} D_x u - D_y u = 0 .$$

D'après les formules du §1.2 ceci équivaut à

$$\alpha c^{\frac{1}{\alpha}} \xi^{-1} D_\xi w - \alpha\gamma c^{\frac{1}{\alpha}} \xi^{-2} w + \alpha c^{\frac{2-\alpha}{\alpha}} \xi^{-2} D_\eta w + c D_\eta w = 0 ,$$

sur $\eta = 1$. Cette condition peut être écrite plus simplement comme

$$D_\eta w + \xi^{-1} M w = 0 ,$$

sur $\eta = 1$, où M est un opérateur différentiel linéaire du premier ordre à coéfficients bornés (pour $\xi \geq 1$ pour fixes les idées) et à dérivées premières bornées.

En suivant les mêmes étapes que pour le problème de Dirichlet on va établir le résultat suivant où Γ_a désigne le segment

$$\{(a, y); 0 < y < \varphi(a)\}$$

de la frontière de Ω.

Théorème 7.1. *Pour tout $f \in L^p(\Omega)$ il existe $u \in W^{2,p}(\Omega)$ unique solution de*

$$\Delta u = f \ \ dans\ \Omega$$
$$u = 0 \ \ sur\ \Gamma_a$$
$$\frac{\partial u}{\partial \nu} = 0 \ \ sur\ \Gamma/\Gamma_a\,.$$

On remarque qu'il faut se limiter à $p \geq 2$ pour être assuré de l'existence d'une solution variationnelle.

Pour éviter des difficultés liées à la non unicité on a conservé la condition de Dirichlet loin du point de rebroussement. Ce qui importe ici c'est que la condition soit du type Neumann sur les deux arcs de Γ qui aboutissent au rebroussement.

Démonstration. Le problème de référence est ici

$$\begin{aligned} \Delta u &= f \ \text{ dans } \Omega_0 \\ u &= 0 \ \text{ pour } x = 0\,, \quad 0 < y < 1 \\ D_y &= 0 \ \text{ pour } x > 0\,, \quad y = 0 \text{ ou } 1\,. \end{aligned}$$

Comme au §1 on montre que pour $f \in L^p(\Omega_0)$ il existe une unique $u \in W^{2,p}(\Omega_0)$ solution de ce problème.

Le morceau nouveau de technique est ici qu'il faut considérer le problème non homogène:

$$\begin{aligned} \Delta u &= f \ \text{ dans } \Omega_0\,, \\ u &= 0 \ \text{ pour } x = 0\,, \quad 0 < y < 1\,, \\ D_y u &= g_0 \ \text{ pour } x > 0\,, \quad y = 0\,, \\ D_y u &= g_1 \ \text{ pour } x > 0\,, \quad y = 1\,. \end{aligned}$$

Ce problème admet une solution unique dans $W^{2,p}(\Omega_0)$ si et seulement si $f \in L^p(\Omega_0)$ et $g_0, g_1 \in W_0^{1-\frac{1}{p},p}(\mathbb{R}_+)$. Ici le symbole zero signifie que $g_0(0) = g_1(0) = 0$. C'est la condition de raccord naturelle avec une donnée de Dirichlet homogène.

En translatant à Ω_a on voit que

$$T_a : u \mapsto \{\Delta u; D_\eta u|_{\eta=0}, D_\eta u|_{\eta=1}\}$$

est un isomorphisme de

$$\{u \in W^{2,p}(\Omega_a); u|_{x=a} = 0\} \ \text{ sur } \ L^p(\Omega_a) \times W_0^{1-\frac{1}{p},p}(]a, +\infty[)^2$$

et que la norme de T_a^{-1} est indépendante de a.

En raisonnant comme au §1.3 on en déduit que pour a assez petit l'opérateur

$$u \mapsto \{\Delta u + \xi^{-1}Lu; D_\eta u|_{\eta=0} \ , \ D_\eta u + \xi^{-1}Mu|_{\eta=1}\}$$

est aussi un isomorphisme dans les mêmes espaces. C'est l'analogue du théorème 3.1.

On conclut en raisonnant comme au §4 ce qui donne le même surcroît de régularité que dans la Proposition 4.1.

Proposition 7.2. *On aurait pu de la même façon considérer un problème mêlé en conservant la condition de Dirichlet sur la courbe $y = \varphi(x)$.*

II: Le problème de Dirichlet pour l'équation biharmonique dans le domaine modèle

2.1. Le problème biharmonique de référence

On établit ici le

Théorème 1.1. *Pour $f \in W^{-1,p}(\Omega_0)$ donné, il existe $u \in W^{3,p}(\Omega_0)$ unique, solution de $\Delta^2 u = f$ dans Ω_0 avec les conditions aux limites*

$$u = 0 \ \textit{sur} \ \partial\Omega_0$$

$$D_y u = 0 \text{ sur } \{y = 0\} \text{ et sur } \{y = 1\}$$
$$D_x^2 u = 0 \text{ sur } \{x = 0\} \, .$$

Démonstration. Les conditions aux limites sur le segment vertical $\{x = 0\}$ sont choisies pour permettre d'effectuer une réflexion impaire par rapport à la variable x. Partant de $F \in W^{-1,p}(]0,1[\times\mathbb{R})$ il revient au même de chercher $U \in W^{3,p}(]0,1[\times\mathbb{R}) \cap W_0^{2,p}(]0,1[\times\mathbb{R})$ solution de $\Delta^2 U = F$ dans $]0,1[\times\mathbb{R}$. L'existence et l'unicité d'une telle U est prouvée dans Grisvard [4]. □

Remarque 1.2. Dans la même référence on prouve que si $F \in L^p(]0,1[\times\mathbb{R})$ alors $U \in W^{4,p}(]0,1[\times\mathbb{R})$ d'où le résultat analogue sur Ω_0: pour $f \in L^p(\Omega_0)$ il existe $u \in W^{4,p}(\Omega_0)$ unique solution de $\Delta^2 u = f$ dans Ω_0 avec les conditions aux limites ci-dessus.

2.2. Quelques préliminaires

On part de $f \in W^{-1,p}(\Omega)$ avec $p \geq 2$. L'inclusion de $W^{-1,p}(\Omega)$ dans $H^{-1}(\Omega)$ implique l'existence et l'unicité d'une solution variationnelle $u \in H_0^2(\Omega)$ de l'équation $\Delta^2 u = f$. On sait de plus, par des résultats classiques, que $u \in W^{3,p}(\omega)$ pour tout ouvert $\omega \subset \Omega$ dont la frontière ne contient pas les trois coins de Ω.

Pour étudier le comportement de u près de l'origine (le seul qui nous importe ici) on introduit une fonction de troncature φ qui dépend de la seule variable x et telle que $\varphi(x) = 1$ pour $x \leq \frac{a}{3}$ et $\varphi(x) = 0$ pour $x \geq \frac{2a}{3}$. D'après ce qui a été rappelé ci-dessus, il est clair que

$$\begin{gathered}\varphi u \in H_0^2(\Omega)\\ \Delta^2(\varphi u) = f_1 \in W^{-1,p}(\Omega)\,.\end{gathered}$$

En outre l'avantage de φu est d'être identiquement nulle près du segment $\{x = a\}$, elle y vérifie donc n'importe quelle condition au bord. C'est ce qui va permettre par changement de variable de se ramener au problème biharmonique de référence introduit plus haut.

Dans les §§ qui suivent on considérera φu au lieu de u, mais on la notera encore u pour ne pas alourdir les notations.

2.3. Changement de variables dans un espace dual

On va étudier l'équation $\Delta^2 u = f$ dans Ω avec f donné dans $W^{-1,p}(\Omega)$. Il faut donc étudier l'image g de f dans le changement de variable du I, §1.2.

Lemme 3.1. *Si $u \in W_0^{1,p}(\Omega)$ alors $x^{-\alpha}u \in L^p(\Omega)$.*

Démonstration. De l'inégalité

$$\int_0^1 |\psi(y)|^p \, dy \leq C \int_0^1 |\psi'(y)|^p \, dy$$

valable pour $\psi \in W_0^{1,p}(]0,1[)$, on en déduit par homothétie l'inégalité

$$L^{-p} \int_0^L |\psi(y)|^p \, dy \leq C \int_0^L |\psi'(y)|^p \, dy$$

pour $\psi \in W_0^{1,p}(]0,L[)$.

On en déduit que

$$\int_0^{x^\alpha} |x^{-\alpha}u(x,y)|^p \, dy \leq C \int_0^{x^\alpha} |D_y u(x,y)|^p \, dy$$

pour $u \in W_0^{1,p}(\Omega)$ d'où le résultat. □

Corollaire 3.2. *Si $f \in W^{-1,p}(\Omega)$ on a*

$$f = f_0 + D_x f_1 + D_y f_2$$

avec $x^\alpha f_0, f_1, f_2 \in L^p(\Omega)$.

Les formules du §1.2 montrent alors que si on pose

$$g_i(\xi,\eta) = f_i(x,y)\,, \quad i = 0,1,2$$

on a

$$\xi^{-\beta(1+\frac{2}{p})}g_0 \in L^p(\Omega_a)$$

puis

$$D_y f_2 = c\xi^\beta D_\eta g_2 = D_\eta h_2$$

où

$$h_2 = c\xi^\beta g_2$$

donc

$$\xi^{-\beta(1+\frac{2}{p})}h_2 \in L^p(\Omega_a)$$

puis enfin

$$D_x f_1 = D_\xi\{-c\xi^\beta g_1\} + D_\eta\left\{-\alpha c^{\frac{1}{\alpha}}\eta\xi^{\beta-1}g_2\right\} + \left\{c\beta + \alpha c^{\frac{1}{\alpha}}\right\}\xi^{\beta-1}g_2$$

d'où, dans des notations évidentes

$$D_x f_1 = D_\xi h_1' + D_\eta h_2' + h_3'$$

où

$$\xi^{-\beta(1+\frac{2}{p})}h_1', \xi^{-\beta(1+\frac{2}{p})}h_2', \xi^{-\beta(1+\frac{2}{p})}h_3' \in L^p(\Omega_a)$$

car ξ^{-1} est borné sur Ω_a.

Au total on a obtenu le

Lemme 3.3. *Si $f \in W^{-1,p}(\Omega)$ alors g défini par $g(\xi,\eta) = f(x,y)$ vérifie*

$$g = k_0 + D_\xi k_1 + D_\eta k_2$$

avec $\xi^{-\beta(1+\frac{2}{p})}k_i \in L^p(\Omega_a)$, $i = 0, 1, 2$.

2.4. Changement de variables dans l'équation biharmonique

On a vu au §1.2 que l'équation $\Delta u = f$ équivaut par changement de variable à l'équation

$$c^2\left\{\Delta v + \frac{1}{\xi}Lv\right\} = \xi^{-2\beta}g$$

dans Ω_a où L est un opérateur différentiel linéaire du second ordre à coefficients bornés (et à dérivées de tous ordres bornées). Il s'ensuit que l'équation $\Delta^2 u = f$ équivaut à

$$c^4 \left\{ \Delta^2 v + \frac{1}{\xi} \mathcal{M} v \right\} = \xi^{-4\beta} g$$

dans Ω_a où $\mathcal{M}$ est un opérateur différentiel linéaire du quatrième ordre à coefficients bornés.

D'après le lemme 3.3 on sait que

$$\xi^{-\beta(1+\frac{2}{p})} g \in W_p^{-1}(\Omega_a) .$$

Il est donc logique de prendre comme nouvelle fonction inconnue une fonction w dont l'équation a un second membre proportionnel à $\xi^{-\beta(1+\frac{2}{p})} g$.

Pour cela on pose

$$w = \xi^\delta v, \delta = \beta + \frac{2\beta}{p'} .$$

On obtient alors

$$c^4 \left\{ \Delta^2 w + \frac{1}{\xi} M w \right\} = \Phi \in W^{-1,p}(\Omega_a)$$

où M est encore un opérateur différentiel linéaire du quatrième ordre à coéfficients bornés (et à dérivées bornées).

D'après le théorème 1.1 du II, on sait que Δ^2 est un isomorphisme de

$$\Big\{ u \in W^{3,p}(\Omega_a); u = 0$$
$$\text{sur } \partial\Omega_a, D_y u = 0 \text{ sur } y = 0, 1, D_x^2 u = 0 \text{ sur } x = a \Big\} .$$

sur $W^{-1,p}(\Omega_a)$ et que la norme de son inverse ne dépend pas de a. Pour a assez petit on en déduit que $\Delta^2 + \frac{1}{\xi} M$ est aussi un isomorphisme. Ceci nous permet d'affirmer que

$$w \in W^{3,p}(\Omega_a) .$$

On a donc $\xi^{-\frac{2\beta}{p}} \left\{ \xi^{3\beta} v \right\} \in L^p(\Omega_a)$ et par le lemme 3.1 on a

$$x^{-3\alpha} u \in L^p(\Omega) .$$

En raisonnant comme dans le §1.4 on voit au total que

$$x^{(j-3)\alpha} D^j u \in L^p(\Omega)$$

pour $j = 1, 2, 3$. En particulier on a $u \in W^{3,p}(\Omega)$.

2.5. Bilan

L'analyse ci-dessus est valable pour φu définie au §2.2 et non pour u. On a donc établi le résultat suivant.

Théorème 5.1. *Soit $u \in H_0^2(\Omega)$ solution de $\Delta^2 u = f$ avec $f \in W_p^{-1}(\Omega)$, alors $u \in W^{3,p}(\omega)$ où $\omega = \Omega \cap \{x < a'\}$ avec $a' < a$ arbitraire.*

Remarque 5.2. On a en plus la précision de décroissance

$$x^{-3\alpha} u, x^{-2\alpha} \nabla u, x^{-\alpha} D^2 u \in L^p(\Omega) .$$

2.6. Application aux équations de Stokes

Dans le cas des équations linéaires de Stokes on obtient, en considérant la fonction de courant, le

Théorème 6.1. *Pour $f \in L^p(\Omega)^2$ donné, l'unique solution $u \in \left[H_0^1(\Omega)\right]^2$ de*

$$-\Delta u = f - \nabla p \ \textit{avec} \ \operatorname{div} u = 0$$

dans Ω appartient á $\left[W^{2,p}(\omega)\right]^2$.

Remarque 6.2. Dans le cas $p = 2$ la régularité H^2 de u est connue au voisinage des deux coins de Ω car ils sont convexes. On en déduit que globalement $u \in H^2(\Omega)^2 \cap H_0^1(\Omega)^2$.

Remarque 6.3. Par transformation de Fourier partielle ce résultat peut être étendu à un domaine tridimensionnel modèle tel que $\Omega \times \mathbf{T}$ où $\mathbf{T}$ est le tore. On obtient que pour

$$f \in L^2(\Omega \times \mathbf{T})^3$$

il existe

$$u \in H^2(\Omega \times \mathbf{T})^3 \cap H_0^1(\Omega \times \mathbf{T})^3$$

unique solution de $-\Delta u = f - \nabla p$ avec $\operatorname{div} u = 0$.

Références

[1] Agmon–Douglis–Nirenberg, Estimates near the boundary for solutions of elliptic partial differential equations satisfying general boundary conditions, I, *Communications Pure Appl. Math.* **12** (1959), 623–627.

[2] Coifman–Weiss, Transference methods in analysis, *Conference board on Mathematical Sciences, Conference series in Mathematics* **31** (1977), AMS, Providence, Rhode Island, 1–59.

[3] Dore–Venni, On the closedness of the sum of two closed operators, *Math Zeitschrift* **196** (1987), 189–201.

[4] Grisvard, *Elliptic problems in nonsmooth domains*, Monographs and Studies in Mathematics **24** (1985), Pitman.

[5] Ibuki, Dirichlet problem for elliptic equations of the second order in a singular domain of $\mathbb{R}^2$, *Journal Math Kyoto Univ.* **14** (1) (1974), 54–71.

[6] Khelif, Problèmes aux limites pour le laplacien dans un domaine à points cuspides, *CRAS, Paris* **287** (1978), 1113–1116.

[7] Lions–Magenes, *Problèmes aux limites non homogènes et applications*, Dunod, Paris, 1968.

Elliptic Problems in Domains with Edges: Anisotropic Regularity and Anisotropic Finite Element Meshes

Thomas Apel and Serge Nicaise

Abstract

This paper is concerned with the anisotropic singular behaviour of the solution of elliptic boundary value problems in domains with edges and its consequences for anisotropic FEM. We first deal with the description of the analytic properties of the solution in newly defined anisotropic weighted Sobolev spaces. The finite element method with anisotropic, graded meshes and piecewise linear shape functions is then investigated for such problems; the schemes exhibit optimal convergence rates with decreasing mesh size. For the proof, new local interpolation error estimates in anisotropically weighted Sobolev spaces are derived.

1 Motivation and main ideas

1.1 The boundary value problem and analytical results

In this paper we want to study the approximation properties of the finite element method with anisotropic meshes for certain elliptic boundary value problems over three-dimensional domains.

Let $\Omega \subset \mathbf{R}^3$ be a bounded domain with non-intersecting edges. Especially we will focus on prismatic domains

$$\Omega = G \times I, \tag{1.1}$$

where $G \subset \mathbf{R}^2$ is a polygonal domain and $I =]0, z_0[\subset \mathbf{R}^1$ is an interval. The domain G may have a corner with interior angle $\omega > \pi$ at the origin; thus Ω has an edge which is part of the x_3-axis. The case of more than one edge can be treated similarly because the edge singularities we are interested in are of local nature only.

Over this domain Ω, we consider the variational form of the Dirichlet problem

$$\begin{aligned} -\Delta u &= f \quad \text{in } \Omega, & (1.2)\\ u &= 0 \quad \text{on } \partial\Omega, & (1.3)\end{aligned}$$

given by:

$$\textit{find } u \in \overset{\circ}{H}{}^1(\Omega) \textit{ such that } a(u,v) = (f,v) \textit{ for all } v \in \overset{\circ}{H}{}^1(\Omega). \tag{1.4}$$

The bilinear form $a(.,.)$ and the linear form $(f,.)$ are defined by

$$a(u,v) := \int_\Omega \sum_{i,j=1}^{3} \partial_i u \partial_j v \, d\underline{x}, \tag{1.5}$$

$$(f,v) := \int_\Omega f v \, d\underline{x}. \tag{1.6}$$

We use the abbreviations ∂_i for $\frac{\partial}{\partial x_i}$ and ∂_{ij} for $\partial_i \partial_j$. The datum f is supposed to be in $L^p(\Omega)$ $(p \geq 2)$.

It is well known that for domains with edges with interior angle $\omega > \pi$ the so-called shift theorem ($u \in H^2(\Omega)$ for $f \in L^2(\Omega)$) does not hold, and there are many papers where the regularity of the solution of these and more general problems are studied. We mention here the papers of Kondrat'ev [8] and Maz'ya-Plamenevskiĭ [10].

In [8], a representation formula for the solution u for $f \in L^2(\Omega)$ is given:

$$u = \xi(r)\gamma(\underline{x})\, r^\lambda \sin(\lambda\varphi) + u_r \quad \text{with} \quad \lambda = \frac{\pi}{\omega}, \; \gamma \in W_\lambda^{2,2}(\Omega), \tag{1.7}$$

where r, φ are polar coordinates in the plane perpendicular to the edge $u_r \in H^2(\Omega)$ and $\xi(r)$ is a smooth cut-off function. $W_\lambda^{2,2}(\Omega) := \{v \in \mathcal{D}'(\Omega) : r^\lambda D^\alpha v \in L^2(\Omega) \; \forall |\alpha| \leq 2\}$, $\alpha = (\alpha_1, \alpha_2, \alpha_3)$ is a multi-index and $D^\alpha := \partial_1^{\alpha_1} \partial_2^{\alpha_2} \partial_3^{\alpha_3}$.

In [10], the solution is described in the framework of another type of weighted Sobolev spaces: let $f \in L^p(\Omega)$ then

$$u \in V_\beta^{2,p}(\Omega) \quad \text{for } \beta > 2 - \frac{2}{p} - \frac{\pi}{\omega}, \tag{1.8}$$

$V_\beta^{2,p}(\Omega) := \{v \in \mathcal{D}'(\Omega) : r^{\beta-2+|\alpha|} D^\alpha v \in L^p(\Omega) \; \forall |\alpha| \leq 2\}$.

The anisotropic structure of the edge is reflected by the factor r^λ in (1.7) and the weights in the definition of the spaces $W_\lambda^{2,2}(\Omega)$ and $V_\beta^{2,2}(\Omega)$, because r is the distance to the edge and is independent of the tangential coordinate of the edge. Using these results it has been possible to justify a mesh refinement strategy near edges [2, 5, 9] in order to improve the approximation order (which is in general low because of the low regularity of the solution) of the standard finite element method. In this strategy, isotropic elements (that are elements whose ratio of the diameters of the smallest circumscribed and the largest inscribed balls is bounded independently of the mesh size h) are used, and the size of the elements is determined by their distance to the edge.

This result is not really satisfactory because it seems to be natural to treat anisotropic structures like edges with anisotropic finite elements. According to [1] an element is called anisotropic when its diameter in different directions has different asymptotics and, consequently, the ratio of the outer and the inner ball is growing to infinity for $h \to 0$ (i.e., it is non regular in Ciarlet's sense [6]). As shown in that paper for problems with smoother data than we assume here, these elements can be applied successfully in the finite element method with graded meshes near edges.

It was an open problem to justify this anisotropic strategy also for problems with $f \in L^p(\Omega)$ ($p \geq 2$). But the analytic results (1.7) and (1.8) have been insufficient because the weighted Sobolev spaces used have the disadvantage that all derivatives of the same order have the same weight. This drawback is removed in Section 2 by using more appropriate, anisotropically weighted Sobolev spaces. For $p < 6$ it is proved that the solution u of (1.4) satisfies

$$u \in A^{2,p}_\beta(\Omega) \text{ with } \begin{cases} \beta > 2 - \frac{2}{p} - \frac{\pi}{\omega} & \text{for } 2 - \frac{2}{p} \geq \frac{\pi}{\omega} > 1 - \frac{2}{p}, \\ \beta = 0 & \text{for } 2 - \frac{2}{p} < \frac{\pi}{\omega}. \end{cases} \tag{1.9}$$

The space $A^{2,p}_\beta(\Omega)$ is defined by $A^{2,p}_\beta(\Omega) := \{v \in \mathcal{D}'(\Omega) : \|v; A^{2,p}_\beta(\Omega)\| < \infty\}$,

$$\begin{aligned} |v; A^{2,p}_\beta(\Omega)|^p &:= \int_\Omega \{r^{\beta p} \sum_{i,j=1}^{2} |\partial_{ij}u|^p + \sum_{i=1}^{3} |\partial_{3i}u|^p\}\, d\underline{x}, \\ \|v; A^{2,p}_\beta(\Omega)\|^p &:= |v; A^{2,p}_\beta(\Omega)|^p \\ &+ \int_\Omega \{r^{(\beta-1)p} \sum_{i=1}^{2} |\partial_i u|^p + r^{-p}|\partial_3 u|^p + r^{(\beta-2)p}|u|^p\} d\underline{x}, \end{aligned}$$

and x_3 is the direction of the edge. Particularly, that means $\partial_3 u \in V^{1,p}_0(\Omega) \hookrightarrow W^{1,p}(\Omega)$. For $\frac{\pi}{\omega} \leq 1 - \frac{2}{p}$ (that means $p \geq 2/(1 - \frac{\pi}{\omega})$) we do not have $\partial_3 u \in W^{1,p}(\Omega)$.

1.2 The class of finite element meshes

Assume that we are given a family $\mathcal{T}$ of finite element partitions $\mathcal{T}_h$ with the usual regularity properties:

(a) $\overline{\Omega} = \bigcup_{i=1}^m \overline{\Omega}_i$, where Ω_i are tetrahedra,

(b) $\Omega_i \cap \Omega_j = \emptyset$ for $i \neq j$ $(i, j = 1, \ldots, m)$,

(c) any edge or face of Ω_i is either a subset of $\partial\Omega$ or an edge or face of another Ω_j $(i, j = 1, \ldots, m)$.

Then we introduce the finite element space V_h of all continuous functions whose restriction to any Ω_i $(i = 1, \ldots, m)$ is a polynomial of first degree. Furthermore, we let V_{0h} be defined by $V_{0h} := \{v_h \in V_h : v_h|_{\partial\Omega} = 0\}$.

The finite element solution of problem (1.4) is defined by:

$$\textit{find } u_h \in V_{0h} \textit{ such that } a(u_h, v_h) = (f, v_h) \textit{ for all } v_h \in V_{0h}. \tag{1.10}$$

The investigation of the finite element error $u - u_h$ in the energy norm (here equivalent to the $W^{1,2}(\Omega)$-norm) is usually reduced via Céa's lemma to a general approximation problem. If we want to take advantage of anisotropic finite element meshes (and this kind of mesh seems to be natural near edges, see above), we need an approximation operator for which error estimates are available that take these different asymptotic mesh sizes of the elements into account. As far as we know, such estimates are only available for the interpolation operator [1]. But in order to use these local estimates and to extend them to weighted Sobolev spaces, the mesh must satisfy two more conditions (d) and (e). Moreover, another assumption (f) is necessary for the extension of these estimates to weighted Sobolev spaces. This extension is done in Subsection 3.1 because it is necessary for our global estimate in Subsection 3.2.

For the explanation and for further use we introduce the following notation: Assume we are given a finite element Ω_i. Let e_i be the longest edge of Ω_i and f_i the larger of the two faces of Ω_i with $e_i \subset \overline{f}_i$. Then we denote by $h_{3,i} := \text{meas}_1(e_i)$ the length of e_i, by $h_{2,i} := 2\,\text{meas}_2(f_i)/h_{3,i}$ the diameter of f_i perpendicularly to e_i and by $h_{1,i} := 6\,\text{meas}_3(\Omega_i)/(h_{2,i}h_{3,i})$ the diameter of Ω_i perpendicularly to f_i. Note that $h_{3,i} \geq h_{2,i} \geq h_{1,i}$.

Introduce further local Cartesian coordinate systems $(x_{1,i}, x_{2,i}, x_{3,i})$ such that $(0,0,0)$ is a vertex of Ω_i, e_i is part of the $x_{3,i}$-axis, and f_i is part of the $x_{2,i}, x_{3,i}$-plane. Note that each coordinate system can be transformed via a translation and three rotations around the $x_{j,i}$-axes by an angle $\psi_{j,i}$ $(j = 1, 2, 3)$ into the original coordinate system (x_1, x_2, x_3). (The angles $\psi_{j,i}$ depend on the order of the three rotations but this influence is of lower order.)

Let the following assumptions also be fulfilled:

(d) all elements Ω_i have to fulfill the maximal angle condition: let $\gamma_{e,i}$ be the maximal angle between faces of Ω_i and $\gamma_{f,i}$ the maximal interior angle of the four triangular faces of Ω_i $(i = 1, \ldots, m)$. Then the relations $\gamma_{e,i} \leq \gamma_0 < \pi$ and $\gamma_{f,i} \leq \gamma_0 < \pi$ have to be fulfilled with γ_0 independent of the element counter i and the mesh size parameter h,

(e) the elements are located such that the angles $\psi_{j,i}$ fulfill the following relations for $i = 1, \ldots, m$:

$$|\tan\psi_{1,i}| \leq C\frac{h_{2,i}}{h_{3,i}}, \; |\tan\psi_{2,i}| \leq C\frac{h_{1,i}}{h_{3,i}}, \; |\tan\psi_{3,i}| \leq C\frac{h_{1,i}}{h_{2,i}},$$

with the exception that the first (respectively the third) inequality is not necessary if $h_{2,i}$ is of order $h_{3,i}$ (respectively $h_{1,i}$ is of order $h_{2,i}$),

(f) all elements Ω_i with distance $r_i = 0$ to the edge (x_3-axis) have two vertices such that the straight line through them is parallel to the x_3-axis.

So we introduce a graded mesh by conditions (a)–(f) and the following choice of the element sizes:

(g) With h being the mesh size parameter, $\mu \in]0,1]$ the grading parameter, r_i the distance of Ω_i to the edge ($r_i := \min_{(x_1,x_2,x_3)\in\overline{\Omega}_i}(x_1^2 + x_2^2)^{1/2}$) and some constant $R > 0$, we define real numbers h_i ($i = 1, \ldots, m$)

$$h_i := \begin{cases} h^{1/\mu} & \text{for } r_i = 0, \\ h r_i^{1-\mu} & \text{for } 0 < r_i \leq R, \\ h & \text{for } r_i > R, \end{cases} \tag{1.11}$$

and assume that there are positive constants C_1 and C_2 such that for the element sizes $h_{1,i}$, $h_{2,i}$, $h_{3,i}$ the relations

$$\begin{array}{rcccl} C_1 h_i & \leq & h_{j,i} & \leq & C_2 h_i, \quad j = 1,2, \\ C_1 h & \leq & h_{3,i} & \leq & C_2 h, \end{array} \tag{1.12}$$

are fulfilled for $i = 1, \ldots, m$.

Let us notice that for domain Ω with the special structure (1.1), it is easy to construct such a mesh, see [4].

1.3 Outline of the paper

In Section 2 we prove the anisotropic regularity (1.9) of the solution u from (1.4). In Section 3 we use the standard way for bounding the finite element error, namely the estimation of the interpolation error. One difficulty is that the anisotropic local interpolation error estimate

$$|v - Iv; W^{1,p}(\Omega_i)| \leq C \sum_{j=1}^{3} h_{j,i} |\partial_j u; W^{1,p}(\Omega_i)| \tag{1.13}$$

does not hold for $p = 2$, but only for $p > 2$. That is why we restrict our consideration to problems with a right hand side $f \in L^p(\Omega)$ with $p > 2$. Another task is to prove an approximation result for elements Ω_i touching the edge, since the solution does not belong to $W^{2,p}(\Omega_i)$ $(p > 2)$ there even if we would assume smooth data. Under the conditions (d)–(f) and certain assumptions on p and β (see Theorem 3.7), we get

$$|v - Iv; W^{1,p}(\Omega_i)| \leq C\,(h_{1,i}^{1-\beta} + h_{3,i})|v; A_\beta^{2,p}(\Omega_i)|, \quad (1.14)$$

$$\|v - Iv; L^p(\Omega_i)\| \leq C\,(h_{1,i}^{2-\beta} + h_{3,i}^2)|v; A_\beta^{2,p}(\Omega_i)|. \quad (1.15)$$

The global result is then formulated in Theorem 3.8 and we get for $p \in]2, (1 - \frac{\pi}{\omega})^{-1}[$

$$\|u - u_h; W^{1,2}(\Omega)\| \leq Ch^s \|f; L^p(\Omega)\| \quad (1.16)$$

with

$$s = \begin{cases} 1 & \text{for } \mu < \frac{\pi}{\omega} \cdot \frac{p}{2p-2}, \\ \frac{2}{p} - 1 + \frac{1}{\mu}\frac{\pi}{\omega} - \varepsilon & \text{for } \mu \geq \frac{\pi}{\omega} \cdot \frac{p}{2p-2}. \end{cases} \quad (1.17)$$

Note that we use the symbol C for a generic positive constant, which means C may be of different value at each occurrence. But C is always independent of the function under consideration and of the finite element mesh.

For a numerical test, we refer to [3], where we derive approximation orders from the finite element errors for different mesh size parameters h and observe a good agreement of the calculated approximation orders with the expected ones (see (1.17)).

This paper is the short version of [3], to which we refer the reader for the details.

2 Anisotropic regularity near the edge

Let $D := C \times \mathbf{R}$ be a dihedral cylinder of $\mathbf{R}^3$, where C is an infinite cone of $\mathbf{R}^2$ of opening ω. As before, we denote by $\underline{x} = (x_1, x_2, x_3)$ the Cartesian coordinates in D, where $x_3 \in \mathbf{R}$ and $(x_1, x_2) \in C$ and by (r, φ) the polar coordinates in C. We are concerned with the edge regularity of the variational solution $v \in \overset{\circ}{H}{}^1(D)$ of the Dirichlet problem

$$-\Delta v = g \in L^p(D), \quad (2.1)$$

for $p \geq 2$. Since we are only interested in the local behaviour of the solution, we suppose that v exists and has a compact support.

For studying the regularity of v, we shall employ some weighted Sobolev spaces of Kondrat'ev type, introduced for instance in [10]. For $l \in \mathbf{N}, \beta \in$

$\mathbf{R}, p \in]1, +\infty[$, we recall that

$$V_\beta^{l,p}(D) := \{v \in \mathcal{D}'(D) : r^{\beta-l+|\alpha|} D^\alpha v \in L^p(D), \forall |\alpha| \le l\}.$$

We start with a weak isotropic regularity result:

Lemma 2.1 *For any $\varepsilon > 0$, we have $v \in V_{1+\varepsilon}^{2,2}(D)$.*

Proof. Using Hardy's inequalities, we find that $v \in V_{-1+\varepsilon}^{0,2}(D)$. This inclusion and Theorem 4.1 of [10] lead to the conclusion. ■

This allows us to use comparison theorems in weighted Sobolev spaces in order to get anisotropic regularity:

Theorem 2.2 *Let $\alpha \ge 0$ be such that*

$$\begin{cases} 2 - \frac{2}{p} - \alpha & < & \frac{\pi}{\omega} & \text{if } 2 - \frac{2}{p} \ge \frac{\pi}{\omega}, \\ \alpha & = & 0 & \text{if } 2 - \frac{2}{p} < \frac{\pi}{\omega}. \end{cases} \tag{2.2}$$

Then

$$v \in V_\alpha^{2,p}(D). \tag{2.3}$$

If moreover

$$1 - \frac{2}{p} < \frac{\pi}{\omega}, \tag{2.4}$$

then

$$\partial_3 v \in V_0^{1,p}(D). \tag{2.5}$$

Proof. The inclusion (2.3) is a direct consequence of Theorem 7.2 of [10], using Lemma 2.1 and since $g \in V_{1+\varepsilon}^{0,2}(D) \cap V_\alpha^{0,p}(D)$.

The inclusion (2.5) follows now from Theorem 3.1 of [11], since the assumption of that theorem is equivalent to (2.4). ■

We shall now improve this theorem using recent results of Grisvard [7]. The obtained inclusions will sometimes recover the above ones, but due to the convenient form of the Laplace operator, their proofs are simpler. Let us first recall Theorem 6.6 of [7]:

Theorem 2.3 *Suppose that* $\frac{j\pi}{\omega} \neq 2 - \frac{2}{p}$ *for all* $j \in \mathbf{Z}$. *Then the solution* $v \in \overset{\circ}{H}{}^1(D)$ *of problem (2.1) admits the decomposition*

$$v = v_r + \sum_{0<\frac{j\pi}{\omega}<2-\frac{2}{p}} (K_j \overset{x_3}{\star} q_j)\, \psi_j, \tag{2.6}$$

where $v_r \in W^{2,p}(D)$ *is the regular part of* v, $q_j \in B^{2-\frac{2}{p}-\frac{j\pi}{\omega},p}(\mathbf{R})$ *(that means in the classical Sobolev space* $W^{2-\frac{2}{p}-\frac{j\pi}{\omega},p}(\mathbf{R})$, *if* $2-\frac{2}{p}-\frac{j\pi}{\omega} \notin \mathbf{Z}$, *otherwise in the Besov space* $B^{2-\frac{2}{p}-\frac{j\pi}{\omega},p}(\mathbf{R})$*),* ψ_j *are the 2D-singular functions of the Laplace operator in* C:

$$\psi_j(r,\varphi) := \xi(r) r^{j\pi/\omega} \sin(j\pi\varphi/\omega), \tag{2.7}$$

and finally K_j *are kernels defined by*

$$K_j(r,x_3) \quad := \quad \frac{r}{\pi(r^2+x_3^2)} \quad \text{if } \frac{j\pi}{\omega} > 1 - \frac{2}{p},$$
$$K_j(r,x_3) \quad := \quad \frac{2r^3}{\pi(r^2+x_3^2)^2} \quad \text{if } \frac{j\pi}{\omega} \leq 1 - \frac{2}{p}.$$

There exists a positive constant C *independent of* g *such that*

$$\|v_r; W^{2,p}(D)\| + \sum_{0<\frac{j\pi}{\omega}<2-\frac{2}{p}} \|q_j; B^{2-\frac{2}{p}-\frac{j\pi}{\omega},p}(\mathbf{R})\| \leq C\, \|g; L^p(\mathbf{R})\|.$$

Here and in the sequel, $K \overset{x_3}{\star} q$ means the convolution with respect to the edge parameter x_3:

$$(K \overset{x_3}{\star} q)(r,x_3) := \int_{\mathbf{R}} K(r,s) q(x_3 - s)\, ds.$$

In view of that theorem, if we want to prove inclusions of type (2.3) or (2.5), it suffices to show that the 3D-singularity function

$$v_j := (K_j \overset{x_3}{\star} q_j)\psi_j \tag{2.8}$$

satisfies such inclusions. Their proofs are based on the next general result concerning convolution with arbitrary kernels, which is inspired from Theorem 6.5 of [7] (notice that this theorem had a different goal).

Theorem 2.4 *Let* $K(r,x_3)$ *be a kernel satisfying*

$$|K(r,x_3)| \leq C \frac{r^\beta}{(r^2+x_3^2)^\gamma}, \forall r > 0, x_3 \in \mathbf{R}, \tag{2.9}$$

with some $C > 0, \beta \in \mathbf{R}, \gamma > \frac{1}{2}$ *and*

$$\int_{\mathbf{R}} K(r,x_3)\, dx_3 = 0. \tag{2.10}$$

For $q \in B^{\sigma,p}(\mathbf{R})$, *with* $\sigma \in]0,1]$, *we set*

$$h(r,x_3) := (K \overset{x_3}{\star} q)(r,x_3). \tag{2.11}$$

If $\sigma < 2\gamma - 1$ *and* $\beta \geq -1 - \frac{2}{p} - \sigma + 2\gamma$, *then there exists a constant* $C_1 > 0$ *(independent of* q*) such that*

$$\left(\int_0^1 \int_{\mathbf{R}} |h(r,x_3)|^p \, r dr dx_3\right)^{1/p} \leq C_1 \|q; B^{\sigma,p}(\mathbf{R})\|. \tag{2.12}$$

Proof. Based on Young's theorem. ■

We are now able to prove some anisotropic regularities:

Theorem 2.5 *If* $0 < \frac{j\pi}{\omega} < 2 - \frac{2}{p}$, *then*

$$\partial_{33} v_j \in L^p(D), \tag{2.13}$$

and there exists a positive constant C *such that*

$$\|\partial_{33} v_j; L^p(D)\| \leq C \, \|g; L^p(\mathbf{R})\|. \tag{2.14}$$

Proof. If $1 - \frac{2}{p} < \frac{j\pi}{\omega} < 2 - \frac{2}{p}$, we use Theorem 2.4, with $K(r,x_3) = r^{j\pi/\omega} \partial_{33} K_j(r,x_3)$, since

$$\partial_{33} v_j = (K \overset{x_3}{\star} q_j)(r,x_3)\, \xi(r) \sin(j\pi\varphi/\omega).$$

Conversely, if $0 < \frac{j\pi}{\omega} \leq 1 - \frac{2}{p}$, then we know that $q_j \in B^{\sigma',p}(\mathbf{R})$, with $\sigma' = 2 - \frac{2}{p} - \frac{j\pi}{\omega} \geq 1$. If $\sigma' = 1$, we use Theorem 2.4 as above with $K(r,x_3) = r^{j\pi/\omega} \partial_{33} K_j(r,x_3)$, $q = q_j \in B^{\sigma,p}(\mathbf{R})$, when $\sigma = \sigma'$. On the contrary, if $\sigma' > 1$, we apply Theorem 2.4 with $K(r,x_3) = r^{j\pi/\omega} \partial_3 K_j(r,x_3)$, $q = \partial_3 q_j \in B^{\sigma,p}(\mathbf{R})$, when $\sigma = \sigma' - 1$. ■

Analogously, we can consider other derivatives:

Theorem 2.6 *If* $0 < \frac{j\pi}{\omega} < 2 - \frac{2}{p}$, *then*

$$\begin{aligned} \partial_3 v_j &\in L^p(D), & (2.15)\\ v_j &\in L^p(D), & (2.16) \end{aligned}$$

with norms depending continuously on the L^p*-norm of* g. *If moreover,* $1 - \frac{2}{p} < \frac{j\pi}{\omega}$, *then*

$$\begin{aligned} \partial_{13} v_j, \partial_{23} v_j &\in L^p(D), & (2.17)\\ r^{\gamma-1}\partial_1 v_j, r^{\gamma-1}\partial_2 v_j &\in L^p(D), & (2.18)\\ r^{\gamma-2} v_j &\in L^p(D), & (2.19)\\ r^{-1}\partial_3 v_j &\in L^p(D), & (2.20) \end{aligned}$$

with $\gamma > 2 - \frac{2}{p} - \frac{j\pi}{\omega}$, *the norms depending continuously on the* L^p*-norm of* g.

Proof. Properties (2.15), (2.17) and (2.20) follow from Theorem 2.4. To establish the other inclusions, we can no more apply Theorem 2.4 because the corresponding kernels do not satisfy (2.10). Therefore, we proceed as follows: for the derivative $\partial_r v_j$, for instance, we must consider a term of the form

$$h(r, x_3) := (\partial_r K_j) \overset{x_3}{\star} q_j\, \psi_j.$$

As $|\partial_r K_j| = \mathcal{O}(\frac{1}{r^2 + x_3^2})$, we get

$$\begin{aligned} \|h(r, x_3); L^p_{x_3}(\mathbf{R})\| &\le Cr^{j\pi/\omega} \int_{\mathbf{R}} |\partial_r K_j(r, s)|\ \|q_j(x_3 - s); L^p_{x_3}(\mathbf{R})\|\, ds\\ &\le Cr^{-1+j\pi/\omega}\|q_j; L^p(\mathbf{R})\|. \end{aligned}$$

Multiplying this estimate by $r^{\gamma-1}$ and integrating with respect to r on $]0, 1[$ and to φ, we obtain the inclusion $r^{\gamma-1}h \in L^p(D)$. Other terms are treated analogously. ■

Corollary 2.7 *Let* $u \in \overset{\circ}{H}{}^1(\Omega)$ *be the solution of* $-\Delta u = f$, *with* $f \in L^p(\Omega)$. *Then*

$$u \in A^{2,p}_\beta(\Omega) \text{ with } \begin{cases} \beta > 2 - \frac{2}{p} - \frac{\pi}{\omega} & \text{for } 2 - \frac{2}{p} \ge \frac{\pi}{\omega} > 1 - \frac{2}{p},\\ \beta = 0 & \text{for } 2 - \frac{2}{p} < \frac{\pi}{\omega}, \end{cases}$$

and

$$\|u; A^{2,p}_\beta(\Omega)\| \le C\|f; L^p(\Omega)\|.$$

3 Interpolation error estimates

3.1 Local error estimates in weighted Sobolev spaces

As introduced in Subsection 1.2, we are interested in local approximation error estimates for anisotropic elements. In [1], interpolation error estimates in classical Sobolev spaces were derived. These are useful far from the edge, but unfortunately, we can not apply them for tetrahedrons along the edge. In this subsection, we shall extend these results of [1] to weighted Sobolev spaces and consider particularly the three-dimensional case. We remark that interpolation error estimates for functions from weighted Sobolev spaces were already proved in [13] for the two-dimensional isotropic case.

We consider first estimates on a reference element $\Omega_0 \in \mathcal{R}$ where $\mathcal{R}$ is a finite set of reference elements, whose elements have the following essential property (P):

(P) For each axis x_i ($i = 1, 2, 3$) of the coordinate system there exists one edge E_i of the reference element, which is parallel to this axis and, for normalization, which has length $\text{meas}_1(E_i) = 1$.

Using a similar notation as in [1, §2] we denote by P a space of polynomials, and since each monomial $x^\alpha = x_1^{\alpha_1} x_2^{\alpha_2} x_3^{\alpha_3}$ can be identified with the multi-index $\alpha \in \mathbf{N}^3$, we also identify P with the corresponding set of multi-indices. The hull $\overline{P}$ of P is the set $\overline{P} := P \cup \{\alpha + e_i : \alpha \in P,\ i = 1, 2, 3\}$ ($\{e_i\}_{i=1}^3$ denotes the canonical basis of $\mathbf{R}^3$) and the boundary $\partial\overline{P}$ of P is the set $\overline{P} \setminus P$. Note that $\max_{\alpha\in\overline{P}} |\alpha| = 1 + \max_{\alpha\in P} |\alpha|$.

We introduce now anisotropic weighted Sobolev spaces on Ω_0: For a finite set $P \subset \mathbf{N}^3$ with $0 \in P$ and for $\beta \in \mathbf{R}$ we set

$$V_\beta^{P,p}(\Omega_0) := \{v \in \mathcal{D}'(\Omega_0) : \|v; V_\beta^{P,p}(\Omega_0)\| < \infty\},$$

where

$$\|v; V_\beta^{P,p}(\Omega_0)\|^p := \sum_{\alpha\in P} \int_{\Omega_0} |r^{\beta-k+|\alpha|} D^\alpha v|^p \, d\underline{x},$$

$k := \max_{\alpha\in P} |\alpha|$, $D^\alpha := \partial_1^{\alpha_1} \partial_2^{\alpha_2} \partial_3^{\alpha_3}$, and $r(\underline{x}) := (x_1^2 + x_2^2)^{1/2}$. For $v \in V_\beta^{\overline{P},p}(\Omega_0)$ we also introduce the seminorm

$$|v; V_\beta^{\overline{P},p}(\Omega_0)|^p := \sum_{\alpha\in\partial\overline{P}} \int_{\Omega_0} |r^{\beta-k-1+|\alpha|} D^\alpha v|^p \, d\underline{x}.$$

Lemma 3.1 *Let $P \in \mathbf{N}^3$, P finite with $0 \in P$. Then we have the compact embedding*

$$V_\beta^{\overline{P},p}(\Omega_0) \overset{c}{\hookrightarrow} V_\beta^{P,p}(\Omega_0).$$

Proof. Use the compact embedding $W^{1,p}(\Omega_0) \overset{c}{\hookrightarrow} L^p(\Omega_0)$ (Rellich–Kondrašov theorem) and the fact that $v \in V_\beta^{\overline{P},p}(\Omega_0)$ satisfies $r^{\beta-k+|\alpha|}D^\alpha v \in W^{1,p}(\Omega_0)$, for any fixed $\alpha \in P$. ∎

With the help of Hölder's inequality, we can show that, under some condition on β, elements of $V_\beta^{P,p}(\Omega_0)$ are in $L^1(\Omega_0)$, as well as all derivatives with respect to P:

Lemma 3.2 *Let $P \subset \mathbf{N}^3$, P finite with $0 \in P$. If $\beta < 2 - \frac{2}{p}$ then for all $v \in V_\beta^{P,p}(\Omega_0)$ and all $\alpha \in P$, $D^\alpha v \in L^1(\Omega_0)$.*

From Lemmas 3.1 and 3.2 and using the same arguments as in [1, Lemma 2], we obtain the following lemma.

Lemma 3.3 *Let $P \in \mathbf{N}^3$ be a finite set of multi-indices with $0 \in P$. If $\beta < 2 - \frac{2}{p}$ then*

$$\|v; V_\beta^{\overline{P},p}(\Omega_0)\| \le C|v; V_\beta^{\overline{P},p}(\Omega_0)| \tag{3.1}$$

for all $v \in V_\beta^{\overline{P},p}(\Omega_0)$ satisfying $\int_{\Omega_0} D^\alpha v \, d\underline{x} = 0$ for $\alpha \in P$.

We are now ready to give the interpolation estimate, first in a very general form, and then especially for our purposes.

Lemma 3.4 *Let $\beta < 2 - \frac{2}{p}$ be a real number, and let $P, Q \subset \mathbf{N}^3$ and $\gamma \in \mathbf{N}^3$ be such that $0 \in Q$ and $Q + \gamma \subset P$. Further, introduce a linear operator $I : C^\mu(\Omega_0) \to P$, $\mu \in \mathbf{N}$, and assume that there are linear functionals $F_i \in \left(V_\beta^{\overline{Q},p}(\Omega_0)\right)'$, $i = 1, \dots, j$, $j = \dim D^\gamma P$, satisfying*

$$F_i(D^\gamma I v) = F_i(D^\gamma v) \quad (i = 1, \dots, j) \text{ for all } v \in C^\mu(\Omega_0) \cap V_\beta^{\overline{Q}+\gamma,p}(\Omega_0),$$
$$F_i(D^\gamma q) = 0 \text{ for all } i = 1, \dots, j \Longrightarrow D^\gamma q = 0 \text{ for all } q \in P.$$

Then for all $v \in C^\mu(\Omega_0) \cap V_\beta^{\overline{Q}+\gamma,p}(\Omega_0)$, there holds

$$\|D^\gamma(v - Iv); V_\beta^{\overline{Q},p}(\Omega_0)\| \le C|D^\gamma v; V_\beta^{\overline{Q},p}(\Omega_0)|.$$

Proof. We follow the proof of Lemma 3 of [1], since Lemma 1 of [1] can be extended to the spaces $V_\beta^{P,p}(\Omega_0)$ (due to Lemma 3.2), while Lemma 2 of [1] is replaced by Lemma 3.3. ∎

Theorem 3.5 *Suppose that $0 \le \beta < 1 - \frac{1}{p}$, $p > 2$, and let Iv be the linear Lagrange interpolant of v with respect to the vertices. Then for all $v \in A^{2,p}_\beta(\Omega_0) \cap C(\Omega_0)$ we have*

$$\begin{aligned}&\|r^{\beta-1}\partial_i(v - Iv); L^p(\Omega_0)\| \le \\ &\le C\{\int_{\Omega_0} \left[r^{p\beta}\left(|\partial_{1i}v|^p + |\partial_{2i}v|^p\right) + |\partial_{3i}v|^p\right] d\underline{x}\}^{1/p},\ i = 1, 2,\end{aligned} \tag{3.2}$$

$$\|r^{-1}\partial_3(v - Iv); L^p(\Omega_0)\| \le C\{\int_{\Omega_0} (\sum_{j=1}^{3} |\partial_{j3}v|^p)\, d\underline{x}\}^{1/p}. \tag{3.3}$$

Proof. We set $Q := \{(0,0,0)\}$, $\overline{Q} := \{(0,0,0)\} \cup \{e_i\}_{i=1,2,3}$ and remark that $v \in A^{2,p}_\beta(\Omega_0)$ implies $\partial_i v \in V^{1,p}_\beta(\Omega_0) = V^{Q,p}_\beta(\Omega_0)$ $(i = 1, 2)$ and $\partial_3 v \in V^{1,p}_0(\Omega_0) = V^{\overline{Q},p}_0(\Omega_0)$. To prove the assertion we apply Lemma 3.4 with $P = \overline{Q}$, $\gamma := e_i$ and $F_1(v) := \int_{E_i} v\, dx_i$, where E_i is that edge of Ω_0 which is parallel to the x_i-axis (see Property (P)). The condition on p yields the continuity of F_1. ■

Lemma 3.4 can also be applied to prove an L^p-estimate:

Theorem 3.6 *Suppose that $0 \le \beta < 2 - \frac{3}{p}$, $p \ge 1$, and let Iv be the linear Lagrange interpolant of v with respect to the vertices. Then for all $v \in A^{2,p}_\beta(\Omega_0)$ we have*

$$\|r^{\beta-2}(v - Iv); L^p(\Omega_0)\| \le C\,|v; V^{2,p}_\beta(\Omega_0)|. \tag{3.4}$$

Now we transform these estimates to the actual finite elements Ω_i. Using an affine linear transformation and the properties (d)-(f) on the mesh, one can prove

Theorem 3.7 *Let $I_h v$ be the linear Lagrange interpolant of $v \in A^{2,p}_\beta(\Omega_i)$ with respect to the vertices. Assume that the element Ω_i has at least one common point with the edge. Then for $0 \le \beta < 2 - \frac{3}{p}$, $p \ge 1$, the following local interpolation error estimate holds:*

$$\|v - I_h v; L^p(\Omega_i)\| \le C\,(h^{2-\beta}_{1,i} + h^2_{3,i})|v; A^{2,p}_\beta(\Omega_i)|. \tag{3.5}$$

Moreover, if $0 \le \beta < 1 - \frac{1}{p}$, $p > 2$, then for all $v \in A^{2,p}_\beta(\Omega_i)$ the norm of the derivatives of the interpolation error can be estimated by

$$|v - I_h v; W^{1,p}(\Omega_i)| \le C\,(h^{1-\beta}_{1,i} + h_{3,i})|v; A^{2,p}_\beta(\Omega_i)|. \tag{3.6}$$

3.2 Global error estimates

In this section, we investigate the global interpolation error, that is the difference between the solution u of our boundary value problem (1.4) and its piecewise linear interpolant $I_h u$ on the family of anisotropic graded meshes introduced in Subsection 1.2. The difficulty is that we are interested on the one hand in an estimate in the energy norm which is equivalent to $\| \,.\,; W^{1,2}(\Omega)\|$, in order to apply Céa's lemma for the finite element error. But on the other hand the local interpolation error estimate (3.6) is valid for $\| \,.\,; W^{1,p}(\Omega_i)\|$ with $p > 2$ only.

Theorem 3.8 *Let u be the solution of the boundary value problem (1.4) and $2 < p < (1 - \frac{\pi}{\omega})^{-1}$. Then for the interpolation error $u - I_h u$, I_h defined on the family of meshes in Subsection 1.2, the following estimate holds:*

$$|u - I_h u; W^{1,2}(\Omega)| \leq Ch^s \|f; L^p(\Omega)|,$$

with s given by (1.17).

Proof. We reduce the estimation of the global error to the evaluation of the local errors and distinguish between the $m_0 = \mathcal{O}(h^{-1})$ elements whose closure has at least one common point with the edge, and the $m - m_0 = \mathcal{O}(h^{-3})$ elements away from the edge:

$$\begin{aligned} |u - I_h u; W^{1,2}(\Omega)|^2 &= \sum_{i=1}^{m_0} |u - I_h u; W^{1,2}(\Omega_i)|^2 \\ &+ \sum_{i=m_0+1}^{m} |u - I_h u; W^{1,2}(\Omega_i)|^2. \end{aligned} \tag{3.7}$$

For the elements in the first sum we apply the local estimate (3.6). Using Hölder's inequality, we have for $i = 1, \ldots, m_0$

$$\begin{aligned} |u - I_h u; W^{1,2}(\Omega_i)|^p &\leq (\operatorname{meas} \Omega_i)^{-1+p/2} |u - I_h u; W^{1,p}(\Omega_i)|^p \\ &\leq C(hh_i^2)^{-1+p/2}(h_i^{1-\beta} + h)^p |u; A_\beta^{2,p}(\Omega_i)|^p. \end{aligned}$$

Summing up these estimates for all $i = 1, \ldots, m_0$, and using again Hölder's inequality, we can conclude

$$\begin{aligned} \sum_{i=1}^{m_0} |u - I_h u; W^{1,2}(\Omega_i)|^2 &\leq m_0^{1-2/p} \left(\sum_{i=1}^{m_0} |u - I_h u; W^{1,p}(\Omega_i)|^p \right)^{2/p} \\ &\leq C \sum_{i=1}^{m_0} h^{-1+2/p} (hh_i^2)^{1-2/p} (h_i^{1-\beta} + h)^2 |u; A_\beta^{2,p}(\Omega_i)|^2 \\ &\leq C \left(h^{(2-\beta-2/p)/\mu} + h^{1+(1-2/p)/\mu} \right)^2 \|f; L^p(\Omega)\|^2. \end{aligned}$$

Since for $\beta = \max\{0;\ 2-\frac{2}{p}-\frac{\pi}{\omega}+\varepsilon'\}$ there holds $\frac{1}{\mu}(2-\frac{2}{p}-\beta) > s$, and we have directly $1+\frac{1}{\mu}(1-\frac{2}{p}) > 1 \geq s$ (with s from (1.17)), we get

$$\sum_{i=1}^{m_0} |u - I_h u; W^{1,2}(\Omega_i)|^2 \leq C h^{2s} \|f; L^p(\Omega)\|^2. \tag{3.8}$$

For the elements in the second sum of (3.7) we can use that $u \in W^{2,p}(\Omega_i)$, $i = m_0+1,\ldots,m$, and thus apply the local estimates (3.2)–(3.3) with $\beta = 0$. Again with Hölder's inequality, we have for $i = m_0+1,\ldots,m$:

$$\begin{aligned} |u - I_h u; W^{1,2}(\Omega_i)|^p &\leq (\operatorname{meas}\Omega_i)^{-1+p/2} |u - I_h u; W^{1,p}(\Omega_i)|^p \\ &\leq C(hh_i^2)^{-1+p/2}(h_i^p \sum_{\substack{|\alpha|=2\\ \alpha_3=0}} \|D^\alpha u; L^p(\Omega_i)\|^p + h^p \sum_{\substack{|\alpha|=2\\ \alpha_3>0}} \|D^\alpha u; L^p(\Omega_i)\|^p). \end{aligned} \tag{3.9}$$

Using the refinement conditions (1.11)–(1.12), (3.9) may be transformed into

$$|u - I_h u; W^{1,2}(\Omega_i)|^p \leq C h^{ps+3(p-2)/2} \|u; A^{2,p}_\beta(\Omega_i)\|^p$$

with s from (1.17). Summing up these estimates for all $i = m_0+1,\ldots,m$, and using again Hölder's inequality, we can conclude with Corollary 2.7

$$\sum_{i=m_0+1}^{m} |u - I_h u; W^{1,2}(\Omega_i)|^2 \tag{3.10}$$

$$\leq (m-m_0)^{1-\frac{2}{p}} (\sum_{i=m_0+1}^{m} |u - I_h u; W^{1,2}(\Omega_i)|^p)^{2/p} \tag{3.11}$$

$$\leq C h^{-3(1-\frac{2}{p})} h^{\frac{2}{p}(ps+\frac{3}{2}(p-2))} \|u; A^{2,p}_\beta(\Omega)\|^2$$

$$= C h^{2s} \|f; L^p(\Omega)\|^2. \tag{3.12}$$

From (3.8) and (3.12) we get

$$|u - I_h u; W^{1,2}(\Omega)| \leq C h^s \|f; L^p(\Omega)\|. \tag{3.13}$$

■

Corollary 3.9 *Let u be the solution of the boundary value problem (1.4), $2 < p < (1-\frac{\pi}{\omega})^{-1}$, and let u_h be the finite element solution of (1.10), using a family of meshes as defined in Subsection 1.2. Then the error estimate*

$$\|u - u_h; W^{1,2}(\Omega)\| \leq C h^s \|f; L^p(\Omega)\|$$

holds, with s from (1.17).

Acknowledgement. The first author was supported by DFG (German Research Foundation), No. La 767-3/1.

References

[1] Th. Apel and M. Dobrowolski, Anisotropic interpolation with applications to the finite element method. *Computing*, 47:277–293, 1992.

[2] Th. Apel and B. Heinrich, Mesh refinement and windowing near edges for some elliptic problem. *SIAM J. Numer. Anal.*, 31:695–708, 1994.

[3] Th. Apel and S. Nicaise., Elliptic problems in domains with edges: Anisotropic regularity and anisotropic finite element meshes. Preprint SPC 94_16, TU Chemnitz-Zwickau, 1994.

[4] Th. Apel, R. Mücke, and J. R. Whiteman, An adaptive finite element technique with a-priori mesh grading. Technical Report 9, BICOM Institute of Computational Mathematics, 1993.

[5] Th. Apel, A.-M. Sändig, and J. R. Whiteman, Graded mesh refinement and error estimates for finite element solutions of elliptic boundary value problems in non-smooth domains. Technical Report 12, BICOM Institute of Computational Mathematics, 1993. To appear in Math. Meth. Appl. Sci.

[6] P. Ciarlet, *The finite element method for elliptic problems.* North-Holland Publishing Company, Amsterdam, 1978.

[7] P. Grisvard, Singular behaviour of elliptic problems in non Hilbertian Sobolev spaces. *J. Math. Pures Appl.*, 74:3–33, 1995.

[8] V. A. Kondrat'ev, Singularities of the solution of the Dirichlet problem for a second order elliptic equation in the neighbourhood of an edge. *Differencial'nye uravnenija*, 13(11):2026–2032, 1977. (Russian).

[9] M. S. Lubuma and S. Nicaise, Dirichlet problems in polyhedral domains II: approximation by FEM and BEM. *J. Comp. Appl. Math.*, 61:13–27, 1995.

[10] V. G. Maz'ya and B. A. Plamenevskiĭ, L_p-estimates of solutions of elliptic boundary value problems in domains with edges. *Trudy Moskov. Mat. Obshch.*, 37:49–93, 1978. (Russian).

[11] V. G. Maz'ya and J. Roßmann, Über die Asymptotik der Lösung elliptischer Randwertaufgaben in der Umgebung von Kanten. *Math. Nachr.*, 138:27–53, 1988.

[12] L. A. Oganesyan and L. A. Rukhovets, *Variational-difference methods for the solution of elliptic equations.* Izd. Akad. Nauk Armyanskoi SSR, Jerevan, 1979. (Russian).

[13] G. Raugel, *Résolution numérique de problèmes elliptiques dans des domaines avec coins.* PhD thesis, Université de Rennes (France), 1978.

Thomas Apel
Technische Universität Chemnitz-Zwickau
Fakultät für Mathematik
D–09107 Chemnitz, Germany
e-mail: na.apel@na-net.ornl.gov

Serge Nicaise
Université de Valenciennes et du Hainaut Cambrésis
LIMAV and URA D 751 CNRS "GAT"
Institut des Sciences et Techniques de Valenciennes
B.P. 311, F–59304 - Valenciennes Cedex, France
e-mail: snicaise@univ-valenciennes.fr

Unique Continuation of Harmonic Functions at Boundary Points and Applications to Problems in Complex Analysis

M.S. Baouendi and Linda Preiss Rothschild

Dedicated to the memory of Pierre Grisvard

This is a report on some recent results obtained by the authors on unique continuation at boundary points for harmonic functions satisfying appropriate local sign conditions on the boundary. The authors' investigation was initially motivated by some questions in several complex variables. After we present the main results we shall indicate the connection to these questions. We will also mention some open related problems. We consider an open set Ω in $\mathbb{R}^n$ and a boundary point x_0. We need to assume that, near x_0, the boundary of Ω is either a piece of a hyperplane or a piece of a sphere. We state our results in the latter case.

Let B be the open unit ball in the Euclidean space $\mathbb{R}^n$ centered at 0, and S its boundary, the unit sphere. Let $\mathcal{O}$ be an open neighborhood in $\mathbb{R}^n$ of a point x_0 in S. Put $\Omega = \mathcal{O} \cap B$ and $V = \mathcal{O} \cap S$. We shall assume that Ω is connected. A continuous function u in Ω *vanishes of infinite order at* x_0 if for every positive integer N

$$\lim_{\substack{x\in\Omega \\ x\to x_0}} \frac{u(x)}{|x-x_0|^N} = 0.$$

Similarly u *vanishes of infinite order in the normal direction at* x_0 if for every N

$$\lim_{\substack{0<t<1 \\ t\to 1}} \frac{u(tx_0)}{(1-t)^N} = 0.$$

We now state our main results.

Theorem 1. *Let $u \in C^0(\bar{\Omega})$ be harmonic in Ω and satisfy the following conditions.*

(1) *There exists $P(x)$, a nontrivial homogeneous polynomial in n variables, such that $P(x + x_0) = P(x)$ for all x in $\mathbb{R}^n$, and $P(x)u(x) \geq 0$ for $x \in V$.*

(2) *For every positive N, the function $x \mapsto |u(x)||x - x_0|^{-N}$ is integrable on V.*

The authors were partially supported by National Science Foundation Grant DMS 9203973

(3) *For every multi-index α in n variables with $|\alpha| \leq d$, where d is the degree of $P(x)$, the function $\partial^\alpha u(x)$ vanishes of infinite order in the normal direction at x_0.*

Then $u(x) \equiv 0$ in a neighborhood of x_0 in V, and hence u extends as a real analytic function in a neighborhood of x_0 in $\mathbb{R}^n$.

The following unique continuation result is an immediate consequence of Theorem 1.

Corollary 1. *Let $u \in C^0(\bar{\Omega})$ be harmonic in Ω and satisfy the following conditions.*

(i) *There exists $P(x)$, a nontrivial homogeneous polynomial in n variables, such that $P(x + x_0) = P(x)$ for all x in $\mathbb{R}^n$, and $P(x)u(x) \geq 0$ for $x \in V$.*

(ii) *For every multi-index α in n variables with $|\alpha| \leq d$, where d is the degree of $P(x)$, the function $\partial^\alpha u(x)$ vanishes of infinite order at x_0.*

Then $u(x) \equiv 0$ in Ω.

Note that in Corollary 1 if we take $P(x) \equiv 1$ and replace condition (i) by the stronger condition $u(x) \geq 0$ for $x \in \overline{\Omega}$ then the conclusion of the corollary follows from the classical Hopf lemma (see e.g. [13]). Similarly if the condition (i) is replaced by $u|_V \equiv 0$, then the conclusion follows immediately from the classical local real analyticity of u.

Theorem 1 was first proved for the case $P(x) \equiv 1$ by the authors in [5]. For boundary points in a piece of a hyperplane, and arbitrary homogeneous polynomial $P(x)$, Theorem 1 can be found in [6]. We indicate here how the result in [6] implies Theorem 1 stated above and conversely. Consider the Kelvin transform defined by

$$v(x) = \frac{1}{|x|^{n-2}} u\left(\frac{x}{|x|^2}\right).$$

If u is harmonic in an open neighborhood of the point $p_0 = (1, 0, \ldots, 0)$ in the ball defined by $(x_1 - 1/2)^2 + x_2^2 + \ldots + x_n^2 \leq 1/4$, then $v(x)$ is harmonic in an open neighborhood of the point p_0 in the half space $\{x_1 \geq 1\}$. Hence local results, similar to those stated in Theorem 1 and Corollary 1, for boundary points contained in an open subset of a sphere can easily be translated to boundary points contained in an open subset of a hyperplane and vice versa. We thank Jean-Michel Bony for suggesting to us the use of the Kelvin transform. The proof in [6] is based on an asymptotic expansion in the normal direction of harmonic functions in a half plane and makes explicit use of the Poisson integral for harmonic functions in a half space. We refer the reader to this work for the details of the proof.

We consider now the case $n = 2$. Let U be an open neighborhood of 0 in the complex plane $\mathbb{C}$. Denote by U_+ (resp. $\overline{U}_+$) the set of points $z = x + iy \in U$ with $y > 0$ (resp. $y \geq 0$) and assume that U_+ is connected. The following results are easy consequences of Theorem 1 and Corollary 1 above.

Corollary 2. *Let $u \in C^0(\overline{U}_+)$ be harmonic in U_+ and vanishing of infinite order at* 0. *Then $u \equiv 0$ in U_+ if one of the following conditions holds.*

(a) *$x \mapsto u(x, 0)$ does not change sign in a neighborhood of* 0 *in* $\mathbb{R}$.
(b) *$x \mapsto xu(x, 0)$ does not change sign in a neighborhood of* 0 *in* $\mathbb{R}$, *and* $y \mapsto \frac{\partial u}{\partial x}(0, y)$ *vanishes of infinite order at* $y = 0$.
(c) *$x \mapsto xu(x, 0)$ does not change sign in a neighborhood of* 0 *in* $\mathbb{R}$ *and* $y \mapsto v(0, y)$ *vanishes of infinite order at* $y = 0$, *where $v(x, y)$ is a harmonic conjugate of u near* 0 *in* U_+.

The following unique continuation result in complex analysis is an immediate consequence of Corollary 2.

Corollary 3. *Let h be a holomorphic function in U_+, vanishing of infinite order at* 0. *If $u = \Re h$ is continuous in $\overline{U}_+$ and the function $x \mapsto u(x, 0)$ does not change sign in a neighborhood of* 0, *except perhaps at* 0, *then* $h \equiv 0$.

We state other applications to problems in one and several complex variables.

Theorem 2. *Let $U \subset \mathbb{C}$ be a connected open neighborhood of* 0. *Suppose that $H : \overline{U}_+ \to \mathbb{C}^n$ is continuous, holomorphic in U_+, and vanishes of infinite order at* 0. *If H satisfies*

$$H(U \cap \mathbb{R}) \subset \{z \in \mathbb{C}^n : |\Im z| \leq \beta |\Re z|\} \tag{1}$$

for some β with $0 \leq \beta < 1$, then $H \equiv 0$. If $n = 1$ the conclusion holds for $\beta = 1$ also.

Proof. We write $H = (H_1, \ldots, H_n)$ with $H_j = u_j + iv_j$, where the u_j and v_j are real. By the hypothesis, we have

$$\sum_j v_j(x, 0)^2 \leq \beta^2 \sum_j u_j(x, 0)^2, \quad x \in U \cap \mathbb{R}.$$

Then the function $h = H_1^2 + \ldots + H_n^2$ satisfies the hypotheses of Corollary 3, since $\Re h = \sum_j u_j(x, 0)^2 - \sum_j v_j(x, 0)^2 \geq 0$. We may then conclude $h \equiv 0$, which implies in particular, $\sum_j u_j(x, 0)^2 = \sum_j v_j(x, 0)^2$. Since $\beta < 1$, we obtain $\sum_j u_j(x, 0)^2 = 0$, which implies $u_j(x, 0) = v_j(x, 0) = 0$ for all j, i.e. $H \equiv 0$.

If $n = 1$ and $\beta = 1$, then we note that $H^2 = u^2 - v^2 + 2iuv$ satisfies the hypothesis of Corollary 1, and we conclude immediately that $H \equiv 0$. ∎

Theorem 3. *If H is as in Theorem 2, but with* (1) *replaced by*

$$H(U \cap \mathbb{R}) \subset M, \tag{2}$$

where M is a C^1 totally real submanifold of $\mathbb{C}^n$, then $H \equiv 0$.

Note that any C^1 curve in $\mathbb{C}^n$ is a totally real submanifold. If M is a totally real submanifold of dimension r containing the origin, we may find local holomorphic coordinates $z = (z', z'') \in \mathbb{C}^r \times \mathbb{C}^{n-r}$ such that M is given near 0 by

$$\Im z' = \phi(\Re z'), \quad z'' = \psi(\Re z'),$$

with ϕ and ψ of class C^1 defined near 0 in $\mathbb{R}^r$, with $\phi(0) = 0$, $\psi(0) = 0$, $d\phi(0) = 0$, $d\psi(0) = 0$. It then follows immediately that near 0, M is contained in a set of the form (1) with $\beta < 1$, hence Theorem 3 follows from Theorem 2.

Theorem 3 was first proved with more regularity assumptions on the manifold M in a joint work with S. Alinhac [3]; the version stated above is in [4]. (See also Bell and Lempert [8], Alexander [1], [2], Huang and Krantz [10].)

Noting that the manifold M in Theorem 3 can be replaced by a finite union of C^1 curves (see [4] for details), we obtain the following generalized Schwarz reflection principle in one complex variable.

Theorem 4. *Let h be a continuous function in $\overline{U}_+$, holomorphic in U_+ and assume that $h(U \cap \mathbb{R})$ is contained in a real analytic subset of $\mathbb{C}$.*

(a) *If h vanishes of infinite order at 0 then, $h \equiv 0$.*
(b) *If h is smooth up to the boundary in $\overline{U}_+$, then h extends holomorphically to an open neighborhood of $\overline{U}_+$ in $\mathbb{C}$.*

Note that the conclusion of Theorem 5 need not hold if h is only assumed to be of class C^k, even if k is taken to be large.

We now describe another type of related problems encountered in several complex variables. A *smooth analytic disc valued in* $\mathbb{C}^n$ is a map $A : \overline{\Delta} \to \mathbb{C}^n$, where Δ denotes the unit disc in $\mathbb{C}$, with A smooth in $\overline{\Delta}$ and holomorphic in Δ. We shall say that the disc A is *attached* to a manifold $M \subset \mathbb{C}^n$ if $A(S) \subset M$, where $S = \partial\Delta$. Similarly, A is *partially attached* to M if $A(I) \subset M$, where $I \subset S$ is an open arc with $1 \in I$. It should be mentioned that starting with the work of Bishop [9] and Lewy [12], analytic discs attached to hypersurfaces and their boundary derivatives have played an important role in local and global CR geometry for such problems as holomorphic extension of CR functions and propagation of holomorphic extension.

It follows from Theorem 3 that if A is partially attached to a totally real manifold M and flat at 1, i.e. $A^{(j)}(1) = 0, j = 1, 2, \ldots$, then A is constant. A natural question is the following: Can one assume M to be a smooth real hypersurface in $\mathbb{C}^n$ and obtain the same unique continuation result? For instance, using the classical Hopf lemma, it is well known that if M is a strictly pseudoconvex hypersurface and A is a sufficiently small nonconstant analytic disc attached to M, then $A'(1) \neq 0$. For simplicity we restrict ourselves to hypersurfaces in $\mathbb{C}^2$ given by

$$\Im z_2 = p(z_1, \bar{z}_1), \tag{3}$$

where

$$p(u,\bar{u}) = \sum_{2\le j+k\le m} a_{jk} u^j \bar{u}^k, \quad a_{jk} = \bar{a}_{kj}, \tag{4}$$

is a real-valued polynomial defined for $u \in \mathbb{C}$.

If $p(z_1, \bar{z}_1)$ given by (3) is non trivial and non negative, then it follows from Corollary 3 that any disc A flat at 1 and attached to M with $A(1) = 0$ must be constant. It has been an open question whether there can exist a smooth nonconstant analytic disc A attached to a hypersurface of finite type in $\mathbb{C}^2$ in the sense of Kohn [11], with A flat at 1. In [7] the following result is proved.

Theorem 5. *Let M be a hypersurface in $\mathbb{C}^2$ given by (3), and assume that $a_{jk} = 0$ whenever $j = k$ in (4). Then given $\epsilon > 0$ there are nontrivial smooth analytic discs A attached to M with $A(1) = 0$, flat at 1, and satisfying $|A(\zeta)| < \epsilon$ for all $\zeta \in \overline{\Delta}$.*

We close with some open problems.

Problem 1. To what extent can Theorem 1 and Corollary 1 be generalized for harmonic functions for more general domains, say with real analytic boundary? Recent work of Shapiro [14] and Shklover [15] are relevant. Can Corollary 1 be extended to solutions of more general elliptic operators, even with $P(x) \equiv 1$?

Problem 2. For what real manifolds M and for what conditions on the mapping H does the conclusion of Theorem 3 hold? A related question is the following. If M and M' are two real analytic hypersurfaces of finite type, say in $\mathbb{C}^2$, and H a holomorphic map defined on one side of M, smooth up to the boundary, and mapping M into M', can H vanish of infinite order at a point in M without being constant?

References

1. H. Alexander, *Boundary behavior of certain holomorphic maps*, Michigan Math. J. **38** (1991), 117-128.
2. ———, *A weak Hopf Lemma for holomorphic mappings*, preprint.
3. S. Alinhac, M.S. Baouendi, and L. P. Rothschild, *Unique continuation and regualrity at the boundary for holomorphic functions*, Duke J. Math. **61** (1990), 635-653.
4. M.S. Baouendi and L. P. Rothschild, *Unique continuation and a Schwarz reflection principle for analytic sets*, Comm. P.D.E. **18** (1993), 1961-1970.
5. ———, *A local Hopf lemma and unique continuation for harmonic functions*, Duke J. Math., Inter. Research Notices **71** (1993), 245-251.
6. ———, *Harmonic functions satisfying weighted sign conditions on the boundary*, Ann. Inst. Fourier, Grenoble **43** (1993), 1311-1318.
7. ———, *Flat analytic discs attached to real hypersurfaces of finite type*, Math. Research Letters **1** (1994), 359-367.
8. S. Bell and L. Lempert, *A C^∞ Schwarz reflection principle in one and several complex variables*, J. Diff. Geom., **32** (1990), 889-915.
9. E. Bishop, *Differentiable manifolds in complex Euclidean space*, Duke Math. J. **32** (1965), 1-22.
10. S. Huang and S. G. Krantz, *A unique continuation problem for holomorphic mappings*, Comm. P.D.E. **18** (1993), 241-263.

11. J.J. Kohn, *Boundary behavior of $\bar{\partial}$ on weakly pseudoconvex manifolds of dimension two*, J. Diff. Geom. **6** (1972), 523-542.
12. H. Lewy, *On the local character of the solution of an atypical differential equation in three variables and a related problem for regular functions of two complex variables*, Ann. of Math. **64** (1956), 514-522.
13. C. Miranda, *Partial differential equations of elliptic type*, Ergeb.Math. Grenzgeb (n.F.), 2, Springer-Verlag, Berlin, 1970.
14. H.S. Shapiro, *Notes on a theorem of Baouendi and Rothschild*, preprint.
15. V. Shklover, *Remarks on the local Hopf's lemma*, preprint.

Department of Mathematics-0112, University of California
San Diego, La Jolla, CA 92093-0112
sbaouendi@ucsd.edu, lrothschild@ucsd.edu

The Wave Shaping Problem*

Claude Bardos and Mikhael Belyshev

Résumé

Cet article contient un certain nombre de résultats classiques et nouveaux sur la structure des solutions des équations d'ondes avec conditions aux limites. Ces résultats sont motivés par les problèmes d'observation de contrôle et d'identification. On essaye d'ailleurs de mettre en évidence les relations existant entre ces différentes questions.

1. Introduction

With the introduction of the HUM method J. L. Lions [Li] completely clarified the relations existing beetween observation, stabilization and control for the solutions of the wave equation. A few years later it was found out by Lebeau, Rauch (and C.B.) [BLR] that the answer to several questions raised in this theory were completely given by the geometrical structure of the problem. Simultaneously M. Belyshev with several coworkers (cf. [BK] and the references) found out the connection beetween problem of boundary controllability and the inverse problem. Futher, he used the control to prove the uniqueness of solutions of inverse problem in the region that could be reached by waves. To achieve this goal he introduced the operator W^T described in section 3. Roughly speaking, this operator associates to the Dirichlet or Neumann boundary data the value of the solution $u(.,T)$ at time T, ignoring the value of the derivative $\partial_t u(.,T)$. Therefore the present paper is mostly devoted to this type of "half" control. In the mean time the analysis produces new results on the structure of the solution of the wave equation.

This contribution is a report on a joint work in progress. However I (Claude Bardos) wrote the present version, expressing my gratitude for the opportunity of including an article in the volume dedicated to memory of Pierre Grisvard. As it has been said, Pierre had a fantastic influence on the French mathematical community, of course on his students that really appreciated him, but also on many mathematicians of his generation. He personally helped me a lot.

---

* Partially supported by the Direction Générale de l'Armement under contract DRET ERS92/1441/A300/DRET/DS/SR, and by the CNRS in the ENS Steklov-Landau agreement

2. Geometry and definition

Recent results on control theory have been obtained analysing the underlining Riemannian-Lagrangian geometry of the wave equation

$$\partial_t^2 u - \nabla_x \cdot A(x) \nabla_x u = 0 \tag{1}$$

in a bounded domain $\Omega \subset \mathbf{R}^n$.

In (1) $A(x)$ is a symmetric positive matrix with smooth (say in $C^2(\mathbf{R}^n)$) coefficients; therefore there exists two strictly positive constants such that one has:

$$\forall \xi \in R^n \quad \alpha ||\xi||^2 \leq (A(x)\xi, \xi) \leq \beta ||\xi||^2. \tag{2}$$

The differential operator $\nabla_x \cdot A(x) \nabla_x$ is denoted by $\mathbf{A}$. Its symbol

$$H(x, \xi, \tau) = \frac{1}{2}\{-\tau^2 + (A(x)\xi, \xi)\} = H(p, q) \tag{3}$$

is a real valued function defined in $T^*(R^{n+1})$. In accordance with the theory of Hamiltonian mechanics the following variables *position* and *impulsion* are introduced:

$$q = (x, t), \quad p = (\xi, \tau). \tag{4}$$

The solutions of the differential system

$$\dot{q} = \nabla_p H(q, p), \quad \dot{p} = -\nabla_q H(q, p) \tag{5}$$

or

$$\dot{x}(s) = \nabla_\xi (H(x, \xi, \tau)) = A(x)\xi \tag{6}$$

$$\dot{t}(s) = \frac{\partial}{\partial \tau}(H(x, \xi, \tau)) = -\tau \tag{7}$$

$$\dot{\xi}(s) = -\nabla_x (H(x, \xi, \tau)) = -\nabla_x (A(x)\xi, \xi) \tag{8}$$

$$\dot{\tau}(s) = 0 \tag{9}$$

are well defined and the quantity $H(q, p)$ remains constant along the trajectories of this flow. This leads to the introduction of the *characteristic manifold.*

$$C = \{(q, p) / H(q, p) = H(x, \xi, \tau) = 0\} \subset T^*(R^d \times R_t)$$

The solutions of (5) define a foliation of C and are called *bicharacteristic*. The projection on $R^n \times R_t$ of these curves $\pi : (q,p) \to q = (x,t)$ are called rays. Along a bicharacteristic one has $\tau = constant$ and

$$\tau^2 = (A(x(s))\xi(s), \xi(s)), \tag{10}$$

By homogeneity one can choose $|\tau| = 1$ and on the corresponding bicharacteristic we have

$$(A(x(s))\xi(s), \xi(s)) = 1 .$$

With (7) and (9) this implies the relation $t = \pm s + t_0$; therefore t can be chosen as a parameter and the rays $t \to x(t)$ turn out to be the geodesics for the metric given by the matrix $A(x)^{-1}$, i.e. the curves which minimise the geodesic length:

$$L = \int_{t_0}^{t_1} \sqrt{(A^{-1}(x(s))\dot{x}(s), \dot{x}(s))}ds . \tag{11}$$

The geodesics and the bicharacteristics are in particular related by the formula:

$$\dot{x}(t) = A(x(t))\xi(t) ; \tag{12}$$

the mapping

$$(x, \xi) \to (x, A(x)\xi) \tag{13}$$

defines a isomorphism beetween $T^*(\mathbf{R}^n)$ and $T(\mathbf{R}^n)$ equipped with the scalar products $((\xi_1, \xi_1)) = (A(x)\xi_1, \xi_1)$ and $((\gamma_1, \gamma_2)) = (A(x)^{-1}\gamma_1, \gamma_2)$.

Ω denotes a bounded open set of $\mathbf{R}^n$, with boundary $\partial\Omega$ and outward unitary normal $\nu(x)$. It will be convenient to denote by M the cylinder $\Omega \times \mathbf{R}_t$ with boundary $\partial M = \partial\Omega \times \mathbf{R}_t$. The tangent space $T(\partial M)$ is identified with the space of the (x, t, γ, s) such that γ is orthogonal to $\nu(x)$ for the euclidean initial scalar product. Restricted on ∂M the Hamiltonian

$$H(x, \xi, \tau) = \frac{1}{2}\{-\tau^2 + (A(x)\xi, \xi)\} = \frac{1}{2}\{-\tau^2 + ((\xi, \xi))\} \tag{14}$$

induces a Hamiltonian flow on the cotangent bundle $T^*(\partial M)$ (identified with the points:

$$\{(x, t, \xi, \tau), x \in \partial\Omega, t \in \mathbf{R}_t, \xi \in \mathbf{R}^n, \quad A(x)\xi.\nu(x) = 0\}). \tag{15}$$

Any vector ξ is uniquely decomposed in its orthogonal projection (for the scalar product $((.,.))$) on $\nu(x)$ and on the orthogonal subspace $\nu(x)^\perp$:

$$\xi = \xi' + \xi_\nu, \quad |\tau|^2 - (A(x)\xi, \xi) = |\tau|^2 - ||\xi'||^2 - ||\xi_\nu||^2 \tag{16}$$

These constructions lead to the extension of the notion of bicharacteristic flows up to the boundary as maps from $\mathbf{R}_s$ with value in $T^*(\bar{M}) \cup T^*(\partial M)$. Introduce the *characteristic variety* as a subspace VC of $T^*(\bar{M}) \cup T^*(\partial M)$ defined by the following formulas:

$$VC = \{(x,t,\xi,\tau) \in T^*(\bar{M}), \quad |\tau|^2 - (A(x)\xi,\xi) = 0\} \cup$$

$$\{(x,t,\xi',\tau) \in T^*(\partial M), \quad |\tau|^2 - (A(x)\xi',\xi') \geq 0\}. \tag{17}$$

This submanifold is itself decomposed as the sum of the two following:

$$H = \{(x,t,\xi',\tau)\,/x \in \partial\Omega, \quad |\tau|^2 - (A(x)\xi',\xi') > 0\} \tag{18}$$

$$G = \{x,t,\xi',\tau/x \in \partial\Omega, \quad |\tau|^2 - (A(x)\xi',\xi') = 0\}. \tag{19}$$

Any point of H is associated with the two following elements of $T^*(\bar{M})$:

$$(x,t,\xi',\xi_\nu = \sqrt{|\tau|^2 - (A(x)\xi',\xi')}.\nu(x), \tau)$$

and

$$(x,t,\xi',\xi_\nu = -\sqrt{|\tau|^2 - (A(x)\xi',\xi')}.\nu(x), \tau) \tag{20}$$

which are identified by an equivalence relation.

Any element of G is identified with an element of $T^*(\bar{M})$. Furthermore G is invariant under the action of the Hamiltonian flow and the projection on ∂M of this flow are the geodesics for the induced metric.

A section of ray $x(s), t(s)$ (or the corresponding bicharateristic) is said to be *creeping* if satisfies the following relations: For $0 < |s - s_0| < \eta$ one has $x(s_0) + (s - s_0)\dot{x}(s_0) \notin \Omega$. A point on the boundary is said to be *incoming on the boundary* if for $0 < |s-s_0| < \eta, s-s_0 < 0$ one has $x(s_0)+(s-s_0)\dot{x}(s_0) \in \Omega$ and for $0 < |s-s_0| < \eta, 0 < s-s_0$ one has $x(s_0)+(s-s_0)\dot{x}(s_0) \notin \Omega$. The points which are *outgoing on the boundary* are defined in a similar way exchanging the values of s before and after s_0. A diffractive ray at the point $x(s_0)$ is a ray contained in Ω in the neighbourhood of a unique point $x_0 \in \partial\Omega$ and which satisfies, for s close to s_0, $x(s_0) + (s - s_0)\dot{x}(s_0) \in \Omega$.

The C^∞ bicharacteristic flow is constructed in the following way:

1 The flow is contained in $\{\dot{T}^*(\bar{M}) \cup T^*(\partial M)\} \cap VC$.
2 In M bicharacteristics propagate according to the Hamiltonian flow.
3 When they meet the boundary in a hyperbolic point they reflect according to the equation (19) which corresponds in the constant coefficients case to the law of geometric optic.
4 If they meet the boundary at a diffractive point, they are extended by continuity in M.
5 If they meet the boundary at an incoming point they are extended by the induced bicharacteristic on $T^*(\partial M)$ and remain on the boundary as long as they creep. They are extended by continuity inside $T^*(M)$ as soon as they meet an outgoing point.

With the riemannian distance denoted by $d(x,y)$ in Ω, one can define the *future* and *past cone* of influence of any subdomain $D \subset \bar{M}$.

$$K^+(D) = \{(x,t) \in M, \quad \exists\,(x_0,t_0) \in D,\ d(x,x_0) < t - t_0\ \} \tag{21}$$

$$K^-(D) = \{(x,t) \in M, \quad \exists\,(x_0,t_0) \in D,\ d(x,x_0) < -(t - t_0)\}\,. \tag{22}$$

For any $T > 0$ one can introduce the "section" of $K^+(D)$ (same constructions can be done for $K^-(D)$) denoted by

$$\Omega^T(D) = \{x \in \Omega, \text{ such that } (x,T) \in K^+(D)\}\,. \tag{23}$$

The sets $\Omega^T(D)$ increase with T; assume that D is strictly contained in $\bar{\Omega}\times]0,\mathbf{R}_t^+[$ then, since Ω is bounded, there exists a unique time $0 < T^*(D) < \infty$ defined by the relation

$$T_*(D) = \inf\{T \in \mathbf{R}_t^+ \ \text{ such that } \Omega^T(D) = \Omega\}\,. \tag{24}$$

This time will be called the *filling time*. For $D = \Gamma \times R_t^+$ the following notations will be used $\Omega^T(\Gamma) = \Omega^T(\Gamma)\times]0,T[$ and $T_*(\Gamma) = T_*(\Gamma\times]0,T[)$.

3. Wave shaping

In this section old and news results are used to describe the shape of the wave u which is a solution of the equation:

$$\partial_t^2 u - \mathbf{A}u = 0 \text{ in } \Omega\times]0,T[\,. \tag{25}$$

First recall the two observations:

1) The boundary of ∂M is not characteristic for the differential operator appearing in (25); therefore for any extendible distribution (restriction to M of a distribution defined in a neighbourhood of M) the restriction to ∂M of u and the normal derivative $\partial_\nu u$ are well defined distributions.

2) The operator $\mathbf{A}$ is time independent therefore for any regularizing sequence

$$\rho_\epsilon(t) = \frac{1}{\epsilon}\rho(\frac{t}{\epsilon})$$

converging to the Dirac mass, $u_\epsilon = \rho_\epsilon * u(.,x)$ is a solution of (25). Since the wave front set of the solution of (24) is contained in the wave cone, $u_\epsilon(x,t) \in C^\infty(\Omega \times \mathbf{R}_t)$. Furthermore if u is in $H^s(\Omega\times]0,T[)$, u_ϵ converges strongly to u in this space and the support of u_ϵ differs from the support of u given terms of the order of ϵ. This implies that almost all the results proven for smooth solutions and independant of the regularity of these solutions can be extended to distributions.

For the sake of simplicity we emphasize the Dirichlet boundary value problem with zero initial data:

$$\partial_t^2 u - \mathbf{A}u = 0 \text{ in } \Omega \times \mathbf{R}_t, u(t,\sigma) = g \text{ on } \partial\Omega \times \mathbf{R}_t,$$

$$\text{supp } g = \Sigma \subset \partial\Omega \times R_t^+, \quad u(.,T) = 0, \text{ for } t < 0. \tag{26}$$

Recall that (26) is, for any distribution g defined on $\partial\Omega \times \mathbf{R}_t^+$, a well posed problem in the class of extended distributions defined in on $\Omega \times \mathbf{R}_t$ (cf. Kreiss [K], Sakamoto [Sa] Hormander [H], Chazarain and Piriou [CP] etc.).

For any $T > 0$ one can introduce the operators:

$$L^T : g \to (u_g(.,T), \partial_t u_g(.,T) \text{ and } W^T : g \to u_g(.,T)\,. \tag{27}$$

For any $s \in \mathbf{R}, s < 3/2$, L^T is continuous from $H^s(\partial\Omega\times]0,T[)$ to $H^s(\Omega) \times H^{s-1}(\Omega)$ and W^T is continuous from $H^s(\partial\Omega\times]0,T[)$ to $H^s(\Omega)$. Same results hold true for $s \geq 3/2$ provided g satisfies some extra compatibility conditions (cf. Rauch and Massey [RM] for details). In particular L^T is continuous from $L^2(\partial\Omega\times]0,T[)$ to $L^2(\Omega) \times H^{-1}(\Omega)$ and W^T is continuous from $L^2(\partial\Omega\times]0,T[)$ to $L^2(\Omega)$ (in this special situation a clever and elementary proof can be found in Lascieka, Lions and Triggiani [LLT]). Finally, with $\Gamma \subset \partial\Omega$, L_Γ^T and W_Γ^T will denote the restriction of the operators L^T and W^T to the space of distributions g with support in $\Gamma\times]0,T[$.

Theorem 1. *The support of the solution $u(x,t)$ of (26) is contained in $K^+(\Sigma)$, the future region of influence of Σ.*

The proof of this result is elementary and can be obtained by integration on the intersection of M with a convenient characteristic cone C of the following type:

$$C = \{(x,t)/c(x,t) > 0 \text{ with } |\partial_t c(x,t)|^2 - (A(x)\nabla c(x,t), \nabla c(x,t)) = 0\}$$

cf. Smirnov [Si].

Complementary results on the support (in the future) of the solution of the wave equation are described below.

Lemme 2. *Let $\omega' \subset \omega$ two open subsets of Ω. Suppose that w is a solution of the equation (25) in $\omega \times]-\infty, T[$ with the following properties:*
1) $w(x,T) = 0$ *for* $x \in \omega$
2) $w(x,t) = 0$ *for* $x \in \omega, \quad t < 0$
3) (*support* w) $\cap \{\omega' \times [0,T]\} = \emptyset$ *i.e. a part of the cylinder* $\omega \times]0,T[$ *is not engaged by the wave* w. *Then* $w = 0$ *everywhere in* $\omega \times]-\infty, T]$.

Proof. Extend w by odd continuation to $\omega \times \mathbf{R}_t$ according to the rule:

$$\tilde{w}(x,t) = -w(x, 2T - t) \text{ for } T \leq t \leq 2T$$

$$\tilde{w}(x,t) = 0 \text{ for } 2T \leq t \tag{28}$$

and obtain a solution of the wave equation

$$\partial_t^2 \tilde{w} - \mathbf{A}\tilde{w} = 0 \text{ in } \omega \times R_t \tag{29}$$

By virtue of 3) one has

$$\forall (x,t) \in \omega' \times \mathbf{R}_t, \quad \tilde{w}(x,t) = 0, \tag{30}$$

The time Fourier transform of $\tilde{w}$:

$$\hat{w}(.,\tau) = \int_{-\infty}^{+\infty} e^{-i\tau t} \tilde{w}(.,t) dt$$

is in ω a solution of the Helmoltz equation:

$$(\tau^2 + A)\hat{w}(.,\tau) = 0 \tag{31}$$

which by 3) and (29) is zero in $\omega' \times \mathbf{R}_\tau$. In accordance with the classical uniqueness theorem for the elliptic equation (cf. Landis [La]) this implies that $\tilde{w}(x,\tau)$ is zero in $\omega \times \mathbf{R}_\tau$. Returning to $\hat{w}$ and w this proves the lemma.

The next basic ingredient is Holmgren theorem.

Theorem 3 [Holmgren]. *Let B be a ball with center $x_0 \in \bar{\Omega}$ and radius $\epsilon > 0$. Then any solution u of the wave equation*

$$\partial_t^2 u - \mathbf{A}u = 0 \tag{32}$$

which is zero on the cylinder $B \times]-T, T[\cap M$ is also zero in the region defined by

$$\{(x,t) \in \Omega \times]-T, T[/ d(x,B) < T - |t|\}. \tag{33}$$

Remark 4. This result is classical for the case of real analytic coefficients [J]. To extend it to a nonanalytical case required more than 40 years. In particular the difficulty of the problem can be illustrated by a basic example given by Alinhac and Baouendi [AB]. They constructed some C^∞ potential $q(x,t)$ such that the perturbed wave equation:

$$\partial_t^2 u - \Delta u + q(x,t)u = 0$$

does not satisfy the unique continuation principle accross non characteristic surfaces.

Recent progress was initated by Robbiano [Ro]. Then Hormander proved the validity of a variant of (33) i.e. for

$$(x,t) \in M, \quad Kd(x,x_0) < T - |t|$$

with some absolute constant $K \in [1, \sqrt{\frac{27}{23}}[$. At last efforts were crowned by D. Tataru [T] who proved the uniqueness of the continuation of the wave equation's solutions across any noncharacteristic surface. Using the standard methods of John one can show that Tataru's result implies $K = 1$, a property which is instrumental for our paper.

Corollary 5. *Let u be a solution of the problem*

$$\partial_t^2 u - \mathbf{A}u = 0 \tag{34}$$

with Cauchy data zero at times $t = 0$. If $u = 0$ in the cylinder $B\times]0,T[$, then u is zero in the region

$$\{(x,t) \in \Omega\times]0,T[/d(x,B) < T - t\}, \tag{35}$$

Proof. Extend the function u by zero for negative times, at least up to the time $-T$, and use the above Holmgren theorem.

$HF(\Sigma)$, the Huyghens front of the wave given by (26) is defined as the closure in $\Omega \times \mathbf{R}_t$ of the boundary $K^+(\Sigma)$ and one has

Theorem 6 [Huyghens Rule]. *Any point $(x,t) \in HF(\Sigma)$ belongs to the support of the solution of the problem (26).*

Proof. The proof is made by reducing to contradiction. Assume the existence of a point $(x_0,t_0) \in HF(\Sigma)$ which does not belong to the support of $u(x,t)$. Then the solution is equal to zero on a cylindrical neighbourhood $\omega\times]t_0+\epsilon, t_0-\epsilon[$ with ω denoting a ball of center x_0 and of radius η small enough. By reducing the size of ϵ and ω, one can assume that the solution is zero in the cylinder $\omega\times]-\infty, t_0+\frac{\epsilon}{2}[$. With the precised form of the Holmgren theorem this implies that u is zero in a small neighbourhood of any point (x_1,t_1) such that

$$t_1 < t, \; d(x_0,x_1) < t_0 - t_1 + \epsilon/2\,.$$

This is in contradiction to the definition of $HF(\Sigma)$ which implies the existence of at least one point $(x_1, t_1) \in \Sigma$ for which $d(x_0, x_1) = t_0 - t_1$ and the definition of Σ itself, because for any neighbourhood $w \subset \partial\Omega \times \mathbf{R}_t$ of any point $(\sigma, t) \in \Sigma \subset \partial\Omega \times \mathbf{R}_t$, one has

$$g_{|w} \neq 0 \text{ and therefore } u_{|w} \neq 0\,. \tag{36}$$

Observe that the mapping $t \to u(.,t)$ defined by the solution of (26) is for any distribution $g \in \mathbf{D}'(\partial M)$ continuous in the space $\mathbf{D}'(\Omega)$. Therefore for any T the support of $u(.,T)$ is well defined and that for T strictly less than the filling time $T_*(\Sigma)$ (defined in Section **2**). One has $HF(\Sigma) \neq \emptyset$. With these considerations one can improve the previous result and obtain

Proposition 7. *For $T < T_*$ any point of the front at time T ie $HF^T(\Sigma) = HF(\Sigma) \cap \{(x,T))\}$ is in the support of $u(.,T)$.*

Proof. Assume that $w(x,T)$ is zero in an open neigbourhood of x_0 such that $(x_0, T) \in HF(\Sigma)$. Then there exists an open set $\omega' \subset \omega$ such that

$$K^+(\Sigma) \cap (\omega' \times]-\infty, T[) = \emptyset \tag{37}$$

and Lemma 2 can be applied showing that u is zero in a neighbourhood in space time of a point $(x,T) \in HF(\Sigma)$. The proofs follow from Theorem 6.

Proposition 7 implies the following

Corollary 8. *For $T < T_*$ the operator W_Γ^T is injective from the space of distributions with support contained in $\Gamma \times]0,T[$ to the space $D'(\Omega)$.*

Proof. An opportunity

$$u(.,T) = W_\Gamma^T g = 0 \tag{38}$$

breaks the Huyghens rule. So a wave generated by a nonzero boundary control cannot disappear at any time $t < T_*$.

4. Approximate and exact controlability in subregions

The construction of Lions for boundary controllability will be systematically used and is recalled below:

For any Cauchy data given at time T, $\Phi = (\phi_0, \phi_1)$, one introduces the solution $\phi(x,t)$ of the wave equation

$$\partial_t^2 \phi - \mathbf{A}\phi = 0 \tag{39}$$

with homogenous Dirichlet boundary data and denotes by u_ϕ the solution of the inhomogenous mixed problem with zero Cauchy data at time $t = 0$ and with boundary data

$$u_\phi(t,\sigma) = \partial_\nu \phi(t,\sigma)_{|\Gamma \times]0,T[}\,.$$

This process defines the operators

$$\Lambda(\Phi) = (\partial_t u_\phi(.,T), -u_\phi(.,T)) = L(\partial_\nu \phi(t,\sigma)_{|\Gamma\times]0,T[}) \quad (40)$$

and

$$\Xi(\Phi) = u_\phi(.,T) = W^T(\partial_\nu \phi(t,\sigma)_{|\Gamma\times]0,T[}). \quad (41)$$

The operator Λ is continuous from $E = H_0^1(\Omega) \times L^2(\Omega)$ to its dual $E^* = L^2(\Omega) \times H^{-1}(\Omega)$ and the operator Ξ is continuous from $H_0^1(\Omega) \times L^2(\Omega)$ to $L^2(\Omega)$.

For any solution $\psi(x,t)$ of the homogenous wave equation with Cauchy data at time T $\Psi = (\psi_0, \psi_1)$, one has, with the Green formula applied to the equations

$$\partial_t^2 u_\phi - \mathbf{A} u_\phi = 0, \quad \partial_t^2 \psi - \mathbf{A}\psi = 0 \quad (42)$$

the relation

$$< \Lambda(\Phi), \Psi > = < \partial_t u_\phi(x,t), \psi_0(x,t) > - < u_\phi(x,t), \psi_1(x,t) >$$

$$= \int_{\Gamma\times]0,T[} \partial_\nu \phi(t,\sigma) \partial_\nu \psi(t,\sigma) d\sigma dt \,. \quad (43)$$

This shows that Λ is a positive self-adjoint operator from E into its dual.

Denoted by $D(\Omega,\Gamma)$ the distance beetween Ω and Γ:

$$D(\Omega,\Gamma) = \sup_{x\in\Omega} d(x,\Gamma). \quad (44)$$

Theorem 9.

i) *Approximate controllability. Assume that* $2D(\Omega,\Gamma) < T$, *then the range of* L *and* Λ *are dense in* $E^* = H^{-1}(\Omega) \times L^2(\Omega)$,

ii) *Exact controllability [BLR]. Assume that any ray of geometric optic meets* $\Gamma\times]0,T[$ *in a non diffractive point then the operator* Λ *is coercive and therefore exact controllability holds* [BLK].

Proof of i) The range of L contains the range of Λ and if this space is not dense in E^*, there exists a non zero element $\Phi \in E$ which is orthogonal to the image of Λ; one has

$$\int_{\Gamma\times]0,T[} |\partial_n u\phi(\sigma,t)|^2 d\sigma dt = < \Lambda(\Phi), \Phi > = 0 \quad (45)$$

and this shows that on $\Gamma\times]0,T[$

$$\phi(t,\sigma) = \partial_\nu \phi(t,\sigma) = 0 \quad \forall(\sigma,t) \in \Gamma\times]0,T[\,. \quad (46)$$

Therefore with the Holmgren theorem, ϕ is identically zero. This contradicts the hypothesis. In fact this proof is basically due to Russel who was the first to observe the connection between the Holmgren theorem and approximate controllability. In the present form it uses the basic improvement of Tataru for the Holmgren theorem.

The proof of ii) is given in [BLR] where it is shown that the geometric condition is also generically necessary.

One of the goals of the present paper is to produce a local version of the above criteria. However observe that in the above form the statement i) cannot be localised; more precisely one has the

Remark 10. The space $L(L^2(\Gamma\times]0,T[)$ is not dense in the subspace of distributions $\Phi \in E^* = H^{-1}(\Omega)\times L^2(\Omega))$ with support in $C = \{x \in \Omega,\ 2d(x,\Gamma) < T\}$.

To produce a counterexample introduce, in one space variable, with $\Omega =]0,1[$, $T<1$, the solution $\phi(x,t) = f(x+t)$ with support $f \subset]T,1[$ and observe that for any control g given on $\{x=0\}\times]0,T[$ one has:

$$< \partial_t u_g(.,T), \phi(.,T) > - < u_g(.,T), \partial_t\phi(.,T) >= 0$$

A weaker version due to Belyshev and including only half of the information (density for the function or time derivative, but not both) will be given in the next section. On the other hand, at present one can produce the following local version of the BLR criteria:

Theorem 11. *Let C be a open subset of Ω, Γ an open subset of $\partial\Omega$ and $T>0$ a positive time.*

i) *Assume that any generalised C^∞ ray $(x(t),t)$, with $x(T)\in C$, meets $\Gamma\times]0,T[$ at a non diffractive point. Then local exact controllability is true i.e.: For any pair $(u_0^d,u_1^d)\in L^2(\Omega)\times H^{-1}(\Omega)$ there exists at least one $g\in L^2(\Gamma\times]0,T[)$ such that*

$$(u_g(.,T),\partial_t u_g(.,T))_{|C} = (u_0^d,u_1^d)_{|C} \tag{47}$$

ii) *Assume the existence of a generalised ray $(x(t),t)$ with $x(T)\in C$ which does not meets $\Gamma\times]\bar{0},T[$ then i) is false.*

Proof. Introduce the space E_C of solutions of $\phi(x,t)$ of the wave equation with Dirichlet boundary data and with finite energy which satisfy the property:

$$\text{supp } \phi(.,T)\cup\text{supp } \partial_t\phi(.,T)\subset\bar{C} \tag{48}$$

E_C is a closed subspace of E. With the lifting lemma of [BLR] (which gives an estimate of the H^1 microlocal norm near any non diffractive direction in term of the norm of the local norm of $\partial_\nu u$ in $L^2(\partial M)$) and the H^1 microlocal theorem of propagation of Melrose and Sjostrand [MS], one obtains the existence of two constants α and β such that

$$||\phi||_E^2 \leq \alpha \int_{\Gamma\times]0,T[} |\partial_\nu \phi(s,\sigma)|^2 d\sigma ds + \beta ||\phi||^2_{L^2(\Omega\times]0,T[)}. \qquad (49)$$

In fact as in [BLR], β can be taken equal to zero. This is proved by absurdum. If this were false, there would exist a sequence ϕ_n of solutions of the wave equation (with homogenous Dirichlet boundary data) in $\Omega \times \mathbf{R}_t$ with

$$\lim_{n\to\infty} \int_{\Gamma\times]0,T[} |\partial_\nu \phi_n(s,\sigma)|^2 d\sigma ds = 0 \text{ and } ||\phi_n||^2_{L^2(\Omega\times]0,T[} = 1 \qquad (50)$$

From (49) one observes that the Cauchy data of this sequence at time T are uniformly bounded in the space E of finite energy and have their support in $\bar{C}$. Since the space of solutions of finite energy are compactly imbedded in the space of solutions equipped with the norm

$$||\phi||^2_{L^2(\Omega\times]0,T[)},$$

any cluster point ϕ (in the weak $L^2(\Omega\times]0,T[)$ topology) of the family ϕ_n is a solution of the wave equation with Dirichlet boundary conditions and with the following properties:

$$\phi \in E_C, \quad \int_{\Gamma\times]0,T[} |\partial_\nu \phi_n(s,\sigma)|^2 d\sigma ds = 0, \qquad (51)$$

$$||\phi||^2_{L^2(\Omega\times]0,T[)} = 1. \qquad (52)$$

Denote by N the space of solutions of the wave equation with Dirichlet boundary data which satisfies the properties (51) and observe, using in particular the relation

$$\partial_t^2 \phi(.,T) = \mathbf{A}\phi(.,T),$$

that, for any $\phi \in N$, one has

$$\text{supp}\partial_t\phi(.,T) \cup \text{supp}\partial_t^2\phi(.,T) \subset \bar{C}, \quad \int_{\Gamma\times]0,T[} |\partial_\nu \partial_t \phi(s,\sigma)|^2 d\sigma ds = 0. \qquad (53)$$

Therefore $\partial_t\phi \in N$. As in [BLR] this implies both that N is invariant under the map $\phi \to \partial_t\phi$ and that N is finite dimensional. If $N \neq 0$ there exists in this space a non trivial eigenvector w for the map ∂_t ; i.e. $\partial_t w(x,t) = \lambda w$ giving

$$\lambda^2 w - \mathbf{A}w = 0. \qquad (54)$$

With (53) this implies that (uses the unique continuation property for second order elliptic equation) one has

$$w(x,t) = 0 \text{ for } 0 \leq t \leq T \tag{55}$$

Therefore w is identically zero.

Finally with the relation:

$$\Lambda(\Phi, \Phi) = \int_{\Gamma\times]0,T[} |\partial_\nu \phi_n(s,\sigma)|^2 d\sigma ds \tag{56}$$

observe that the operator Λ, when restricted to E_C, is a continuous and coercive map from E_C into its dual E_C^*. Therefore for any $l \in E_C^*$ and in particular for any pair

$$(u_0, u_1) \in E^* = L^2(\Omega) \times H^{-1}(\Omega)$$

there exists $\Phi \in E_C$ such that one has

$$\forall \Psi \in E_C, \quad < u_1, \psi_0(x,t) > - < u_0, \psi_1(x,t) > = < \Lambda(\Phi), \Psi >$$

$$= < \partial_t u_\Phi(.T), \psi_0(x,t) > - < u_\Phi(.T), \psi_1(x,t) > \tag{57}$$

or

$$\forall x \in C \quad u_\Phi(x,T) = u_0(x), \quad \partial_t u_\Phi(x,T) = u_1(x) \tag{58}$$

which completes the proof of the statement i). The proof of the statement ii) follows the lines of [BLR] (inspired by the pioneering remark of Ralston [Ra1], [Ra2]) and needs no modification. It is left to the reader.

Remark 12. The above theorem gives no information on the structure of the solution obtained by the control process outside the region C.

5. Partial local controllability

In this section sufficient conditions to obtain a given value of a wave at given time T and in a subregion C will be produced. These conditions are weaker than in the previous section and therefore the price to pay is the following: no control is in this situation available for the solution outside the subdomain and no control is available the derivative of the solution. However this construction is at the basis of the BC method for inverse problems introduced by Belyshev.

Theorem 13. *Let $C \subset \bar{\Omega}$ be a compact contained in $\bar{\Omega}$, Γ a closed subregion of $\partial\Omega$ and $T > 0$ a positive time.*

i) *If for any point $x \in C$ one has $d(x,\Gamma) < T$, then $W^T(L^2(\Gamma\times]0,T[)$ is dense in $L^2(C)$.*

ii) *Assume that any generalized ray $(x(t), t)$ with $x(T) \in C$ meets $\Gamma\times]0,T[$ or $\Gamma\times]T,2T[$ at a non diffractive point then exact partial local controlability holds, i.e:*

For any function $u_0 \in L^2(\Omega)$ there exists at least one $g \in L^2(\Gamma\times]0,T[)$ such that the solution of (26) satisfies the relation

$$\forall x \in C, \quad W^T(g)(x) = u_g(x,T) = u_0(x)\,. \tag{59}$$

Remark 14. The meaning of the hypothesis of the point ii) of Theorem 13 is that for any generalized ray ρ passing through a point $x \in C$ there is at least one point $x_0 \in \Gamma \cap \rho$ such that $d(x, x_0) < T$.

Proof. If i) is not true there exists a non zero $\psi_1 \in L^2(\Omega)$ with support in C orthogonal to $W^T(L^2(\Gamma]\times]0,T[)$. Introduce the solution $\psi(x,t)$ of the wave equation with zero Dirichlet boundary data and Cauchy data at time T given by

$$\psi(.,T) = 0, \quad \partial_t\psi(.,T) = \psi_1.$$

From the formula

$$\forall g \in L^2(\Gamma\times]0,T[), \quad 0 =< \partial_t u_g(T), \psi_1(T) > - < u_g(T), \partial_t\psi(T) >$$

$$= \int_{\Gamma\times]0,T[} g(t,\sigma)\partial_\nu\psi(t,\sigma)d\sigma dt \tag{60}$$

one deduces that

$$\partial_\nu\psi(t,\sigma) = 0 \text{ on } \Gamma\times]0,T[\,. \tag{61}$$

Notice that ψ can be extended to the time interval $]0,2T[$ by odd continuation around the point $t = T$, remain a solution of the wave equation with zero Dirichlet boundary datas therefore (61) remains valid on the interval $]0,2T[$. Then the **Holmgren Theorem** shows that $\psi(x,T)$ is zero for $d(x,\Gamma) < T$ and the proof of i) is completed. To prove ii) one introduces the space $\phi(x,t) \in E^1_C$ of finite energy solutions of the wave equation, with Dirichlet boundary conditions and Cauchy data at time $t = T$ equal to $(0, \phi_1(x))$ with $\text{supp}\phi_1 \subset C$. Due to the energy conservation this space can be equipped by any of the equivalent norms

$$||\phi||_{H^1(\Omega\times]T_1,T_2[)} \text{ or } (E(\psi))^{1/2} = (\int_\Omega |\phi_1(x)|^2 dx)^{1/2}. \tag{62}$$

These functions are odd with respect to T:

$$\forall t \in]T,2T[, \quad \phi(.,T) = -\phi(2T-t,.)$$

Therefore with the geometric hypothesis one has (cf. [BLR])

$$||\phi||_E^2 \leq \alpha \int_{\Gamma\times]0,2T[} |\partial_\nu \phi(s,\sigma)|^2 d\sigma ds + \beta ||\phi||^2_{L^2(\Omega\times]0,2T[}. \tag{63}$$

Proceeding as in the previous section one can show that the constant β appearing in (63) can be taken equal to 0. Therefore for any $\phi \in E_C^1$ one has

$$||\phi||_E^2 \leq \alpha \int_{\Gamma\times]0,2T[} |\partial_\nu \phi(s,\sigma)|^2 d\sigma ds \tag{64}$$

or with the odd symmetry of ϕ:

$$||\phi||_E^2 \leq \alpha \int_{\Gamma\times]T,2T[} |\partial_\nu \phi(s,\sigma)|^2 d\sigma ds = 2\alpha \int_{\Gamma\times]0,T[} |\partial_\nu \phi(s,\sigma)|^2 d\sigma ds \tag{65}$$

Therefore,

$$\Lambda(\Phi,\Phi) = \int_{\Gamma\times]0,T[} |\partial_\nu \phi_n(s,\sigma)|^2 d\sigma ds \tag{66}$$

one observes that the operator Λ when restricted to E_C^1 is a continuous and coercive map from E_C^1 into its dual E_C^{1*}. For any $l \in E_C^{1*}$ and in particular for any function $u_0 \in L^2(\Omega)$ there exists $\Phi \in E_C^1$ such that one has

$$\forall \Psi \in E_C^1, \quad < \Lambda(\Phi), \Psi > = < \partial_t u_{\partial_\nu \phi}(.,T), \psi(.,T) > - < u_{\partial_\nu \phi}(.,T), \partial_t \psi(.,T) >$$

$$= < u_{\partial_\nu \phi}(.,T), \partial_t \psi(.,T) > = < u_0, \psi_1(x,t) > \tag{67}$$

or $\forall x \in C, \quad u_{\partial_\nu \phi}(x) = u_0(x)$ and the proof is completed.

In seems rather interesting to complete the statement i) of Theorem 13 by the following

Proposition 15. *Let Γ be an open subset of $\partial\Omega$ and $T < T_*(\Gamma)$. Then no non zero function $u(.,T)$ with support $\bar{\omega}$ strictly contained in $\Omega^T(\Gamma)$ belongs to the space*

$$W^T(L^2(\Gamma\times]0,T[)).$$

In conjunction with the point i) of Theorem 13 this implies the relation:

$$W^T(L^2(\Gamma\times]0,T[)) \neq clos W^T(L^2(\Gamma\times]0,T[)) \tag{68}$$

i.e. the image $W^T(L^2(\Gamma\times]0,T[))$ is not closed in $L^2(\Omega^T)$.

Proof. Extend $u(.,t)$ by even continuation around the point $t = T$, this produces a new function still denoted $u(x,t)$ which is zero for $t \notin [0, 2T]$. Its Fourier transform $\tilde{w}$ is a solution of the following equation:

$$\omega^2\tilde{w}(x,\omega) + A(x,\omega)\tilde{w}(x,o) = e^{-i\omega T}\partial_t u(x,T). \tag{69}$$

For any ω the right hand side of (69) is zero on $\Omega - \Omega^T$ non empty open set which is connected to any point of $\partial\Omega$ by a continuous curve γ such that

$$\gamma \cap \text{ support } \partial_t u(x,T) = \emptyset .$$

Therefore using the unique continuation property for the solution of the elliptic equation (69) one obtains that

$$\forall x \in \partial\Omega, \quad \tilde{u}(x,\omega) = 0 \tag{70}$$

With the inverse Fourier transform, one obtains

$$\forall x \in \partial\Omega \quad u(x,t) = 0$$

which implies that u is identically zero.

6. Conclusion and connection with inverse problems

In the previous section a subset of the boundary Γ and a time $T < T^*$ have been introduced, an operator W^T from $L^2(\Gamma\times]0,T[)$ with value in $L^2(\Omega)$ has been constructed. Their properties have been studied in some detail. It has been shown that this operator is injective, that its image is dense in $L^2(\Omega^T)$ but never coincides with this space and that nevertheless, it is surjective, when restricted to a convenient subspace $L^2(C)$ with C satisfying the hypothesis of the point ii) of Theorem 13.

All these properties remain valid when the Neumann control replaces the Dirichlet control and when the operator W is defined by the formulas

$$u(.,0) = \partial_t u(.,0) = 0, \qquad \text{support } g \subset \Gamma\times]0,T[\tag{71}$$

$$\partial_t^2 u - \mathbf{A}u = 0, \quad \partial_\nu u = g, \tag{72}$$

In the meantime one can introduce the Dirichlet-Neumann operator defined by (71), (72) and

$$R^T g = u_{\partial\Omega\times]0,T[}\,. \tag{73}$$

The inverse problem has several aspects: First prove that the knowledge of the Neumann-Dirichlet operator determines the operator itself; prove that an error in the knowledge of this operator produces an error of the same order of magnitude for the coefficients of the operator and finally give formulas for the recontruction of the operator. It turns out that the point i) of Theorem 3 is related to the first question while point 2 is related to the second and the third.

In fact Belyshev and Kurylev (cf. [BK] and other references in this paper) have for any $T < T^*$ introduced an operator C^T with the formula

$$(C^T g_1, g_2)_{L^2(\partial\Omega\times]0,T[)} = (W^T g_1, W^T g_2)_{L^2(\Omega^T)} \tag{74}$$

and shown mostly with integration by part that C^T was completely and explicitly determined by the Neumann-Dirichlet operator R^{2T}. Using the injectivity of W^T (cf. Corollary 8) they introduce the closure Φ^T of the space $L^2(\partial\Omega\times]0,T[)$ for the scalar product

$$(W^T g_1, W^T g_2)_{L^2(\Omega^T)}$$

as expected this space is in general much bigger than the space $L^2(\partial\Omega\times]0,T[)$ and in fact much bigger than a space of distributions. However it is isomorphic to $L^2(\Omega^T)$. Using the density of $W^T(L^2(\partial\Omega\times]0,T[))$ they produce a dense set of known functions for which $\mathbf{A}u$ can be computed in terms of the operator C, obtaining therefore the uniqueness of the coefficients of $\mathbf{A}$ in Ω^T.

An example of the connection beetween the point ii) of Theorem 3 and the stability of the inverse problem can be found in the work of Isakov and Sun [IS]. They considered the inverse problem for the two operators.

$$\partial_t^2 - \Delta - q_i(x), \quad i = 1, 2. \text{ support } q_i \subset \bar{\Omega}.$$

Their goal is the recovery of the potential in a subdomain C by the knowledge of the restriction on $\Gamma\times]0,T[$ of the Neumann-Dirichlet operator. The hypothesis made on the subdomain C is similar to the hypothesis made in point ii) of Theorem 3 and instead of a general result on propagation of singularities they use special test functions of the form

$$u(x,t) = \phi(x + t\omega)\beta(x,t)exp[i\tau(x.\omega + t)] + r(x,t) \tag{75}$$

with $\tau \to \infty$ (i.e. localised plane waves) to determine

$$\int_{-\infty}^{\infty} q_1(x + t\omega) - q_2(x + t\omega)dt$$

in terms of the difference of the two Dirichlet-Neumann operators. They complete the proof with an inverse Radon transform. In particular in three space variables they obtain an estimate of the following type:

$$\sup_{x\in C} |q_1(x) - q_2(x)| \le \alpha\big(||R_1^{2T} - R_2^{2T}||_{L^2(\Gamma\times]0,2T),H^{1/6}(\Gamma\times]0,2T)}\big)^{1/4}. \tag{77}$$

Remark 16. The phony Sobolev exponent appearing in the right hand side of (77) is due to the fact that the Neumann problem is not well posed in the Lopatinsky sense (cf. [Lo]) and therefore that the trace of the solution on the boundary does not belong in general to the space $H^1(\partial\Omega\times]0,T[)$. A careful analysis of the proof would show that the full knowledge of the Neumann-Dirichlet operator is not needed; what is needed is the action of this operator on a special class of solutions, say localised plane waves. This corresponds both to the notion of propagation singularities and to the intuitive description of X rays analysis in Non Destructive Control Theory.

References

[AB] S. Alinhac and S. M. Baouendi, Construction de solutions nulles et singulières pour des opérateurs de type principal, Sem. Goulaouic Schwartz **exposé 22**, (1978-1979). Detailed article to appear in *Math. Zeitschrift* (1995).

[BK] M. Belyshev and Y. Kurylev, Boundary control, wave field continuation and inverse problems for the wave equation, *Computers Math. Applic.* **22** No 3/5 (1991), pp. 27–52.

[BLR] C. Bardos, G. Lebeau, J. Rauch, Sharp sufficient conditions for the observation, control and stabilization of waves from the boundary, *SIAM Journal on Control Theory and Applications* **30** (1992), 1024–1065.

[CP] J. Chazarain and A. Piriou, *Introduction à la théorie des équations aux dérivées partielles linéaires*, Gauthiers Villars, Paris, 1981.

[H] L. Hörmander, *The Analysis of Linear Partial Differential Operators*, Vol. III, Springer-Verlag, Berlin, 1984.

[IS] V. Isakov and Z. Sun, Stability estimates for hyperbolic inverse problems with local boundary data, *Inverse problems* **8** (1992), 193–206.

[J] F. John, On linear partial differential equation with analytic coefficients, *Comm. Pure Appl. Math.* **2** (1949), 209–253.

[K] H.O. Kreiss, Initial boundary value problems for hyperbolic systems, *Comm. Pure Appl. Math.* **23** (1970), 277–298.

[Li] J.L. Lions, *Contrôlabilité exacte, perturbation et stabilisation des systèmes distribués*, Masson, Paris, (1988).

[LLT] I. Lasiecka, J.L. Lions and R. Triggiani, Non homogenous boundary value problems for second order hyperbolic operators, *J. Math. Pures et Appliquées* **65** (1986), 149–192.

[La] E. M. Landis, On some properties of solutions for elliptic equations, *Dokl. Akad. Nauk SSSR* **107** (1956), 640–643.

[Lo] Ya. B. Lopatinsky, On a method of reducing boundary problems for a system of differential equations of elliptic type to regular integral equations, *Ukrain. Math. Z.* **5** (1953), pp. 123–151 and *Amer. Math. Soc. Transl.* **89** (2) (1970), 149–183.

[MS] R. Melrose and J. Sjöstrand, Singularities of boundary value problems, I, II, *Comm. on Pure and Appl. Math.* **31** (1978), 593–617 and *Comm. on Pure and Appl. Math.* **35** (1982), 129–168.

[Ra1] J. Ralston, Solutions of the wave equation with localized energy, *Comm. Pure. Appl. Math.* **22** (1969), 807–824.

[Ra2] J. Ralston, Gaussian beams and the propagation of singularities, *MAA Studies in Mathematics* **23**, W. Littman, ed., 206–248.

[Ro] L. Robbiano, Fonction de coût et contrôle des solutions des équations hyperboliques, to appear in *Asymptotic Analysis.*

[RM] J. Rauch and F. Massey, Differentiability of solutions to hyperbolic initial-boundary value problems, *Trans. A.M.S.* **189** (1974), 303–318.

[Sa] R. Sakamoto, Mixed problems for hyperbolic equations, *I. J. Math.*, Kyoto University, **10** (1970), 375–401; II. *J. Math.*, Kyoto University, **10** (1970), 403–411.

[Si] V.I. Smirnov, *A Course of Higher Mathematics*, **4**, Fitzmatgiz, Moscow 1958 [Addison Wesley, Reading, MA 1964].

[T] D. Tataru, Carleman estimates and unique continuation for solutions to boundary value problems, to appear in *Comm. in Partial Differential Equations.*

Claude Bardos
Université de Paris 7 et CMLA ENS Cachan
61, Av. Pres. Wilson, 94235 Cachan Cedex, France
bardos@cmla.ens-cachan.fr

Mikhael Belyshev
San Petersbourg Branch of the Steklov Mathematical Institute (LOMI)
Fontanka 27 D-11, 191011, Russia
email: belishev@pdmi.ras.ru

Modélisation mathématique des coques linéairement élastiques

Philippe G. Ciarlet

Cet article est dédié à la mémoire de Pierre Grisvard

Résumé

On analyse le comportement asymptotique du champ du déplacement tri-dimensionnel d'une coque linéairement élastique lorsque l'épaisseur tend vers zéro. Sous deux classes distinctes d'hypothèses, portant sur la géométrie de la surface moyenne, sur les conditions aux limites, et sur l'ordre de grandeur des forces appliquées, des théorèmes de convergence peuvent être établis, qui justifient soit les équations bi-dimensionnelles d'une "coque en flexion," soit celles d'une "coque membranaire." On discute également les mérites du modèle bi-dimensionnel de coques de W.T. Koiter sous les mêmes classes d'hypothèses.

Mathematical Modeling of Linearly Elastic Shells

Abstract

We analyze the asymptotic behavior of the three-dimensional displacement field of a linearly elastic shell as the thickness approaches zero. Under two distinct sets of assumptions on the geometry of the middle surface, on the boundary conditions, and on the order of magnitude of the applied forces, convergence theorems can be established, which justify either the two-dimensional equations of a "flexural shell," or those of a "membrane shell." We also discuss the merits of the two-dimensional shell model of W.T. Koiter under the same sets of assumptions.

1. Introduction

De nombreuses structures élastiques comprennent des *coques*: Tours de refroidissement de centrales nucléaires, fuselages d'avions, carosseries et pare-brise d'automobiles, coques de navires, voiles de voiliers, pales de turbines, etc. Bien que chacune de ces coques soit une structure *tri-dimensionnelle*, sa "petite" épaisseur conduit naturellement à la modéliser par un problème *bi-dimensionnel*, c'est-à-dire posé sur la surface moyenne de la coque, donc par un problème *a priori* "plus simple" à traiter numériquement.

En effet, approcher *directement* la "solution tri-dimensonnelle" offre *actuellement* des difficultés considérables, qui néanmoins ne sont pas forcément insurmontables (voir [6], [55], [56] dans le cas des plaques). A l'inverse, on dispose de méthodes numériques performantes pour approcher les "solutions bi-dimensionnelles" (voir en particulier [7]), bien qu'il reste des problèmes délicats liés au "verrouillage numérique," comme d'ailleurs dans le cas des plaques (voir à ce suject [4], [5], [13], [14], [18], [19], [51]).

Les modèles bi-dimensionnels étant universellement utilisés, deux questions essentielles, en réalité intimement liées, se posent:

Une "géométrie" de la surface moyenne et des conditions aux limites étant données, comment choisir parmi les nombreux modèles bi-dimensionnels "disponibles"? Cette première question est fondamentale: En effet, *il ne sert à rien de concevoir et d'utiliser une méthode numérique performante pour approcher la solution d'un "mauvais" modèle!* Autrement dit, la solution du modèle bi-dimensionnel finalement retenu est-elle "suffisamment proche" de la solution du modèle tri-dimensionnel si l'épaisseur est "suffisamment petite?" Autrement dit encore:

Comment justifier d'une manière rationnelle les modèles bi-dimensionnels à partir du modèle de l'élasticité tri-dimensionnelle? Cette seconde question relève de l'*analyse asymptotique*: Elle consiste à *analyser le comportement de la solution tri-dimensionnelle lorsque l'épaisseur tend vers zéro, la surface moyenne étant donnée.*

Le but de cet article est de faire le point sur la seconde question. De façon plus précise, on se borne à l'*élasticité linéarisée*, et on étudie du point de vue de l'*analyse asymptotique* trois *modèles bi-dimensionnels* de coques bien connus, celui "*en flexion*," celui "*en membrane*," et celui "*de W.T. Koiter*." On reprend ici les résultats annoncés dans trois Notes récentes [24], [25], [29] (on en trouvera les démonstrations détaillées dans [27], [28], [30]), en suivant la présentation de [21].

Dans ce qui suit, on note ε le "petit" paramètre (destiné à tendre vers zéro), et on appelle 2ε "l'épaisseur" de la coque. Il s'agit la d'un *abus de langage*, 2ε mesurant en réalité le *rapport* entre l'épaisseur effective de la coque et une "dimension caractéristique" de la coque. Ainsi, pour une tour de refroidissement, 2ε est-il de l'ordre de $1/500$, l'épaisseur moyenne étant de l'ordre de 30 cm et la hauteur de la tour de l'ordre de 150 m. C'est un *ordre de grandeur* qu'il est utile d'avoir présent à l'esprit.

2. Le problème tri-dimensionnel d'une coque linéairement élastique

Les indices ou exposants grecs (sauf ε) prennent leurs valeurs dans l'ensemble $\{1, 2\}$, et les indices ou exposants latins dans l'ensemble $\{1, 2, 3\}$. La convention de la sommation par rapport aux indices et exposants répétés est utilisée. Le produit scalaire euclidien et la produit vectoriel de $\mathbf{u}, \mathbf{v} \in \mathbf{R}^3$ sont notés $\mathbf{u} \cdot \mathbf{v}$

et $\mathbf{u} \times \mathbf{v}$; la norme euclidienne de $\mathbf{u} \in \mathbf{R}^3$ est notée $|\mathbf{u}|$.

Soit ω un ouvert borné connexe de $\mathbf{R}^2$, de frontière $\partial\omega$ lipschitzienne, ω étant localement d'un même côté de $\partial\omega$. On note $y = (y^\alpha)$ un point courant de $\bar{\omega}$, $\partial_\alpha = \partial/\partial y^\alpha$, et $\partial_{\alpha\beta} = \partial_\alpha\partial_\beta$. Soit $\varphi : \bar{\omega} \to \mathbf{R}^3$ une application injective, de classe C^3, telle que les deux vecteurs $\mathbf{a}_\alpha = \partial_\alpha\varphi$ sont linéairement indépendants en tout point de $\bar{\omega}$. En tout point de S, on définit les vecteurs $\mathbf{a}^\beta$ du plan tangent par les relations $\mathbf{a}^\beta \cdot \mathbf{a}_\alpha = \delta_\alpha^\beta$, la *vecteur normal* $\mathbf{a}^3 = (\mathbf{a}_1 \times \mathbf{a}_2)/|\mathbf{a}_1 \times \mathbf{a}_2|$, les *symboles de Christoffel* $\Gamma_{\alpha\beta}^\sigma = \mathbf{a}^\sigma \cdot \partial_\alpha\mathbf{a}_\beta$, le *tenseur métrique* par ses composantes covariantes $a_{\alpha\beta} = \mathbf{a}_\alpha \cdot \mathbf{a}_\beta$ ou contravariantes $a^{\alpha\beta} = \mathbf{a}^\alpha \cdot \mathbf{a}^\beta$, l'*élément d'aire* $\sqrt{a}dy$ où $a = \text{dét}(a_{\alpha\beta})$, et le *tenseur de courbure* par ses composantes covariantes $b_{\alpha\beta} = -\mathbf{a}_\alpha \cdot \partial_\beta\mathbf{a}_3$ ou mixtes $b_\alpha^\beta = a^{\beta\sigma}b_{\sigma\alpha}$. Enfin, on pose $c_{\alpha\beta} = b_\alpha^\sigma b_{\sigma\beta}$. On notera les symétries:

$$\Gamma_{\alpha\beta}^\sigma = \Gamma_{\beta\alpha}^\sigma, \quad b_{\alpha\beta} = b_{\beta\alpha}, \quad c_{\alpha\beta} = c_{\beta\alpha}\,.$$

Pour tout $\varepsilon > 0$, on définit les ensembles $\Omega^\varepsilon = \omega \times]-\varepsilon, \varepsilon[$ et $\Gamma_0^\varepsilon = \gamma_0 \times [-\varepsilon, \varepsilon]$, où $\gamma_0 \subset \partial\omega$ est de *longueur* > 0, on note $x^\varepsilon = (x_i^\varepsilon)$ un point courant de $\bar{\Omega}^\varepsilon$, et on pose $\partial_i^\varepsilon = \partial/\partial x_i^\varepsilon$; on a donc $x_\alpha^\varepsilon = y_\alpha$ et $\partial_\alpha^\varepsilon = \partial_\alpha$. On considère une *coque élastique* de *surface moyenne* S et d'*épaisseur* 2ε, c'est-à-dire dont la *configuration de référence* est $\mathbf{\Phi}(\bar{\Omega}^\varepsilon)$, où l'application $\mathbf{\Phi} : \bar{\Omega}^\varepsilon \to \mathbf{R}^3$ est définie par $\mathbf{\Phi}(x^\varepsilon) = \varphi(y) + x_3^\varepsilon\mathbf{a}^3(y)$ pour tout $x^\varepsilon = (y, x_3^\varepsilon) \in \bar{\Omega}^\varepsilon$. Pour ε suffisamment petit, les vecteurs $\mathbf{g}_i^\varepsilon = \partial_i^\varepsilon\mathbf{\Phi}$ sont linéairement indépendants. On définit alors les vecteurs $\mathbf{g}^{j,\varepsilon}$ par les relations $\mathbf{g}^{j,\varepsilon} \cdot \mathbf{g}_i^\varepsilon = \delta_i^j$, puis les composantes covariantes $g_{ij}^\varepsilon = \mathbf{g}_i^\varepsilon \cdot \mathbf{g}_j^\varepsilon$ et contravariantes $g^{ij,\varepsilon} = \mathbf{g}^{i,\varepsilon} \cdot \mathbf{g}^{j,\varepsilon}$ du *tenseur métrique*, l'élément de volume $\sqrt{g^\varepsilon}dx^\varepsilon$ où $g^\varepsilon = \text{dét}(g_{ij}^\varepsilon)$, et les *symboles de Christoffel* $\Gamma_{ij}^{p,\varepsilon} = \mathbf{g}^{p,\varepsilon} \cdot \partial_j^\varepsilon\mathbf{g}_i^\varepsilon$.

Les inconnues du problème sont les trois *composantes covariantes* $u_i^\varepsilon : \bar{\Omega}^\varepsilon \to \mathbf{R}$ du *déplacement* $u_i^\varepsilon\mathbf{g}^{i,\varepsilon}$ des points de la coque, le vecteur $u_i^\varepsilon(x^\varepsilon)\mathbf{g}^{i,\varepsilon}(x^\varepsilon)$ représentant pour tout $x^\varepsilon \in \bar{\Omega}^\varepsilon$ le déplacement du point $\mathbf{\Phi}(x^\varepsilon)$. On suppose la coque *partiellement encastrée*, en ce sens que le déplacement s'annule sur la partie $\mathbf{\Phi}(\Gamma_0^\varepsilon)$ de la frontière latérale de la coque. En élasticité linéarisée, l'*inconnue* $\mathbf{u}^\varepsilon = (u_i^\varepsilon)$ résout le problème variationnel:

$$\mathbf{u}^\varepsilon \in \mathbf{V}(\Omega^\varepsilon) = \left\{\mathbf{v}^\varepsilon = (v_i^\varepsilon) \in \mathbf{H}^1(\Omega^\varepsilon); \mathbf{v}^\varepsilon = \mathbf{0} \text{ sur } \Gamma_0^\varepsilon\right\}, \tag{1}$$

$$\int_{\Omega^\varepsilon} A^{ijkl,\varepsilon} e_{k\|l}^\varepsilon(\mathbf{u}^\varepsilon) e_{i\|j}^\varepsilon(\mathbf{v}^\varepsilon)\sqrt{g^\varepsilon}dx^\varepsilon = \int_{\Omega^\varepsilon} f^{i,\varepsilon}v_i^\varepsilon\sqrt{g^\varepsilon}dx^\varepsilon \tag{2}$$

pour tout $\mathbf{v}^\varepsilon \in \mathbf{V}(\Omega^\varepsilon)$, où

$$A^{ijkl,\varepsilon} = \lambda g^{ij,\varepsilon}g^{kl,\varepsilon} + \mu\left(g^{ik,\varepsilon}g^{jl,\varepsilon} + g^{il,\varepsilon}g^{jk,\varepsilon}\right) \tag{3}$$

désignent les composantes contravariantes du *tenseur de l'élasticité tridimensionnelle*, $\lambda > 0$ et $\mu > 0$ sont les *constantes de Lamé*, supposées *indépendantes de* ε, du matériau élastique constituant la coque,

$$e^{\varepsilon}_{i\|j}(\mathbf{v}^{\varepsilon}) = \frac{1}{2}\left(\partial^{\varepsilon}_i v^{\varepsilon}_j + \partial^{\varepsilon}_j v^{\varepsilon}_i\right) - \Gamma^{p,\varepsilon}_{ij} v^{\varepsilon}_p \tag{4}$$

désignent les composantes covariantes du *tenseur linéarisé de déformation* associé à un champ de déplacements $v^{\varepsilon}_i \mathbf{g}^{i,\varepsilon}$ des points de $\Phi(\Omega^{\varepsilon})$, et $f^{i,\varepsilon} \in L^2(\Omega^{\varepsilon})$ sont les composantes covariantes de la *densité des forces de volume appliquées* (des forces de surface sur les faces supérieure et inférieure de la coque peuvent être également prises en compte).

Pour tout $\varepsilon > 0$, *le problème* (1)–(2) *a une solution et une seule.* Pour le voir, il suffit de ré-écrire ce problème en coordonnées cartésiennes, et d'utiliser l'inégalité de Korn "usuelle," comme dans [37, p. 115 et suivantes]; on peut aussi, comme dans [22], établir directement une *inégalité de Korn en coordonnées curvilignes* (combinée ensuite avec l'uniforme définie positivité du tenseur $(A^{ijkl,\varepsilon})$ de (3)), dont la précédente est un cas particulier. Pour établir cette inégalité, un usage essentiel est fait d'un *lemme de J.-L. Lions*, mentionné la première fois dans la Note (27), p. 320 de [48], établi ensuite dans [37, p. 111 et suivantes] pour des ouverts à frontière régulière, étendu enfin à des ouverts à frontière "seulement" lipschitzienne (comme l'ouvert Ω^{ε}) dans [2] et [12].

Dans l'analyse asymptotique qui va suivre, les tenseurs "tri-dimensionnels" $(A^{ijkl,\varepsilon})$ et $\left(e^{\varepsilon}_{i\|j}(\mathbf{v}^{\varepsilon})\right)$ définis en (3) – (4) vont être "remplacés à la limite" par leurs analogues "bi-dimensionnels" $(a^{\alpha\beta\sigma\tau})$, $(\gamma_{\alpha\beta}(\boldsymbol{\eta}))$ et $(\rho_{\alpha\beta}(\boldsymbol{\eta}))$ définis par:

$$a^{\alpha\beta\sigma\tau} = \frac{4\lambda\mu}{\lambda+2\mu} a^{\alpha\beta}a^{\sigma\tau} + 2\mu(a^{\alpha\sigma}a^{\beta\tau} + a^{\alpha\tau}a^{\beta\tau})\,, \tag{5}$$

$$\gamma_{\alpha\beta}(\boldsymbol{\eta}) = \frac{1}{2}\left(\partial_\alpha\eta_\beta + \partial_\beta\eta_\alpha\right) - \Gamma^{\sigma}_{\alpha\beta}\eta_\sigma - b_{\alpha\beta}\eta_3\,, \tag{6}$$

$$\begin{aligned}\rho_{\alpha\beta}(\boldsymbol{\eta}) = {} & \partial_{\alpha\beta}\eta_3 - \Gamma^{\sigma}_{\alpha\beta}\partial_\sigma\eta_3 + b^{\sigma}_{\beta}\left(\partial_\alpha\eta_\sigma - \Gamma^{\tau}_{\alpha\sigma}\eta_\tau\right) + b^{\sigma}_{\alpha}\left(\partial_\beta\eta_\sigma - \Gamma^{\tau}_{\beta\sigma}\eta_\tau\right) \\ & + \left(\partial_\beta b^{\sigma}_{\alpha} + \Gamma^{\sigma}_{\beta\tau}b^{\tau}_{\alpha} - \Gamma^{\tau}_{\alpha\beta}b^{\sigma}_{\tau}\right)\eta_\sigma - c_{\alpha\beta}\eta_3\,.\end{aligned} \tag{7}$$

Ces fonctions représentent respectivement les composantes contravariantes du *tenseur d'élasticité de la surface* S, et les composantes covariantes des *tenseurs linéarisés de déformation* et *de changement de courbure* associés à un champ de déplacements $\eta_i\mathbf{a}^i$ des points de S.

3. Analyse asymptotique des coques "en flexion"

3.1. Le problème tri-dimensionnel sur un ouvert fixe; hypothèses sur les forces

On pose $\Omega = \omega\times]-1,1[$ et $\Gamma_0 = \gamma_0 \times [-1,1]$. Au point $x = (x_i) \in \bar{\Omega}$, on associe le point $x^\varepsilon = (x_i^\varepsilon) \in \bar{\Omega}^\varepsilon$ défini par $x_\alpha^\varepsilon = x_\alpha$, $x_3^\varepsilon = \varepsilon x_3$, en particulier dans (8) – (9), (12) et (16). À l'inconnue $\mathbf{u}^\varepsilon$ et au champ $\mathbf{v}^\varepsilon$ apparaissant dans les équations (2) , on associe l'*inconnue mise à l'échelle* $\mathbf{u}(\varepsilon) = (u_i(\varepsilon))$ et le *champ mis à l'échelle* $\mathbf{v} = (v_i)$ en posant

$$u_i(\varepsilon)(x) = u_i^\varepsilon(x^\varepsilon) \text{ et } v_i(x) = v_i^\varepsilon(x^\varepsilon) \tag{8}$$

pour tout $x \in \bar{\Omega}$.

On *suppose* enfin qu'il existe des foncions $f^i \in L^2(\Omega)$ telles que

$$f^{i,\varepsilon}(x^\varepsilon) = \varepsilon^2 f^i(x) \text{ pour tout } x \in \Omega\,. \tag{9}$$

Alors l'*inconnue mise à l'échelle* $\mathbf{u}(\varepsilon)$ *résout le problème variationnel suivant, posé maintenant sur l'ouvert* Ω *indépendant de* ε:

$$\mathbf{u}(\varepsilon) \in \mathbf{V}(\Omega) = \{\mathbf{v} = (v_i) \in \mathbf{H}^1(\Omega); \mathbf{v} = \mathbf{0} \text{ sur } \Gamma_0\}\,, \tag{10}$$

$$\int_\Omega A^{ijkl}(\varepsilon) e_{k\|l}(\varepsilon)(\mathbf{u}(\varepsilon)) e_{i\|j}(\varepsilon)(\mathbf{v}) \sqrt{g(\varepsilon)}dx = \varepsilon^2 \int_\Omega f^i v_i \sqrt{g(\varepsilon)}dx \tag{11}$$

pour tout $\mathbf{v} \in \mathbf{V}(\Omega)$, où

$$A^{ijkl}(\varepsilon)(x) = A^{ijkl,\varepsilon}(x^\varepsilon) \text{ pour tout } x \in \Omega\,, \tag{12}$$

$$e_{\alpha\|\beta}(\varepsilon)(\mathbf{v}) = \frac{1}{2}(\partial_\alpha v_\beta + \partial_\beta v_\alpha) - \Gamma^p_{\alpha\beta}(\varepsilon) v_p\,, \tag{13}$$

$$e_{\alpha\|3}(\varepsilon)(\mathbf{v}) = \frac{1}{2}\left(\partial_\alpha v_3 + \frac{1}{\varepsilon}\partial_3 v_\alpha\right) - \Gamma^\sigma_{\alpha\beta} v_\sigma\,, \tag{14}$$

$$e_{3\|3}(\varepsilon)(\mathbf{v}) = \frac{1}{\varepsilon}\partial_3 v_3\,, \tag{15}$$

$$\Gamma^p_{ij}(\varepsilon)(x) = \Gamma^{p,\varepsilon}_{ij}(x^\varepsilon) \text{ et } g(\varepsilon)(x) = g^\varepsilon(x^\varepsilon) \text{ pour tout } x \in \Omega\,. \tag{16}$$

Pour tout $\varepsilon > 0$, le problème (10)–(11) a, comme le problème (1)–(2), une solution et un seule $\mathbf{u}(\varepsilon)$. On se propose d'étudier le *comportement asymptotique* de $\mathbf{u}(\varepsilon)$ lorsque $\varepsilon \to 0$. On notera que *les équations* (11) *ne sont pas définies pour* $\varepsilon = 0$, et qu'il s'agit là d'un *problème de pertubations singulières*, au sens de J.-L. Lions [44].

3.2. Une inégalité de Korn généralisée sur l'ouvert Ω pour une coque à surface moyenne "générale"

La clef de voûte de la démonstration de la convergence de $\mathbf{u}(\varepsilon)$ lorsque $\varepsilon \to 0$ (Théorème 2) est l'*inégalité de Korn "généralisée"* (17), qui fait intervenir les fonctions $e_{i\|j}(\varepsilon)(\mathbf{v})$ définies en (13)–(15), au lieu des fonctions "traditionnelles" $\frac{1}{2}(\partial_i v_j + \partial_j v_i)$. C'est la *constante "en* $\frac{1}{\varepsilon}$" qui y apparaît qui permet en effet d'obtenir, avec l'hypothèse (9), les *majorations a priori* fondamentales sur la famille $(\mathbf{u}(\varepsilon))_{\varepsilon>0}$ (voir l'Etape 1 de la démonstration du Théorème 2). On notera que l'inégalité (17) est valable *sans restriction* sur l'application φ introduite au Par. 1; c'est dans ces sens qu'elle s'applique à une coque à surface moyenne "générale."

Remarque. On verra que, pour *certaines* "géométries" de la surface S et pour *certaines* conditions aux limites, on peut établir une inégalité de Korn "plus forte," où la constante est *indépendante de* ε (cf. Théorème 3).

On note $\|\cdot\|_{0,\Omega}$ et $\|\cdot\|_{1,\Omega}$ les normes des espaces $L^2(\Omega)$ et $H^1(\Omega)$.

Théorème 1. *Il existe des constantes $\varepsilon_1 > 0$ et $C_1 > 0$ telles que*

$$\|\mathbf{v}\|_{1,\Omega} \leq \frac{C_1}{\varepsilon} \left\{ \sum_{i,j} \|e_{i\|j}(\varepsilon)(\mathbf{v})\|_{0,\Omega}^2 \right\}^{1/2} \tag{17}$$

pour tout $0 < \varepsilon \leq \varepsilon_1$ et pour tout $\mathbf{v} \in \mathbf{V}(\Omega)$, où $\mathbf{V}(\Omega)$ est l'espace défini en (10).

La démonstration, dont on trouvera les détails dans [30], est longue et technique. Elle procède par contradiction, et repose de façon essentielle sur le *lemme du mouvement rigide infinitésimal* établi en [8] pour une surface "générale" (voir aussi [9], [10]): Si un champ de déplacement $\eta_i \mathbf{a}^i$ de la surface S, où $\boldsymbol{\eta} = (\eta_i) \in H^1(\omega) \times H^1(\omega) \times H^2(\omega)$ et $\eta_i = \partial_\nu \eta_3 = 0$ sur γ_0 supposé comme ici de longueur > 0, vérifie $\gamma_{\alpha\beta}(\boldsymbol{\eta}) = \rho_{\alpha\beta}(\boldsymbol{\eta}) = 0$ (cf. (6)–(7)), alors $\boldsymbol{\eta} = \mathbf{0}$.

Remarque. L'inégalité (17) n'est pas sans rappeler l'inégalité de Korn établie dans [39, Prop. 4.1] (voir aussi [3]) sur l'ouvert "variable" Ω^ε, mais avec les fonctions "traditionnelles" $\frac{1}{2}(\partial_i^\varepsilon v_j^\varepsilon + \partial_j^\varepsilon v_i^\varepsilon)$; en effet, il y apparaît également une constante "en $\frac{1}{\varepsilon}$."

3.3. Limite de la solution tri-dimensionnelle lorsque l'épaisseur tend vers zéro

On note ∂_ν la dérivée normale extérieure le long de $\partial\omega$. Suivant Sanchez–Palencia [54], on définit l'espace

$$\begin{aligned}\mathbf{V}_f(\omega) = \{&\boldsymbol{\eta} = (\eta_i) \in H^1(\omega) \times H^1(\omega) \times H^2(\omega);\ \eta_i = \partial_\nu \eta_3 = 0 \\ &\text{sur } \gamma_0, \gamma_{\alpha\beta}(\boldsymbol{\eta}) = 0 \text{ dans } \omega\},\end{aligned} \tag{18}$$

qui est donc un *espace de "déplacements inextensionnels"* (au premier ordre) de la surface S.

Dans le théorème qui suit, on montre que, *lorsque $\varepsilon \to 0$, l'inconnue mise à l'échelle $\mathbf{u}(\varepsilon) \in \mathbf{V}(\Omega)$ définie en (8) converge dans $\mathbf{H}^1(\Omega)$ vers une limite, qui peut être identifiée* (d'après (20)) *à la solution d'un problème bi-dimensionnel* (cf. (22)), c'est-à-dire posé sur l'ouvert ω. On justifie ainsi par un résultat de *convergence* l'analyse asymptotique formelle de [54] (voir aussi [50]) pour les coques dites "*en flexion*," selon la définition donnée plus loin.

Théorème 2. *Il existe* $\mathbf{u} = (u_i) \in \mathbf{V}(\Omega)$ *tel que:*

$$\mathbf{u}(\varepsilon) \to \mathbf{u} \ \textit{dans}\ \mathbf{H}^1(\Omega)\ \textit{lorsque}\ \varepsilon \to 0\,, \tag{19}$$

$$\mathbf{u}\ \textit{est indépendant de la variable "transverse"}\ x_3\,. \tag{20}$$

Par ailleurs, la "moyenne"

$$\boldsymbol{\zeta} = \frac{1}{2}\int_{-1}^{1} \mathbf{u}\, dx_3\,, \tag{21}$$

à laquelle la limite $\mathbf{u}$ *trouvée en (19) peut être identifiée d'après (21), appartient à l'espace* $\mathbf{V}_f(\omega)$ *défini en (18) et vérifie*

$$\frac{1}{3}\int_\omega a^{\alpha\beta\sigma\tau}\rho_{\sigma\tau}(\boldsymbol{\zeta})\rho_{\alpha\beta}(\boldsymbol{\eta})\sqrt{a}\, dy = \int_\omega \left\{\int_{-1}^{1} f^i\, dx_3\right\}\eta_i\sqrt{a}\, dy \tag{22}$$

pour tout $\boldsymbol{\eta} = (\eta_i) \in \mathbf{V}_f(\omega)$*, les fonctions* $a^{\alpha\beta\sigma\tau}$ *et* $\rho_{\alpha\beta}(\boldsymbol{\eta})$ *étant définies en (5) et (7).*

Démonstration. Elle est analogue *dans son principe* à celle suivie pour les *plaques* (voir [20], th. 3.3-1). Elle comporte quatre étapes principales, très brièvement esquissée ci-dessous; on en trouvera les détails dans [30].

Étape 1. La forme particulière de l'application $\boldsymbol{\Phi}$ qui définit les coques entraîne qu'il existe des constants $\varepsilon_0 > 0$ et $\alpha > 0$ telles que [cf. (12) et (16) pour les définitions]

$$\alpha \le g(\varepsilon)(x) \le \alpha^{-1} \ \text{ et }\ \alpha t_{ij}t_{ij} \le A^{ijkl}(\varepsilon)(x)t_{kl}t_{ij} \tag{23}$$

pour tout $0 < \varepsilon \le \varepsilon_0$, pour tout $x \in \bar{\Omega}$ et pour tout tenseur symétrique (t_{ij}). Faisant $\mathbf{v} = \mathbf{u}(\varepsilon)$ dans les équations (11) et utilisant l'inégalité de Korn généralisée (17) et les inégalités (23) on déduit que *les normes* $\|u_i(\varepsilon)\|_{1,\Omega}$ *et* $\left\|\frac{1}{\varepsilon}e_{i\|j}(\varepsilon)(\mathbf{u}(\varepsilon))\right\|_{0,\Omega}$ *sont bornées indépendamment de* ε. Il existe donc une *suite extraite*, encore notée $(\mathbf{u}(\varepsilon))_{\varepsilon>0}$, telle que (on note $\rightharpoonup$ la convergence faible):

$$u_i(\varepsilon) \rightharpoonup u_i \ \text{ dans }\ H^1(\Omega)\,,\ u_i(\varepsilon) \to u_i \ \text{ dans }\ L^2(\Omega)\,,$$
$$\frac{1}{\varepsilon}e_{i\|j}(\varepsilon)(\mathbf{u}(\varepsilon)) \rightharpoonup e^1_{i\|j} \ \text{ dans }\ L^2(\Omega)\,.$$

Étape 2. On établit que *les limites u_i et $e^1_{i\|j}$ trouvées à l'étape 1 vérifient les relations suivantes:*

$$\mathbf{u} \text{ est indépendant de } x_3 \,,$$
$$\boldsymbol{\zeta} = \frac{1}{2}\int_{-1}^{1} \mathbf{u}\, dx_3 \text{ appartient à l'espace } \mathbf{V}_f(\omega)\,,$$
$$\rho_{\alpha\beta}(\boldsymbol{\xi}) = -\partial_3 e^1_{\alpha\|\beta}\,,\quad e^1_{\alpha\|3} = 0\,,\quad e^1_{3\|3} = -\frac{\lambda}{\lambda+2\rho} a^{\alpha\beta} e^1_{\alpha\|\beta}\,.$$

Alors qu'il est aisé de voir que $\partial_3\mathbf{u}(\varepsilon) \to 0$ dans $\mathbf{L}^2(\Omega)$, (et donc que $\partial_3\mathbf{u} = 0$), que $\gamma_{\alpha\beta}(\mathbf{u}(\varepsilon)) \to 0$ dans $L^2(\Omega)$ (et donc que $\gamma_{\alpha\beta}(\boldsymbol{\xi}) = 0$), et que les fonctions $e^1_{i\|3}$ vérifient les deux dernières relations indiquées (on fixe une fonction $\mathbf{u} = \mathbf{V}(\Omega)$ dans les équations (11) et on fait tendre ε vers zéro); les autres relations ($\xi_3 \in H^2(\omega)$, $\xi_i = \partial_\nu \xi_3 = 0$ sur γ_0, et $\rho_{\alpha\beta}(\bar{u}) = -\partial_3 e^1_{\alpha\|\beta}$), sont de démonstrations très "techniques," beaucoup plus délicates à obtenir.

Étape 3. On montre que *la "moyenne" $\boldsymbol{\zeta} \in \mathbf{V}_f(\omega)$ vérifie effectivement les équations variationnelles (22).* Pour cela, on utilise dans les équations variationnelles (11) les fonctions "particulières" $\mathbf{v} = \mathbf{v}(\varepsilon) = (v_i(\varepsilon))$ proposées dans [50], définies par

$$v_\alpha(\varepsilon) = \eta_\alpha - \varepsilon x_3(\partial_\alpha \eta_3 + 2b^\sigma_\alpha \eta_\sigma)\,,\quad v_3(\varepsilon) = \eta_3\,,$$

où $\boldsymbol{\eta} = (\eta_i)$ est un élément arbitraire de l'espace $\mathbf{V}_f(\omega)$. Ces fonctions sont en effet telles que $\frac{1}{\varepsilon} e_{\alpha\|\beta}(\varepsilon)(\mathbf{v}(\varepsilon)) \to \{-x_3\rho_{\alpha\beta}(\boldsymbol{\eta})\}$ dans $L^2(\Omega)$, la suite $\left(\frac{1}{\varepsilon} e_{\alpha\|3}(\varepsilon)(\mathbf{v}(\varepsilon))\right)_{\varepsilon>0}$ est bornée dans $L^2(\Omega)$, et $e_{3\|3}(\varepsilon)(\mathbf{v}(\varepsilon)) = 0$, ces propriétés permettant de passer à la limite dans les équations (11).

Étape 4. Combinant les résultats des étapes précédentes avec les équations variationnelles (11) écrites avec $\mathbf{v} = \mathbf{u}(\varepsilon)$, on établit enfin les *convergences fortes:*

$$\frac{1}{\varepsilon} e_{i\|j}(\varepsilon)(\mathbf{u}(\varepsilon)) \to e^1_{i\|j} \text{ dans } L^2(\Omega)\,,\quad u_i(\varepsilon) \to u_i \text{ dans } H^1(\Omega)$$

pour *toute* la famille $(\mathbf{u}(\varepsilon))_{\varepsilon>0}$. □

Prenant en compte les hypothèses (9), on voit d'après (22) que la moyenne $\boldsymbol{\xi} \in \mathbf{V}_f(\omega)$ définie en (21) vérifie aussi les équations

$$\frac{\varepsilon^3}{3}\int_\omega a^{\alpha\beta\sigma\tau}\rho_{\sigma\tau}(\boldsymbol{\zeta})\rho_{\alpha\beta}(\boldsymbol{\eta})\sqrt{a}\, dy = \int_\omega \left\{\int_{-\varepsilon}^{\varepsilon} f^{i,\varepsilon} dx^\varepsilon_3\right\} \eta_i \sqrt{a}\, dy \qquad (24)$$

pour tout $\boldsymbol{\eta} = (\eta_i) \in \mathbf{V}_f(\omega)$. Les équations variationnelles (24) sont celles du *modèle bi-dimensionnel d'une coque "en flexion,"* en élasticité linéarisée. Comme il a été observé dans [33], l'existence et l'unicité de la solution des équations variationnelles (24) est un corollaire immédiat de l'existence et l'unicité de la solution du modèle de W.T. Koiter (voir Par. 5) et de l'uniforme définie positivité du tenseur $(a^{\alpha\beta\sigma\tau})$.

On dit qu'*une coque*, ou qu'*une famille de coques* indicée par $\varepsilon > 0$ comme au Par. 1, est "en flexion" si l'espace $\mathbf{V}_f(\omega)$ correspondant (défini en (18)) contient d'*autres* fonctions que la fonction nulle. *C'est naturellement le cas "intéressant" couvert par le Théorème 2.* On dit aussi dans ce cas que la surface S est *"à flexion pure non inhibée" pour les déplacements cinématiquement admissibles*, selon la terminolgie proposée par Sanchez–Palencia [52]. Par exemple, il en est ainsi (comme il a été également noté dans [52]) si la surface S est une portion de cylindre et l'ensemble $\varphi(\gamma_0)$ est contenu dans une ou plusieurs génératrices du cylindre; l'espace $\mathbf{V}_f(\omega)$ est d'ailleurs de dimension infinie dans ce cas.

Si l'espace $\mathbf{V}_f(\omega)$ *est réduit à* $\{\mathbf{0}\}$, c'est-à-dire si la surface S est *"à flexion pure inhibée" pour les déplacements cinématiquement admissibles*, selon la terminologie de [53], le Théorème 2 est encore applicable. Il montre seulement que la suite $(\mathbf{u}(\varepsilon))$ converge vers $\mathbf{0}$ dans $\mathbf{H}^1(\Omega)$. Autrement dit, si $\mathbf{V}_f(\omega) = \{\mathbf{0}\}$ et si les forces sont en $O(\varepsilon^2)$ (au sens de (9)) la coque correspondante devient "rigide" (les déplacements sont nuls) "à la limite;" on vérifie naturellement que ce dernier résultat est compatible avec ce qui suit.

4. Analyse asymptotique des coques "membranaires"

4.1. Le problème tri-dimensionnel sur un ouvert fixe; hypothèses sur les conditions aux limites et sur les forces

On suppose *dans tout le Par. 4* que

$$\gamma_0 = \partial\omega \,. \tag{25}$$

Autrement dit, pour tout $\varepsilon > 0$, le déplacement "tri-dimensionnel" $u_i^\varepsilon \mathbf{g}^{i,\varepsilon}$ s'annule sur *toute* la frontière latérale $\mathbf{\Phi}(\Gamma_0^\varepsilon)$ de la coque, Γ_0^ε désignant l'ensemble $\partial\omega \times [-\varepsilon, \varepsilon]$ *dans tout le Par. 4.* La coque est alors dite *totalement encastrée.*

On pose $\Omega = \omega\times]-1, 1[$, $\Gamma_0 = \partial\omega \times [-1, 1]$, on note $x_0 = (x_i)$ un point courant de $\bar{\Omega}$, et on pose $\partial_i = \partial/\partial x_i$. Comme au Par. 3.1, on associe au point $x = (x_i) \in \bar{\Omega}$ le point $x^\varepsilon = (x_i^\varepsilon) \in \bar{\Omega}^\varepsilon$ défini par $x_\alpha^\varepsilon = x_\alpha$ et $x_3^\varepsilon = \varepsilon x_3$. À l'inconnue $\mathbf{u}^\varepsilon$ et au champ $\mathbf{v}^\varepsilon$ apparaissant dans les équations (2), on associe l'*inconnue mise à l'échelle* $\mathbf{u}(\varepsilon) = (u_i(\varepsilon))$ et le *champ mis à l'échelle* $\mathbf{v} = (v_i)$ comme en (8).

Par contre, on suppose maintenant *au lieu de* (9) qu'il existe des fonctions $f^i \in L^2(\Omega)$ *indépendantes de* ε telles que

$$f^{i,\varepsilon}(x^\varepsilon) = f^i(x) \text{ pour tout } x \in \Omega \,. \tag{26}$$

Alors *l'inconnue mise à l'échelle* $\mathbf{u}(\varepsilon)$ *résout le problème variationnel sui-*

vant, posé maintenant sur l'ouvert Ω indépendant de ε (comparer avec (11)):

$$\mathbf{u}(\varepsilon) \in \mathbf{V}(\Omega) = \{\mathbf{v} = (v_i) \in \mathbf{H}^1(\Omega);\ \mathbf{v} = \mathbf{0} \text{ sur } \Gamma_0 = \partial\omega \times [-1,1]\}\ , \tag{27}$$

$$\int_\Omega A^{ijkl}(\varepsilon) e_{k\|l}(\varepsilon)(\mathbf{u}(\varepsilon)) e_{i\|j}(\varepsilon)(\mathbf{v}) \sqrt{g(\varepsilon}\, dx = \int_\Omega f^i v_i \sqrt{g(\varepsilon)}\, dx \tag{28}$$

pour tout $\mathbf{v} \in \mathbf{V}(\Omega)$, les fonctions $A^{ijkl}(\varepsilon)$, $e_{i\|j}(\varepsilon)(\mathbf{v})$, et $g(\varepsilon)$, étant définies comme en (12)–(16).

Pour tout $\varepsilon > 0$, le problème (27)–(28) a, comme le problème (1)–(2), une solution et une seule $\mathbf{u}(\varepsilon)$. On se propose d'étudier le *comportement asymptotique de* $\mathbf{u}(\varepsilon)$ *lorsque* $\varepsilon \to 0$.

4.2. Une inégalité de Korn généralisée sur l'ouvert Ω pour une coque totalement encastrée à surface moyenne "uniformément elliptique"

La clef de voûte de la démonstration de la convergence de $\mathbf{u}(\varepsilon)$ est une "autre" *inégalité de Korn généralisée* (31). Il est en effet remarquable que, *dans certains cas* (cf. Théorèmes 3 et 4) la "constante en $\frac{1}{\varepsilon}$" apparaissant dans l'inégalité de Korn généralisée (17) peut être remplacée par une constante *indépendante de* ε, au prix il est vrai du "remplacement de $\|v_3\|_{1,\Omega}$ par $\|v_3\|_{0,\Omega}$ dans le membre de gauche de l'inégalité":

Théorème 3. *On définit l'espace*

$$V_m(\omega) = \{\boldsymbol{\eta} = (\eta_i);\ \eta_\alpha \in H_0^1(\omega),\ \eta_3 \in L^2(\omega)\} = H_0^1(\omega) \times H_0^1(\omega) \times L^2(\omega)\ . \tag{29}$$

On suppose qu'il existe une constante c telle que

$$\left\{\sum_\alpha \|\eta_\alpha\|_{1,\omega}^2 + \|\eta_3\|_{0,\omega}^2\right\}^{1/2} \le c \left\{\sum_{\alpha,\beta} \|\gamma_{\alpha\beta}(\boldsymbol{\eta})\|_{0,\omega}^2\right\}^{1/2} \tag{30}$$

pour tout $\boldsymbol{\eta} \in V_m(\omega)$, *où les fonctions* $\gamma_{\alpha\beta}(\boldsymbol{\eta})$ *sont définies en (6). Alors il existe des constantes* $\varepsilon_2 > 0$ *et* $C_2 > 0$ *telles que*

$$\left\{\sum_\alpha \|v_\alpha\|_{1,\Omega}^2 + \|v_3\|_{0,\Omega}^2\right\}^{1/2} \le C_2 \left\{\sum_{i,j} \|e_{i\|j}(\varepsilon)(\mathbf{v})\|_{0,\Omega}^2\right\}^{1/2} \tag{31}$$

pour tout $0 < \varepsilon \le \varepsilon_2$ *et pour tout* $\mathbf{v} \in \mathbf{V}(\Omega)$ *où* $\mathbf{V}(\Omega)$ *est l'espace défini en (27).*

La démonstration, dont on trouvera les détails dans [27], procède par contradiction; les *moyennes* $\frac{1}{2}\int_{-1}^{1} \mathbf{v}\, dx_3$ des fonctions $\mathbf{v} \in \mathbf{V}(\Omega)$ y jouent un rôle essentiel, en permettant notamment de passer de l'inégalité "bi-dimensionnelle" (30) à l'inégalité "tri-dimensionnelle" (31).

On dit que la surface S est *uniformément elliptique* s'il existe une constante b telle que

$$b > 0 \text{ et } b_{\alpha\beta}(y)\xi^\alpha\xi^\beta \geq b\xi^\alpha\xi^\alpha \text{ pour tout } y \in \bar{\omega} \text{ et } (\xi^\alpha) \in \mathbf{R}^2 . \tag{32}$$

Cette hypothèse signifie que les deux rayons de courbure principaux $R_1(y)$ et $R_2(y)$ sont du même signe, en tous les points $\varphi(y) \in S, y \in \bar{\omega}$, et qu'il existe une constante ρ telle que $0 < \rho \leq |R_\alpha(y)| \leq \rho^{-1}$ pour tout $y \in \bar{\omega}$. Le résultat suivant, dont on trouvera l'annonce dans les Notes [23] et [32], et les démonstrations détaillées dans [26] et [34], donne des conditions suffisantes pour que l'hypothèse (30) soit satisfaite.

Théorème 4. *Ou bien on suppose la frontière $\partial\omega$ de ω de classe C^3 et l'application φ analytique dans un ouvert contenant $\bar{\omega}$; ou bien on suppose $\partial\omega$ de classe C^4 et φ de classe C^5 sur $\bar{\omega}$. Alors l'équivalence de normes (30) est satisfaite si la surface $S = \varphi(\bar{\omega})$ est uniformément elliptique au sens de (32).*

4.3. Limite de la solution tri-dimensionelle lorsque l'épaisseur tend vers zéro

On fait l'hypothèse (25) et on suppose l'équivalence de normes (30) satisfaite. Dans le théorème qui suit, on montre que, *lorsque $\varepsilon \to 0$, l'inconnue mise l'échelle $\mathbf{u}(\varepsilon)$ définie en (8) converge dans $H^1(\Omega) \times H^1(\Omega) \times L^2(\Omega)$ vers une limite qui peut être identifiée* (d'après (34)) *à la solution d'un problème bi-dimensionnel* (cf. (36)), c'est-à-dire posé sur l'ouvert ω. On justifie ainsi par un résultat de *convergence* l'analyse asymptotique formelle de [54] (voir aussi [50]) pour les coques dites "*membranaires,*" selon la définition donnée plus loin.

Théorème 5. *On suppose que $\gamma_0 = \partial\omega$ et l'équivalence de normes (30) est satisfaite. Il existe des fonctions $u_\alpha \in H^1(\Omega)$ vérifiant $u_\alpha = 0$ sur Γ_0 et une fonction $u_3 \in L^2(\Omega)$ telles que*

$$u_\alpha(\varepsilon) \to u_\alpha \text{ dans } H^1(\Omega),\ u_3(\varepsilon) \to u_3 \text{ dans } L^2(\Omega) , \tag{33}$$

$$\mathbf{u} = (u_i) \text{ est indépendant de la variable ``transverse'' } x_3 . \tag{34}$$

Par ailleurs, la "moyenne"

$$\boldsymbol{\zeta} = \frac{1}{2}\int_{-1}^{1} \mathbf{u}\, dx_3 , \tag{35}$$

à laquelle la limite $\mathbf{u}$ trouvée en (33) peut être identifiée d'après (34), appartient à l'espace $\mathbf{V}_m(\omega)$ défini en (29) et vérifie

$$\int_\omega a^{\alpha\beta\sigma\tau}\gamma_{\sigma\tau}(\boldsymbol{\zeta})\gamma_{\alpha\beta}(\boldsymbol{\eta})\sqrt{a}\, dy = \int_\omega \left\{\int_{-1}^{1} f^i\, dx_3\right\} \eta_i\sqrt{a}\, dy \tag{36}$$

pour tout $\boldsymbol{\eta} \in \mathbf{V}_m(\omega)$, les fonctions $a^{\alpha\beta\sigma\tau}$ et $\gamma_{\alpha\beta}(\boldsymbol{\eta})$ étant définies en (5)-(6).

Démonstration. Comme celle du Théorème 2, elle est analogue *dans son principe* à celle suivie pour les *plaques*. Elle comporte quatre étapes principales, très brièvement esquissées ci-dessous; on en trouvera les détails dans [27].

Étape 1. Faisant $\mathbf{v} = \mathbf{u}(\varepsilon)$ dans les équations (28), et utilisant l'inégalité de Korn généralisée (31) et les inégalités (23), on déduit que *les normes* $\|u_\alpha(\varepsilon)\|_{1,\Omega}$, $\|u_3(\varepsilon)\|_{0,\Omega}$, $\|e_{i\|j}(\varepsilon)(\mathbf{u}(\varepsilon))\|_{0,\Omega}$ *sont bornées indépendamment de* ε. Il existe donc une *suite extraite*, encore notée $(\mathbf{u}(\varepsilon))_{\varepsilon>0}$, telle que

$$\begin{aligned} &u_\alpha(\varepsilon) \rightharpoonup u_\alpha \text{ dans } H^1(\Omega)\,,\quad u_\alpha(\varepsilon) \to u_\alpha \text{ dans } L^2(\Omega)\,,\\ &u_3(\varepsilon) \rightharpoonup u_3 \text{ dans } L^2(\Omega)\,,\quad e_{i\|j}(\varepsilon)(\mathbf{u}(\varepsilon)) \rightharpoonup e_{i\|j} \text{ dans } L^2(\Omega)\,. \end{aligned}$$

Étape 2. Combinant les convergences ci-dessus avec le comportement des fonctions $\Gamma^p_{ij}(\varepsilon)$ définies en (16) lorsque $\varepsilon \to 0\big(\Gamma^\sigma_{\alpha\beta}(\varepsilon) \to \Gamma^\sigma_{\alpha\beta}$, $\Gamma^3_{\alpha\beta}(\varepsilon) \to b_{\alpha\beta}$ et $\Gamma^\sigma_{\alpha 3}(\varepsilon) \to -b^\sigma_\alpha$ dans $C^0(\bar{\Omega})\big)$, on montre que *les fonctions* u_i *et* $e_{i\|j}$ *sont indépendantes de* x_3 et qu'*elles sont liées par les relations*

$$\begin{aligned} &e_{\alpha\|\beta} = \frac{1}{2}(\partial_\alpha u_\beta + \partial_\beta u_\alpha) - \Gamma^\sigma_{\alpha\beta}u_\sigma - b_{\alpha\beta}u_3\,,\\ &e_{\alpha\|\beta} = 0\,,\quad e_{3\|3} = -\frac{\lambda}{\lambda+2\mu}a^{\alpha\beta}e_{\alpha\|\beta}\,. \end{aligned}$$

Étape 3. Dans les équations variationnelles (28), on *fixe* une fonction $\mathbf{v} \in \mathbf{V}(\Omega)$ *indépendante de* x_3, puis on fait tendre ε vers 0. On déduit alors des résultats précédents que *la fonction* $\boldsymbol{\xi}$ *définie en* (35) *vérifie effectivement les équations variationnelles* (36).

Étape 4. Combinant les résultats des étapes précédentes avec les équations variationnelles (28) écrites avec $\mathbf{v} = \mathbf{u}(\varepsilon)$, on établit successivement les *convergences fortes*:

$$\begin{aligned} &e_{i\|j}(\varepsilon)(\mathbf{u}(\varepsilon)) \to e_{i\|j} \text{ dans } L^2(\Omega)\,,\\ &u_3(\varepsilon) \to u_3 \text{ dans } L^2(\Omega)\,,\quad u_\alpha(\varepsilon) \to u_\alpha \text{ dans } H^1(\Omega)\,, \end{aligned}$$

pour *toute* la famille $(\mathbf{u}(\varepsilon))_{\varepsilon>0}$. □

Prenant en compte les hypothèses (26), on voit d'après (36) que la moyenne $\boldsymbol{\xi} \in \mathbf{V}_m(\omega)$ définie en (35) vérifie aussi les équations

$$\varepsilon \int_\omega a^{\alpha\beta\sigma\tau}\gamma_{\sigma\tau}(\boldsymbol{\zeta})\gamma_{\alpha\beta}(\boldsymbol{\eta})\sqrt{a}\,dy = \int_\omega \left\{\int_{-\varepsilon}^{\varepsilon} f^{i,\varepsilon}dx_3^\varepsilon\right\}\eta_i\sqrt{a}\,dy \qquad (37)$$

pour tout $\boldsymbol{\eta} \in \mathbf{V}_m(\omega)$. Les équations variationelles (37) sont celles du *modèle bi-dimensionnel d'une coque "en membrane,"* en élasticité linéarisée. L'existence et l'unicité de la solution des équations variationnelles (37) est un corollaire immédiat du Théorème 4 (établi dans [23], [26] et [32], [34]) et de l'uniforme définie positivité du tenseur $(a^{\alpha\beta\sigma\tau})$.

On dit qu'*une coque*, ou qu'une *famille de coques* indicée par $\varepsilon > 0$ comme au Par. 1, est "*membranaire*" si $\gamma_0 = \partial\omega$ et si l'inégalité (30) est satisfaite.

5. Justification du modèle de W.T. Koiter

5.1. Le modèle bi-dimensionnel de W.T. Koiter

Pour tout $\varepsilon > 0$, on considère les "mêmes" coques qu'au Par. 2, donc de mêmes surface moyenne $S = \varphi(\bar{\omega})$ et épaisseur 2ε, de mêmes constantes de Lamé λ et μ, et soumises aux mêmes forces de volume, de densité $(f^{i,\varepsilon})$.

A la suite d'un travail fondamental de F. John [38], W.T. Koiter [40] a proposé un modèle de coques *bi-dimensionnel*, c'est-à-dire posé sur l'ouvert ω, dont voici la version *linéarisée*: Les inconnues sont les trois *composantes covariantes* $\zeta_i^\varepsilon : \bar{\omega} \to \mathbf{R}$ du *déplacement* $\zeta_i^\varepsilon \mathbf{a}^i$ des points de la *surface moyenne* $S = \varphi(\bar{\omega})$ de la coque, le vecteur $\zeta_i^\varepsilon(y)\mathbf{a}^i(y)$ représentant pour tout $y \in \bar{\omega}$ le déplacement du point $\varphi(y)$. L'*inconnue* $\boldsymbol{\zeta}^\varepsilon = (\zeta_i^\varepsilon)$ résout alors le problème variationnel:

$$\boldsymbol{\zeta}^\varepsilon \in \mathbf{V}_K(\omega) = \left\{\boldsymbol{\eta} = (\eta_i) \in H^1(\omega) \times H^1(\omega) \times H^2(\omega),\ \eta = \partial_\nu \eta_3 = 0 \text{ sur } \gamma_0\right\}, \tag{38}$$

$$\int_\omega \left\{\varepsilon a^{\alpha\beta\sigma\tau}\gamma_{\sigma\tau}(\boldsymbol{\zeta}^\varepsilon)\gamma_{\alpha\beta}(\boldsymbol{\eta}) + \frac{\varepsilon^3}{3}a^{\alpha\beta\sigma\tau}\rho_{\sigma\tau}(\boldsymbol{\zeta}^\varepsilon)\rho_{\alpha\beta}(\boldsymbol{\eta})\right\}\sqrt{a}\,dy =$$

$$= \int_\omega \left\{\int_{-\varepsilon}^{\varepsilon} f^{i,\varepsilon} dx_3^\varepsilon\right\}\eta_i\sqrt{a}\,dy \text{ pour tout } \boldsymbol{\eta} = (\eta_i) \in \mathbf{V}_K(\omega), \tag{39}$$

les tenseurs $(a^{\alpha\beta\sigma\tau})$, $(\gamma_{\alpha\beta}(\boldsymbol{\eta}))$, $(\rho_{\alpha\beta}(\boldsymbol{\eta}))$ étant définis comme en (5)–(7), et ∂_ν désignant la dérivée normale extérieure le long de $\partial\omega$.

Pour tout $\varepsilon > 0$, *le problème* (38)–(39) *a une solution et une seule* (cf. [8], [9], [11]). On va montrer que, *pour une coque "en flexion"* au sens entendu au Par. 3, et *pour une coque "membranaire"* au sens entendu au Par. 4, *la solution du modèle bi-dimensionnel* (38)–(39) *approche effectivement* (dans un sens qui va être précisé) *la solution du modèle tri-dimensionnel* (1)–(2) *lorsque* $\varepsilon \to 0$. On trouvera les démonstrations détaillées des résultats qui suivent dans [28].

5.2. Justification du modèle de W.T. Koiter pour les coques "en flexion"

On suppose dans ce Par. 5.2 que *l'espace " de déplacements inextensionnels"* $\mathbf{V}_f(\omega)$ défini en (18), qui peut aussi s'écrire sous la forme

$$\mathbf{V}_f(\omega) = \{\boldsymbol{\eta} \in \mathbf{V}_K(\omega); \gamma_{\alpha\beta}(\boldsymbol{\eta}) = 0 \text{ dans } \omega\}, \tag{40}$$

contient d'autres fonctions que la fonction nulle (l'espace $\mathbf{V}_K(\omega)$ est défini en (38)). Suivant la terminologie introduite au Par. 3.3, la coque correspondante est alors *"en flexion."*

Comme au Par. 3, on fait les hypothèses (9) sur les forces appliquées et on définit l'inconnue mise à l'échelle $\mathbf{u}(\varepsilon)$ comme en (8). Il résulte du Théorème 2 que

$$\frac{1}{2}\int_{-1}^{1} \mathbf{u}(\varepsilon) dx_3 \to \boldsymbol{\zeta} \text{ dans } \mathbf{H}^1(\omega) , \tag{41}$$

où $\boldsymbol{\zeta} \in \mathbf{V}_f(\omega)$ vérifie les équations (22). La solution $\boldsymbol{\zeta}^\varepsilon \in \mathbf{V}_K(\omega)$ des équations (39) vérifie par ailleurs

$$\frac{1}{\varepsilon^2}\int_\omega a^{\alpha\beta\sigma\tau}\gamma_{\sigma\tau}(\boldsymbol{\zeta}^\varepsilon)\gamma_{\alpha\beta}(\boldsymbol{\eta})\sqrt{a}\, dy + \int_\omega a^{\alpha\beta\sigma\tau}\rho_{\sigma\tau}(\boldsymbol{\zeta}^\varepsilon)\rho_{\alpha\beta}(\boldsymbol{\eta})\sqrt{a}\, dy$$
$$= \int_\omega \left\{\int_{-1}^{1} f^i \, dx_3\right\} \eta_i\sqrt{a}\, dy \text{ pour tout } \boldsymbol{\eta} \in \mathbf{V}_K(\omega) . \tag{42}$$

Comme il a été noté dans [52, Th. 2.1], il résulte de la relation (40) entre les espaces $\mathbf{V}_K(\omega)$ et $\mathbf{V}_f(\omega)$ et d'un théorème général de perturbations singulières appliqués aux équations (42) que $\boldsymbol{\zeta}^\varepsilon \rightharpoonup \boldsymbol{\zeta}$ dans $\mathbf{V}_K(\omega)$ "faible," où $\boldsymbol{\zeta} \in \mathbf{V}_f(\omega)$ désigne la solution des équations (22), ou (24); on peut en fait établir [28] que cette convergence est *forte*. Revenant, à l'aide de la relation (8), à l'inconnue "physique" $\mathbf{u}^\varepsilon$ dans la relation (41), on a établi le résultat suivant:

Théorème 6. *On suppose que l'espace* $\mathbf{V}_f(\omega)$ *défini en* (18) *n'est pas réduit à* $\{\mathbf{0}\}$ *et que les forces appliquées vérifient les relations (9). Alors*

$$\frac{1}{2\varepsilon}\int_{-\varepsilon}^{\varepsilon} \mathbf{u}^\varepsilon dx_3^\varepsilon = \boldsymbol{\zeta} + o(1) \text{ dans } \mathbf{H}^1(\omega) , \tag{43}$$

$$\boldsymbol{\zeta}^\varepsilon = \boldsymbol{\zeta} + o(1) \text{ dans } \mathbf{H}^1(\omega) , \tag{44}$$

où $\mathbf{u}^\varepsilon$ *est la solution du problème tri-dimensionnel* (1)–(2), $\boldsymbol{\zeta}^\varepsilon$ *celle du modèle bidimensionnel de W.T. Koiter* (38)–(39), *et* $\boldsymbol{\zeta}$ *celle du modèle bi-dimensionnel d'une coque "en flexion"* (24).

Les relations (43) et (44) constituent une *justification du modèle de W.T. Koiter lorsqu'il est appliqué à une coque "en flexion,"* au sens du Par. 3. Elles montrent en effet que, pour une telle coque, les moyennes $\frac{1}{2\varepsilon}\int_{-\varepsilon}^{\varepsilon} \mathbf{u}^\varepsilon \, dx_3^\varepsilon$ données par l'élasticité tri-dimensionnelle et la solution $\boldsymbol{\zeta}^\varepsilon$ du modèle de W.T. Koiter ont *même partie principale* par rapport à ε, dans l'espace $\mathbf{H}^1(\omega)$.

5.3. Justification du modèle de W.T. Koiter pour les coques "membranaires"

On suppose dans ce Par. 5.3 que $\gamma_0 = \partial\omega$ et que l'équivalence de normes (30) est satisfaite. Suivant la terminologie introduite au Par. 4.3, la coque correspondante est alors "membranaire."

Comme au Par. 4, on fait les hypothèses (26) sur les forces appliquées et on définit l'inconnue mise à l'échelle $\mathbf{u}(\varepsilon)$ comme en (8). Il résulte du Théorème 5 que

$$\frac{1}{2}\int_{-1}^{1} \mathbf{u}(\varepsilon)dx_3 \to \boldsymbol{\zeta} \text{ dans } H^1(\omega)\times H^1(\omega)\times L^2(\omega)\,, \tag{45}$$

où $\boldsymbol{\zeta} \in \mathbf{V}_m(\omega)$ vérifie les equations (36). La solution $\boldsymbol{\zeta}^\varepsilon \in \mathbf{V}_K(\omega)$ des équations (39) vérifie par ailleurs

$$\int_\omega a^{\alpha\beta\sigma\tau}\gamma_{\sigma\tau}(\boldsymbol{\zeta}^\varepsilon)\gamma_{\alpha\beta}(\boldsymbol{\eta})\sqrt{a}\,dy + \varepsilon^2\int_\omega a^{\alpha\beta\sigma\tau}\rho_{\sigma\tau}(\boldsymbol{\zeta}^\varepsilon)\rho_{\alpha\beta}(\boldsymbol{\eta})\sqrt{a}\,dy$$
$$= \int_\omega \left\{\int_{-1}^{1} f^i\,dx_3\right\}\eta_i\sqrt{a}\,dy \text{ pour tout } \boldsymbol{\eta}\in \mathbf{V}_K(\omega)\,. \tag{46}$$

Comme il a été démontré dans [36, Th. 7.1], puis noté dans [53, Th. 4.1], il résulte de l'hypothèse (30) et d'un théorème général de perturbations signulières appliqué aux équations (46) que $\boldsymbol{\zeta}^\varepsilon \to \boldsymbol{\zeta}$ dans $H^1(\omega)\times H^1(\omega)\times L^2(\omega)$, où $\boldsymbol{\zeta}\in \mathbf{V}_m(\omega)$ désigne la solution des équations (36), ou (37). Revenant, à l'aide de la relation (8), à l'inconnue "physique" $\mathbf{u}^\varepsilon$ dans la relation (45), on a établi le résultat suivant:

Théorème 7. *On suppose que $\gamma = \partial\omega$, que l'équivalence de normes (30) a lieu, et que les forces appliquées vérifient les relations (26). Alors*

$$\frac{1}{2\varepsilon}\int_{-\varepsilon}^{\varepsilon} \mathbf{u}^\varepsilon dx_3^\varepsilon = \boldsymbol{\zeta} + o(1) \textit{ dans } H^1(\omega)\times H^1(\omega)\times L^2(\omega)\,, \tag{47}$$

$$\boldsymbol{\zeta}^\varepsilon = \boldsymbol{\zeta} + o(1) \textit{ dans } H^1(\omega)\times H^1(\omega)\times L^2(\omega)\,, \tag{48}$$

où $\mathbf{u}^\varepsilon$ est la solution du problème tri-dimensionnel (1)–(2), *$\boldsymbol{\zeta}^\varepsilon$ celle du modèle bi-dimensionnel de W.T. Koiter* (38)–(39) *et $\boldsymbol{\zeta}$ celle du modèle bi-dimensionnel d'une coque "membranaire"* (37).

Les relations (47) et (48) constituent une *justification du modèle de W.T. Koiter lorsqu'il est appliqué à une coque "membranaire,"* au sens du Par. 4. Elles montrent en effet que, pour une telle coque, les moyennes $\frac{1}{2\varepsilon}\int_{-\varepsilon}^{\varepsilon}\mathbf{u}^\varepsilon dx_3^\varepsilon$ données par l'élasticité tri-dimensionnelle et la solution $\boldsymbol{\zeta}^\varepsilon$ du modèle de W.T. Koiter ont *même partie principale* par rapport à ε, dans l'espace $H^1(\omega)\times H^1(\omega)\times L^2(\omega)$.

6. Commentaires

6.1. Pour une *plaque* (cf. [20]), ainsi que pour une *coque "faiblement courbée"* (cf. [15], [31]), l'analyse asymptotique conduit *simultanément* aux modèles en membrane *et* en flexion. Pour une *coque "générale,"* l'analyse asymptotique *peut* conduire au contraire *soit* au modèle "en flexion" (cf. Par. 3), *soit* au modèle "membranaire" (cf. Par. 4), *selon la géométrie de la surface moyenne et les conditions aux limites.* C'est là une *première différence*, encore qu'il soit peut-être possible de l'éliminer par une *autre* démarche, suivie avec succès dans l'approche *formelle* proposée par [16].

Il y a d'*autres différences.* Ainsi les convergences sont-elles obtenues dans des *topologies différentes*; ainsi, la limite est-elle ici *indépendante de* x_3, alors qu'elle est un champ *de Kirchhoff-Love* pour une plaque; etc.

En utilisant les techniques de la Γ-*convergence*, Acerbi, Buttazzo et Percivale [1] ont également obtenu des théorèmes de convergence, mais dans un sens plus "faible" qu'ici, et seulement pour les moyennes sur l'épaisseur. Par ailleurs, la distinction entre le cas "membranaire" et le case "en flexion" n'y apparaît pas liée, comme ici à des conditions géométriques et cinématiques.

6.2. Les *premiers* résultats de *convergence* dans le cas "membranaire" ont été obtenus, semble-t-il, par Ph. Destuynder dans sa thèse d'Etat [35]. En particulier, les convergences qui y sont établies au Théorème 7.9 (p. 305) sous l'hypothèse d'uniforme elipticité de la surface moyenne (p. 280) sont "presque" celles du Théorème 5 pour les composantes $u_\alpha(\varepsilon)$, mais "seulement" $\varepsilon u_3(\varepsilon) \to 0$ dans $L^2(\omega)$ pour la troisième composante; le lien avec les équations membranaires limites restait ainsi "partiellement formel," puisqu'il reposait sur l'existence supposée d'un développement asymptotique formel de $u_3(\varepsilon)$ en puissances de ε.

6.3. En ce qui concerne la "geométrie" de la surface S et les conditions aux limites, on sait donc traiter *deux cas*: Celui où $\gamma_0 = \partial\omega$ et l'équivalence de normes (30) est satisfaite (alors, pour des forces en $O(1)$, le comportement limite est celui d'une coque "membranaire"), et celui où $\mathbf{V}_f(\omega) \neq \{\mathbf{O}\}$ (alors, pour des forces en $O(\varepsilon^2)$, la comportement limite est celui d'une coque "en flexion"). Il reste donc à étudier les *cas "intermédiaires,"* correspondant aux surfaces S *"à flexion pure mal inhibée" pour les déplacements cinématiquement admissibles*, selon la terminologie de [53]. Ainsi, l'espace

$$\mathbf{R}(\omega) = \{\boldsymbol{\eta} = (\eta_i) \in H^1(\omega) \times H^1(\omega) \times L^2(\omega);\ \eta_\alpha = 0 \\ sur \gamma_0, \gamma_{\alpha\beta}(\boldsymbol{\eta}) = 0 \text{ dans } \omega\}$$

peut-il être réduit à $\{\mathbf{0}\}$ sans que l'on ait une équivalence de normes analogue à celle de (30); ainsi, l'espace $\mathbf{V}_f(\omega)$ de (18) peut-il être réduit à $\{\mathbf{0}\}$ sans que l'espace $\mathbf{R}(\omega)$ défini ci-dessus le soit.

6.4. *Dans ces mêmes cas intermédiaires*, l'analyse asymptotique de la solution du *modèle de W.T. Koiter* vient d'être faite [17]. Elle conduit à des problèmes limites "*sensitifs,*" faisant intervenir des espaces fonctionnels tout à fait "inhabituels" (non contenus dans des espaces de distributions!), étudiés dans [45], [46].

6.5. Pour les coques *non linéairement élastiques*, l'analyse asymptotique *formelle* conduit également à des *modèles bi-dimensionnels non linéaires*, soit "membranaires" (cf. [49]), soit "en flexion" (cf. [47]). Un théorème de *convergence* dans le cas "membranaire" vient même d'être établi par Le Dret & Raoult [42] à l'aide des mêmes techniques de Γ-*convergence* qu'ils avaient déjà utilisées avec succès [41], [43] dans le cas des plaques non linéairement élastiques.

References

1. E. Acerbi, G. Buttazzo et D. Percivale, Thin inclusions in linear elasticity: a variational approach, *J. reine angew. Math.* **386** (1988), 99–115.
2. C. Amrouche et V. Girault, Propriétés fonctionnelles d'opérateurs; applications au problème de Stokes en dimension quelconque, Rapport R90025, Laboratoire d'Analyse Numérique, Université Pierre et Marie Curie, Paris, 1990.
3. G. Anzellotti, S. Baldo et D. Percivale, Dimension reduction in variational problems, asymptotic development in Γ-convergence and thin structures in elasticity, *Asymptotic Anal.* **9**, 61–100.
4. D.N. Arnold et F. Brezzi, Some new elements for the Reissner-Mindlin plate model, in *Boundary Value Problems for Partial Differential Equations and Applications*, J.L. Lions and C. Baiocchi, eds., Masson, Paris, (1993), 287–292.
5. D.N. Arnold et F. Brezzi, Locking free finite elements for shells, *Math. Comp.* (à paraître).
6. I. Babuška et L. Li, The problem of plate modeling: Theoretical and computational results, *Comput. Meth. Appl. Mech. Engrg.* **100** (1992), 249–273.
7. M. Bernadou, *Méthodes d'éléments finis pour les problèmes de coques minces*, Masson, Paris (1994).
8. M. Bernadou et P.G. Ciarlet, Sur l'ellipticité du modèle linéaire de coques de W.T. Koiter, *Computing Methods in Applied Sciences and Engineering*, R. Glowinski and J.-L. Lions, eds., Springer-Verlag, Berlin (1976), 89–136.
9. M. Bernadou, P.G. Ciarlet et B. Miara, Existence theorems for two-dimensional linear shell theories, *J. Elasticity* **34** (1994), 111–138.
10. A. Blouza et H. Le Dret, Sur le lemme du mouvement rigide, *C. R. Acad. Sci. Paris*, Sér. I **319** (1994), 1015–1020.
11. A. Blouza et H. Le Dret, Existence et unicité pour le modèle de Koiter pour une coque peu régulière, *C. R. Acad. Sci. Paris*, Sér. I **319** (1994), 1127–1132.
12. W. Borchers et H. Sohr, On the equations rot $v = g$ and div $u = f$ with zero boundary conditions, *Hokkaido Math. J.* **19** (1990), 67–87.
13. F. Brezzi, M. Fortin et R. Stenberg, Error analysis of mixed-interpolated elements for Reissner-Mindlin plates, *Math. Models Meth. Applied Sci.* **1** (1991), 125–151.

14. F. Brezzi et M. Fortin, Numerical approximation of Mindlin-Reissner plates, *Math. Comp.* **47** (1986), 151–158.

15. S. Busse, P.G. Ciarlet et B. Miara, Asymptotic analysis of linearly elastic shallow shells in curvilinear coordinates (à paraître).

16. D. Caillerie et E. Sanchez–Palencia, Elastic thin shells: asymptotic theory in the anisotropic and heterogeneous cases, *Math. Model. Meth. Appl. Sci.* **5** (1995), 473–496.

17. D. Caillerie et E. Sanchez–Palencia, A new kind of singular-stiff problems and application to thin elastic shells, *Math. Model. Meth. Appl. Sci.* **5** (1995), 47–66.

18. D. Chenais et J.-C. Paumier, On the locking phenomenon for a class of elliptic problems, *Numer. Math.* **67** (1994), 427–440.

19. D. Chenais et M. Zerner, Conditions nécessaires pour éviter le verrouillage numérique. Application aux arches, *C. R. Acad. Sci. Paris*, Sér. I **316** (1993), 1097–1102.

20. P.G. Ciarlet, *Plates and Junctions in Elastic Multi-Structures: An Asymptotic Analysis*, Masson, Paris, 1990.

21. P.G. Ciarlet, Mathematical modeling and numerical analysis of linearly elastic shells, in *Proceedings, International Congress of Mathematics '94 (Zürich, August 3–14, 1994)*, Birkhaüser, Basel (à paraître).

22. P.G. Ciarlet, *Mathematical Elasticity, Vol. II: Plates and Shells*, North-Holland, Amsterdam, 1996.

23. P.G. Ciarlet et V. Lods, Ellipticité des équations membranaires d'une coque uniformément elliptique, *C. R. Acad. Sci. Paris*, Sér. I **318** (1994), 195–200.

24. P.G. Ciarlet et V. Lods, Analyse asymptotique des coques linéairement élastiques. I. Coques "membranaires," *C. R. Acad. Sci. Paris*, Sér. I **318** (1994), 863–868.

25. P.G. Ciarlet et V. Lods, Analyse asymptotique des coques linéairement élastiques. III. Une justification du modèle de W.T. Koiter, *C. R. Acad. Sci. Paris*, Sér. I **319** (1994), 299–304.

26. P.G. Ciarlet et V. Lods, On the ellipticity of linear membrane shell equations, *J. Math. Pures Appl.*, (à paraître).

27. P.G. Ciarlet et V. Lods, Asymptotic analysis of linearly elastic shells. I. Membrane shells, *Arch. Rational Mech. Anal.* (à paraître).

28. P.G. Ciarlet et V. Lods, Asymptotic analysis of linearly elastic shells. III. A justification of W.T. Koiter's model, *Arch. Rational Mech. Anal.* (à paraître).

29. P.G. Ciarlet, V. Lods et B. Miara, Analyse asymptotique des coques linéairement élastiques. II. Coques "en flexion," *C. R. Acad. Sci. Paris*, Sér. I **319** (1994), 95–100.

30. P.G. Ciarlet, V. Lods et B. Miara, Asymptotic analysis of linearly elastic shells. II. Flexural shells, *Arch. Rational Mech. Anal.* (à parître).

31. P.G. Ciarlet et B. Miara, Justification of the two-dimensional equations of a linearly elastic shallow shell, *Comm. Pure Appl. Math.* **XLV** (1992), 327–360.

32. P.G. Ciarlet et E. Sanchez–Palencia, Un théorème d'existence et d'unicité pour les équations des coques membranaires, *C. R. Acad. Sci. Paris*, Sér. I **317** (1993), 801–805.

33. P.G. Ciarlet et E. Sanchez–Palencia, Ellipticity of bending and membrane shell equations, *Asymptotic Methods for Elastic Structures*, P.G. Ciarlet, L. Tralricho, and J.M. Viaño, eds., de Gruyter, Berlin (1995), 31–39.

34. P.G. Ciarlet et E. Sanchez–Palencia, An existence and uniqueness theorem for the two-dimensional linear membrane shell equations, *J. Math. Pures Appl.*, (à paraître).

35. P. Destuynder, Sur une justification des modèles de plaques et de coques par les méthodes asymptotiques, Thèse d'Etat, Université Pierre et Marie Curie, Paris, 1980.

36. P. Destuynder, A classification of thin shell theories, *Acta Applic. Math.* **4** (1985), 15–63.

37. G. Duvaut et J.-L. Lions, *Les inéquations en mécanique et en physique*, Dunod, Paris, 1972.

38. F. John, Estimates for the derivatives of the stresses in a thin shell and interior shell equations, *Comm. Pure Appl. Math.* **18** (1965), 235–267.

39. R.V. Kohn et M. Vogelius, A new model for thin plates with rapidly varying thickness. II. A convergence proof, *Quart. Appl. Math.* **43** (1985), 1–22.

40. W.T. Koiter, On the foundation of the linear theory of thin elastic shells, *Proc. Kon. Nederl. Akad. Wetensch.* **B73** (1970), 169–195.

41. H. Le Dret et A. Raoult, Le modèle de membrane non linéaire comme limite variationnelle de l'élasticité non linéaire tridimensionnelle, *C. R. Acad. Sci. Paris*, Sér. I **317** (1993), 221–226.

42. H. Le Dret et A. Raoult, Dérivation variationnelle du modèle non linéaire de coque membranaire, *C. R. Acad. Sci. Paris*, Sér. I, **320** (1995), 511–516.

43. H. Le Dret et A. Raoult, The nonlinear membrane model as variational limit of nonlinear three-dimensional elasticity, *J. Math. Pures Appl.*, (à paraître).

44. J.-L. Lions, *Perturbations singulières dans les problèmes aux limites et en contrôle optimal*, Springer-Verlag, Heidelberg, 1973.

45. J.-L. Lions et E. Sanchez–Palencia, Problèmes aux limites sensitifs, *C. R. Acad. Sci. Paris*, Sér. I **319** (1994), 1021–1026.

46. J.-L. Lions et E. Sanchez–Palencia, Problèmes sensitifs et coques élastiques minces, *Actes du Colloque à la Mémoire de Pierre Grisvard* (Paris, 24–26 nov. 1994), (à paraître), this volume.

47. V. Lods et B. Miara, Analyse asymptotique des coques “en flexion” non linéairèment élastiques, *C. R. Acad. Sci. Paris*, Sér. I, **321** (1995), 1097–1102.

48. E. Magenes et G. Stampacchia, I problemi al contorno per le equazioni differenziali di tipo ellitico, *Ann. Scuola Norm. Sup. Pisa* **12** (1958), 247–358.

49. B. Miara, Analyse asymptotique des coques membranaires non linéairement élastiques, *C. R. Acad. Sci. Paris*, Sér. I **318** (1994), 689–694.

50. B. Miara et E. Sanchez–Palencia, Asymptotic analysis of linearly elastic shells, *Asymptotic Analysis*, (à paraître).

51. J.-C. Paumier, On the locking phenomenon for a linearly elastic three-dimensional clamped plate, (à paraître).

52. E. Sanchez–Palencia, Statique et dynamique des coques minces. I. Cas de flexion pure non inhibée, *C. R. Acad. Sci. Paris*, Sér. I **309** (1989), 411–417.

53. E. Sanchez–Palencia, Statique et dynamique des coques minces. II. Cas de flexion pure inhibée, *C. R. Acad. Sci. Paris*, Sér. I **309** (1989), 531–537.

54. E. Sanchez–Palencia, Passage à la limite de l'élasticité tri-dimensionnelle à la théorie asymptotique des coques minces, *C. R. Acad. Sci. Paris*, Sér. II **311** (1990), 909–916.

55. C. Schwab, Dimensional Reduction for Elliptic Boundary Value Problems, Doctoral Dissertation, University of Maryland, College Park, 1989.

56. C. Schwab, A-posteriori modeling error estimation for hierarchic plate models, *Numer. Math.*, (à paraître).

Laboratoire d'Analyse Numérique, Tour 55, Université Pierre et Marie Curie, 4 Place Jussieu, 75005 Paris

Fully Nonlinear Equations by Linearization and Maximal Regularity, and Applications

Giuseppe Da Prato

1. Fully nonlinear equations

Let X and D be Banach spaces, with D continuously and densely embedded in X, and let $\mathcal{O}$ be an open set in D. We are concerned with the following initial value problem:

$$\begin{cases} u'(t) = F(t, u(t)), \; t \geq 0, \\ u(0) = u_0, \end{cases} \tag{1.1}$$

where $u_0 \in \mathcal{O}$, $F : [0, +\infty[\times\mathcal{O} \to X$, $(t, x) \to F(t, x)$, is a suitable mapping.

We assume

Hypothesis 1

(i) $F \in C([0, +\infty[\times\mathcal{O}; X)$.

(ii) *For any* $t \geq 0$, $F(t, \cdot) \in C^1(\mathcal{O}; X)$ *and for any* $x \in \mathcal{O}$, $F_x(t, x)$ *generates an analytic semigroup in* X *with domain* D. *Moreover, the graph norm of* $F_x(t, x)$ *is equivalent to the norm of* D.

(iii) $F_x \in C([0, +\infty[\times\mathcal{O}; \mathcal{L}(D; X))$.

We want to solve problem (1.1) for t near 0. In the following we set $A = F_x(0, u_0)$.

Typical examples are second order nonlinear parabolic equations, when the nonlinearity involves second order space derivatives, as for instance the following equation arising in the study of the shock displacement in a detonation phenomenon, see [3]:

$$\begin{cases} g_t(t,x) = \log\left(\dfrac{\exp(kgg_{xx}) - 1}{kg_{xx}}\right) - \dfrac{1}{2}g_x^2, \ t \geq 0, \ x \in [-L, L] \\ g_x(t,0) = g_x(t,L), \ t \geq 0, \\ g(0,x) = g_0(x), \ x \in [-L, L], \end{cases} \tag{1.2}$$

where k and L are positive constants. Here the problem is to prove the instability of the detonation solution $g_0 = 1$.

In this case we set

$$F(t,g) = \log\left(\frac{\exp(kgg_{xx}) - 1}{kg_{xx}}\right) - \frac{1}{2}g_x^2.$$

A natural choice of the spaces X and D is

$$X = C([-L,L]), \ \ D = \{\varphi \in C^2([-L,L]) : \ \varphi'(0) = \varphi'(L) = 0\}, \tag{1.3}$$

and $\mathcal{O} = \{\varphi \in X : \ \varphi(x) > 0, \ \forall\, x \in [-L,L]\}$. Moreover

$$A = F_x(t,1) = k^2 D_x^2 + I.$$

We recall a method, introduced by G. Da Prato and P. Grisvard [8] to solve problem (1.1), based on a linearization argument. From now on we assume $\mathcal{O} = D$ for simplicity. We set $A = F_x(0, u_0)$. We fix $T > 0$, u in $C([0,T]; D)$ and we consider the solution v to the problem

$$\begin{cases} v'(t) &= Av(t) + [F(t,u(t) - Au(t)], \ t \geq 0, \\ &:= Av(t) + G(t,u(t)), \ t \geq 0, \\ v(0) &= u_0. \end{cases} \tag{1.4}$$

Solving problem (1.1) is equivalent to finding a fixed point of the mapping

$$\Gamma : u \in C([0,T]; D) \to v.$$

However knowing that the function $t \to G(t,u(t))$ belongs to $C([0,T]; X)$, we cannot conclude that the solution v belongs to $C([0,T]; D)$. Consider in fact the linear equation

$$\begin{cases} z'(t) = Az(t) + f(t), \ t \in [0,T], \\ z(0) = x. \end{cases} \tag{1.5}$$

It is known that if $f \in C([0,T];X)$ then in general z does not belong to $C([0,T];D)$; a counterexample is given in [8], see also [4].

To overcome this problem we change slightly the spaces where we work. Namely we replace X and D with the *continuous* interpolation spaces $D_A(\alpha)$ and $D_A(\alpha+1)$, respectively, for some $\alpha \in]0,1[$. We recall, see [8], that $x \in D_A(\alpha)$ if and only if

$$\lim_{t\to 0} t^{1-\alpha} A e^{tA} x = 0,$$

where e^{tA}, $t \geq 0$, is the analytic semigroup generated by A. Moreover the space $D_A(\alpha+1)$ is defined by

$$D_A(\alpha+1) = \{x \in D(A): \; Ax \in D_A(\alpha)\}.$$

The spaces $D_A(\alpha)$ and $D_A(\alpha+1)$ are endowed respectively with the norms

$$\|x\|_{D_A(\alpha)} = \|x\| + \sup_{t\in]0,1]} \|t^{1-\alpha} A e^{tA} x\|,$$

$$\|x\|_{D_A(\alpha+1)} = \|x\| + \|Ax\|_{D_A(\alpha)}.$$

We remark that, for all $\alpha \in]0,1[$, $D_A(\alpha)$ is the closure of $D(A)$ in $D_A(\alpha,\infty)$, the real interpolation space introduced in [13].

Example 2.1 Let X and D be given by (1.3) and let $A: D \to X$ be defined by

$$A\varphi = \varphi'', \; \forall\, \varphi \in D = D(A).$$

Then A generates an analytic semigroup in X and for all $\alpha \in]0,1/2[$ we have

$$D_A(\alpha) = h^{2\alpha}([0,L]),$$

where $h^{2\alpha}([0,L])$ is the space of 2α–*little Hölder* continuous functions on $[0,L]$, that can be defined as the closure of $C^2([0,L])$ in $C^{2\alpha}([0,L])$.

In [8] the following maximal regularity result is proved.

Proposition 2.2 *Assume that A generates an analytic semigroup in X, and let $x \in D_A(\alpha+1)$ and $f \in C([0,T];D_A(\alpha))$. Then there exists a unique function z*

$$z \in C([0,T];D_A(\alpha+1)) \cap C^1([0,T];D_A(\alpha)),$$

fulfilling (1.5). *Moreover there exists a constant* $C_T > 0$ *such that*

$$\begin{aligned} \|z\|_{C([0,T];D_A(\alpha+1))} &+ \|z\|_{C^1([0,T];D_A(\alpha))} \\ &\leq C_T(\|x\|_{D_A(\alpha+1)} + \|f\|_{C([0,T];D_A(\alpha))}). \end{aligned} \tag{1.6}$$

Another regularity result is obtained when $f \in C^\alpha([0,T];X)$, $x \in D(A)$ and $Ax + f(0) \in D_A(\alpha)$, see the paper by E. Sinestrari, [17]. For a review on maximal regularity results in the literature, see the monograph by A. Lunardi [16].

Remark 2.3 Clearly C_T can be chosen non decreasing in T. However $\liminf_{T\to 0} C_T \neq 0$, in general. This is one of the main differences one finds when dealing with linearization of semilinear and quasilinear problems.■

To solve problem (1.1) we need additional assumptions, roughly that Hypothesis 1 holds when X is replaced by $D_A(\alpha)$ and Y is replaced by $D_A(\alpha+1)$. We remark that, by a general property of interpolation spaces, one has, in view of Hypothesis 1

$$D_A(\alpha) = D_{F_x(t_0,x_0)}(\alpha), \text{ for all } x_0 \in D,\ t_0 \in [0,+\infty[,\ \alpha \in]0,1[, \tag{1.7}$$

whereas, given $\alpha \in]0,1[$, the identity

$$D_A(\alpha+1) = D_{F_x(t_0,x_0)}(\alpha+1), \text{ for all } x_0 \in D,\ t_0 \in [0,+\infty[, \tag{1.8}$$

does not hold in general. So we shall assume, besides Hypothesis 1, that

Hypothesis 2

(i) *There exists* $\alpha \in]0,1[$ *such that condition* (1.8) *holds.*

(ii) $F \in C([0,+\infty[\times D_A(\alpha+1); D_A(\alpha))$.

(iii) *For any* $t \geq 0$, $F(t,\cdot) \in C^1(D_A(\alpha+1); D_A(\alpha))$.

(iv) $F_x \in C([0,+\infty[\times D_A(\alpha+1); \mathcal{L}(D_A(\alpha+1); D_A(\alpha)))$.

Using linearization the following result can be proved, see [8]

Theorem 2.4 *Assume that Hypothesis* 1 *and Hypothesis* 2 *hold. Let* $\alpha \in]0,1[$ *and let* $u_0 \in D_A(\alpha+1)$. *Then there is a maximal positive number* τ_{u_0} *and a unique solution* u *of* (1.1) *such that*

$$u \in C([0,\tau_{u_0}[; D_A(\alpha+1)) \cap C^1([0,\tau_{u_0}[; D_A(\alpha)).$$

Assume moreover that for any $T > 0$, F maps bounded subsets of $[0,T]\times D_A(\beta+1)$ into bounded subsets of $D_A(\beta)$, with $\beta \in]\alpha,1[$. If there exists $M_\beta > 0$ such that $u(t) \in D_A(\beta+1)$ for $t \in [0,\tau_{u_0}[$ and

$$\|u(t)\|_{D_A(\beta+1)} \le M_\beta,\ \forall\, t \in [0,\tau_{u_0}[, \tag{1.9}$$

then $\tau_{u_0} = +\infty$.

Sketch of the proof. For any $T > 0$ we consider the Banach space $Z_T = C([0,T]; D_A(\alpha+1))$, and the mapping Γ

$$Z_T \to Z_T, u \to v = \Gamma(u),$$

where v is the solution to problem (1.4). The mapping Γ is well defined in view of Proposition 1.2. Now we fix $T_0 > 0$ and we estimate $\|\Gamma(u) - \Gamma(\overline{u})\|_{Z_T}$ for any $u, \overline{u} \in Z_T$, and any $T \le T_0$. Setting $v = \Gamma(u), \overline{v} = \Gamma(\overline{u})$, we have

$$(v-\overline{v})' = A(v-\overline{v}) + \int_0^1 G_x(t, \xi u + (1-\xi)\overline{u}) \cdot (u-\overline{u}) d\xi.$$

Let u_0 be the constant function $u_0(t) = u_0,\ t \ge 0$. Now, if $\|u - u_0\|_{Z_T}, \|\overline{u} - u_0\|_{Z_T} \le R$, by (1.6) it follows

$$\|\Gamma(u) - \Gamma(\overline{u})\|_{Z_T} \le C_{T_0} M_1(R, T_0) \|u - \overline{u}\|_{Z_T},$$

where

$$M_1(R,T) = \sup_{t\in[0,T], \|x-u_0\|_{Z_T} \le R} \|G_x(t,x)\|.$$

Since $G_x(0, u_0) = 0$ there exists $T_1 \in]0, T_0]$ and $R_1 > 0$ such that

$$\|\Gamma(u) - \Gamma(\overline{u})\|_{Z_T} \le \frac{1}{2}\|u - \overline{u}\|_{Z_T},\ \forall\, T \in]0, T_1]. \tag{1.10}$$

We conclude the proof by showing that there exists $T_2 \le T_1$ such that Γ maps the ball

$$P(T, R_1) = \{u \in Z_T :\ \|u - u_0\|_{Z_T}, \|\overline{u} - u_0\|_{Z_T} \le R_1\},$$

into itself provided $T \le T_2$. In fact, if $T \le T_2$ and $u \in P(T, R_1)$, we have

$$\begin{aligned}\|\Gamma(u) - u_0\|_{Z_T} &\le \|\Gamma(u) - \Gamma(u_0)\|_{Z_T} + \|\Gamma(u_0) - u_0\|_{Z_T} \\ &\le \frac{1}{2}R_1 + \|\Gamma(u_0) - u_0\|_{Z_T}.\end{aligned} \tag{1.11}$$

Now we remark that $\Gamma(u_0) = z$ is the solution to the problem

$$\begin{cases} z'(t) = Az(t) + G(t, u_0), \ t \geq 0 \\ z(0) = u_0. \end{cases}$$

Thus $\Gamma(u_0) \in C([0,T]; D_A(\alpha+1))$ and $\Gamma(u_0)(0) = u_0$. Therefore, there exists $T_2 \in]0, T_1]$ such that

$$\|\Gamma(u_0) - u_0\|_{Z_{T_1}} \leq \frac{1}{2} R_1.$$

By (1.10) and (1.11) it follows that Γ is a contraction in $P(T_2, R_1)$. Thus Γ has a unique fixed point u in $P(T_2, R_1)$ and u is the solution of (1.1) in $[0, T_1]$. The remainder of the proof: existence and uniqueness of the maximal solution and the last statement of the theorem are based on standard arguments, see [8]. ■

We remark that it is not easy in general to find global solutions of a fully nonlinear problem; for a case where this is possible see [16].

We conclude this section by studying the case when the nonlinearity is *small*. More precisely we consider the problem

$$\begin{cases} u'(t) = Au(t) + \varepsilon F(t, u(t)), \ t \geq 0 \\ u(0) = u_0, \end{cases} \tag{1.12}$$

with $\varepsilon > 0$.

Theorem 2.5 *Assume that Hypothesis* 1 *and Hypothesis* 2 *hold and that* $A = F_x(0, u_0)$. *Then, for any* $T > 0$ *there exists* $\varepsilon_T > 0$ *and* $\lambda_T > 0$ *such that*

(i) *Problem* (1.12) *has a unique solution*

$$u_\varepsilon \in C^1([0,T]; D_A(\alpha)) \cap C([0,T]; D_A(\alpha+1)),$$

for all $\varepsilon \leq \varepsilon_T$ *and any* u_0 *such that* $\|u_0\|_{D_A(\alpha+1)} \leq \lambda_T$.

(ii) *We have*

$$\lim_{\varepsilon \to 0} u_\varepsilon = e^{tA} u_0 \ \text{ in } C^1([0,T]; D_A(\alpha)) \cap C([0,T]; D_A(\alpha+1)). \tag{1.13}$$

Proof. We fix $T > 0$ and consider as before the Banach space

$$Z_T = C([0,T]; D_A(\alpha+1)).$$

Moreover we denote by Γ_ε the mapping

$$Z_T \to Z_T, u \to v = \Gamma_\varepsilon(u),$$

where v is the solution to the problem

$$\begin{cases} v'(t) = Av(t) + \varepsilon F(t, u(t)), \ t \geq 0 \\ v(0) = u_0. \end{cases}$$

Setting $v = \Gamma(u), \overline{v} = \Gamma(\overline{u})$, we have

$$(v - \overline{v})' = A(v - \overline{v}) + \varepsilon \int_0^1 F_x(t, \xi u + (1-\xi)\overline{u}) \cdot (u - \overline{u}) d\xi.$$

Now, if $\|u\|_{Z_T}, \|\overline{u}\|_{Z_T} \leq R$, by (1.6) it follows

$$\|\Gamma_\varepsilon(u) - \Gamma_\varepsilon(\overline{u})\|_{Z_T} \leq \varepsilon C_T M_R \|u - \overline{u}\|_{Z_T}, \tag{1.14}$$

where

$$M_R = \sup_{t\in[0,T], \|x\|_{D_A(\alpha+1)} \leq R} \|F_x(t,x)\|.$$

Moreover, since

$$v_\varepsilon(t) = e^{tA} u_0 + \varepsilon \int_0^T e^{(t-s)A} F(s, u(s)) ds,$$

we have, for all u such that $\|u\|_{Z_T} \leq R$,

$$\|v_\varepsilon\|_{Z_T} \leq L_T \|u_0\|_{D_A(\alpha+1)} + \varepsilon T L_T M_R, \tag{1.15}$$

where

$$L_T = \sup_{t\in[0,T]} \|e^{tA}\|_{\mathcal{L}(D_A(\alpha+1))}.$$

The conclusion follows now from (1.14) and (1.15), and the contraction principle depending on a parameter ■

Fully nonlinear equations arise in several fields of applications, see [16], where a general theory of parabolic equations, including fully nonlinear equations, is developed in several functional spaces.

We recall two other approaches to fully nonlinear equations: the *continuity method* and the *nonlinear semigroups method.* The first one is based on sharp a priori estimates and, under strong assumptions, always gives global solutions. This method has been applied to Hamilton–Jacobi–Bellman equations by N. V. Krylov [11] and P. L. Lions [14]. The second method, based on the Crandall–Liggett Theorem [6], has been used by several people starting from P. Benilan and K.S. Ha [2]. It gives weaker solutions but it may work also for certain degenerate problems.

In the applications both methods require the Maximum Principle, so they work only for second order parabolic equations. On the contrary, the theory based on linearization allows us to study local solutions for any order equations and systems, and to consider problems as stability or Hopf bifurcation. A general qualitative theory for fully nonlinear equations is contained in [16].

2. An application to a parabolic nonlinear equation, perturbed by noise

We are concerned with the problem

$$\begin{cases} du(t,\xi) = D^2_\xi \varphi(u(t,\xi))dt + \varepsilon dW(t,\xi), \ t \geq 0, \ \xi \in [0,\pi] \\ \\ u(t,0) = u(t,\pi) = 0, \ t \geq 0, \\ \\ u(0,\xi) = u_0(\xi), \ \xi \in [0,\pi], \end{cases} \tag{2.1}$$

where $\alpha, \varepsilon > 0$ and W is a cylindrical Wiener process in a probability space $(\Omega, \mathcal{F}, \mathbb{P})$, taking values in the Hilbert space $H = L^2(0,\pi)$, see e.g. [10].

About φ we shall assume

Hypothesis 3 *$\varphi : \mathbb{R} \to \mathbb{R}$ is C^2–mapping such that $\varphi(0) = 0$. Moreover there exists $\nu > 0$ such that*

$$\varphi(0) = 0, \ \varphi'(\alpha) \geq \nu, \ \forall \, \alpha \in \mathbb{R}.$$

Remark 2.6 We are not able to consider the case $\nu = 0$ that arises, for instance, in the porous medium equation, see [1]. Other nonlinear parabolic equations perturbed by multiplicative noise are studied in [9].

In the sequel we put $\nu = 1$ for simplicity and we set

$$\varphi(\alpha) = \alpha + \psi(\alpha),\ \alpha \in \mathbb{R}.$$

We denote by A the linear operator in H defined by

$$\begin{cases} Ax(\xi) = D_\xi^2 x(\xi),\ \forall\, x \in D(A) \\ D(A) = H^2(0,\pi) \cap H_0^1(0,\pi). \end{cases} \tag{2.2}$$

As is well known, A is a self–adjoint operator on H with compact resolvent. Setting

$$e_k(\xi) = \sqrt{\frac{2}{\pi}} \sin k\xi,\ \xi \in [0,\pi],\ k \in \mathbb{N},$$

then $\{e_k\}$ is a complete orthonormal set of eigenfunctions of A.

We recall that the cylindrical Wiener process W can be defined as

$$W(t) = \sum_{k=1}^\infty e_k \beta_k(t),\ t \ge 0, \tag{2.3}$$

where $\{\beta_k\}$ is a sequence of real standard Brownian motions mutually independent. This definition is formal because the series above is almost surely not convergent in $L^2(0,\pi)$. However we will deal in the sequel with the so-called *stochastic convolution*

$$W_A(t) = \int_0^t e^{(t-s)A} dW(s) = \sum_{k=1}^\infty \int_0^t e^{(t-s)A} e_k d\beta_k(s), \tag{2.4}$$

which is well defined in $L^2(0,\pi)$. Moreover for any $\alpha \in]0,1/4[$, we have

$$W_A \in C^\alpha([0,T] \times [0,\pi]),\ \text{ a.s.}, \tag{2.5}$$

see [10].

Now we set $u(t) = u(t,\cdot)$, $t \ge 0$, and we write problem (2.1) in the integral form

$$u(t) = e^{tA} u_0 + \int_0^t e^{(t-s)A} D_\xi^2 \psi(u(s)) ds + \varepsilon W_A(t). \tag{2.6}$$

We reduce problem (2.1) to a deterministic problem by setting

$$v(t) = u(t) - \varepsilon W_A(t),\ t \ge 0 \tag{2.7}$$

Then, for almost $\omega \in \Omega$, v is the solution of the equation

$$v(t) = e^{tA}u_0 + \int_0^t e^{(t-s)A} D_\xi^2 \psi(v(s) + W_A(s))ds + \varepsilon W_A(t),$$

that can be considered as the integral, or mild form of the problem

$$\begin{cases} v'(t) = Av(t) + A\psi\left(v(t) + \varepsilon W_A(t)\right), \; t \geq 0, \\ \\ v(0) = u_0. \end{cases} \tag{2.8}$$

We remark that (2.8) is a family of deterministic problems, one for almost every $\omega \in \Omega$.

If W_A were regular, then equation (2.8) could be solved by using the theory of quasilinear equations, see [12]. Since W_A is only Hölder continuous, we have to look for weak solutions of (2.8). This can be done by setting $z = A^{-1}v$, and transforming the problem in the fully nonlinear equation

$$\begin{cases} z'(t) = Az(t) + \psi\left(Az(t) + \varepsilon W_A(t)\right), \; t \geq 0, \\ \\ z(0) = A^{-1}u_0 := z_0. \end{cases} \tag{2.9}$$

Definition 2.7 *We say that u is a solution of problem* (2.1) *if and only if $z(t) = A^{-1}(u - \varepsilon W_A(t))$ is a solution of problem* (2.9).

We are going to solve problem (2.9) by using Theorem 2.4. We set

$$\begin{cases} X = C_0([0,\pi]) = \{u \in C([0,\pi]) : u(0) = u(\pi) = 0\}, \\ \\ D = \{u \in C^2([0,\pi]) : u(0) = u(\pi) = u''(0) = u''(\pi) = 0\}, \end{cases}$$

and we still denote by A the linear operator in $C_0([0,\pi])$

$$Au = D_\xi^2 u, \; u \in D.$$

A is the infinitesimal generator of an analytic semigroup on $C([0,\pi])$, which we still denote by A. Moreover the interpolation spaces $D_A(\alpha)$ and $D_A(\alpha+1)$ are given by

$$D_A(\alpha) = h_0^{2\alpha}([0,\pi]), \; \alpha \in]0,1/2[,$$

$$D_A(\alpha+1) = \{u \in h_0^{2\alpha}([0,\pi]) : \; D_\xi^2 u \in h_0^{2\alpha}([0,\pi])\}, \; \alpha \in]0,1/2[.$$

Here $h^{2\alpha}([0,\pi])$ is the closure of $C^\infty([0,\pi])$, in the topology of $C^{2\alpha}$, and $h_0^{2\alpha}([0,\pi])$ is the subset of $h_0^{2\alpha}([0,\pi])$ consisting of the functions that vanish at $x=0$ and $x=\pi$.

We prove now the result

Theorem 2.8 *Assume that Hypothesis 3 holds. Let $u_0 \in h_0^{2\alpha}([0,\pi])$ with $\alpha \in]0,1/8[$ and let $\varepsilon \geq 0$. Then there exists a maximal positive stopping time $\tau(u_0)$ such that problem (2.1) has a unique solution $u^\varepsilon \in C([0,\tau_{u_0}[;h_0^{2\alpha}([0,\pi]))$, almost surely. Moreover*

$$\lim_{\varepsilon\to 0} u^\varepsilon = u^0, \quad \text{in } C([0,T];h_0^{2\alpha}([0,\pi])), \quad \text{a.s.,} \tag{2.10}$$

for any $T<\tau_{u_0}$.

Proof. We first solve problem (2.9). We define a mapping

$$F : [0,+\infty[\times D \to X$$

by setting

$$F(t,z) = Az + \psi(Az + \varepsilon W_A(t)),\ t \geq 0,\ z \in D. \tag{2.11}$$

Obviously $F(t,z)$ is differentiable with respect to z and we have

$$F_z(t,z)\zeta = A\zeta + \psi'(Az + \varepsilon W_A(t))\, A\zeta,\ t \geq 0,\ z,\zeta \in D. \tag{2.12}$$

In view of [18], for all $t \geq 0$ and for all $z \in D$, $F_z(t,z)$ is the infinitesimal generator of an analytic semigroup in X with domain D. Moreover the function $F(t,z)$ fulfills Hypothesis 2 almost surely since, in view of (2.5) the trajectories of the process $W_A(t)$ are α–Hölder continuous for any $\alpha \in]0,1/8[$.

We now apply Theorem 2.4 with $\alpha \in]0,1/8[$. By the proof of Theorem 2.4 it follows that there exists $\tau(u_0) > 0$, independent of ε, such that problem (2.9) has a unique solution z^ε with

$$z^\varepsilon \in C([0,\tau_{u_0}[;h^{2\alpha+2} \cap h_0^{2\alpha}([0,\pi])) \cap C^1([0,\tau_{u_0}[;h_0^{2\alpha}([0,\pi])).$$

Then

$$u^\varepsilon = Az^\varepsilon + \varepsilon W_A(t),$$

belongs to $C([0,\tau_{u_0}[;h_0^{2\alpha}([0,\pi]))$, almost surely and solves problem (2.1).

■

Finally, from Theorem 2.5 follows

Theorem 2.9 *Assume that Hypothesis* 3 *holds and let* $\alpha \in]0, 1/8[$. *Then, for any* $T > 0$ *there exists* $\varepsilon_T > 0$ *and* $\lambda_T > 0$ *such that*

(i) *Problem* (2.1) *has a unique solution*

$$u_\varepsilon \in C([0,T]; h^{2\alpha}([0,\pi])),$$

for all $\varepsilon \leq \varepsilon_T$ *and any* u_0 *such that* $\|u_0\|_{h^{2\alpha}([0,\pi])} \leq \lambda_T$.

(ii) *We have*

$$\lim_{\varepsilon \to 0} u_\varepsilon = u_0 \text{ in } C([0,T]; h^{2\alpha}([0,\pi])). \tag{2.13}$$

References

[1] D.G. Aronson, *The Porous Medium Equation,* Lecture Notes in Mathematics No. 1224, A. Fasano and M. Primicerio, eds., Springer-Verlag, 1–46, 1986.

[2] P. Benilan and K.S. Ha, Équations d'évolution du type $du/dt + \beta\partial\varphi(u) \ni 0$ dans $L^\infty(\Omega)$, *C.R. Acad. Sci. Paris* **281** (1975), 947–950.

[3] C.M. Brauner, J. Buckmaster , J.V. Dold, and C. Schmidt–Lainé, "On an evolution equation arising in Detonation Theory," in: *Fluid Dynamical Aspects of Combustion Theory,* M. Onofri and A. Tesei, eds., *Pitman Research Notes Math.* **223**, 196–210, 1991.

[4] J.B. Baillon, Caractère borné de certains générateurs de semigroupes linéaires dans les espaces de Banach, *C.R. Acad. Sci. Paris* **290** (1980), 757–760.

[5] C.M. Brauner, A. Lunardi, and C. Schmidt–Lainé, Stability of travelling waves with interface conditions, *Nonlinear Analysis T.M.A.*, **19** (1992), 455–474.

[6] M.G. Crandall and T.M. Liggett, Generation of semi–groups of nonlinear transformations on general Banach spaces, *Amer. J. Math.* **93** (1971), 265–298.

[7] M.G. Crandall and A. Pazy, Nonlinear evolution equations in *B*–spaces, *Israel J. Math.* **11** (1972), 57–94.

[8] G. Da Prato and P. Grisvard, Équations d'évolution abstraites non-linéaires de type parabolique, *Annali di Matematica Pura ed Applicata* **120** (1979), 329–396.

[9] G. Da Prato and C. Tubaro, Fully nonlinear stochastic partial differential equations, *SIAM J. Control in Mathematical Analysis*, (to appear).

[10] G. Da Prato and J. Zabczyk, Stochastic equations in infinite dimensions, in: *Encyclopedia of Mathematics and its Applications,* Cambridge University Press, 1992.

[11] N.V. Krylov, *Nonlinear Elliptic and Parabolic Equations of the Second Order,* Reidel Publishing, Dordrecht, 1987.

[12] O.A. Ladizhenskaya , V.A. Solonnikov, and N.N. Ural'ceva, Linear and quasilinear equations of parabolic type, *Transl. Math. Monographs* **23** (1968), Amer. Math. Soc., Providence.

[13] J.L. Lions and J. Peetre, Sur une classe d'espaces d'interpolation, Publ. Math de l'I.H.E.S. **19** (1964), 5–68.

[14] P. L. Lions, *Control of diffusion processes in* $\mathbb{R}^N$, Communications in Pure and Applied Mathematics, **XXXIV** (1981), 121–147.

[15] A. Lunardi, Interpolation spaces between domains of elliptic operators and spaces of continuous functions with applications to nonlinear parabolic equations, *Math. Nachr.* **121** (1985), 323–349.

[16] A. Lunardi, *Analytic Semigroups and Optimal Regularity in Parabolic Problems*, Birkäuser, Basel, 1995.

[17] E. Sinestrari, On the abstract Cauchy problem of parabolic type in spaces of continuous functions, *J. of Math. Anal. and Appl.* **107**(1) (1985), 16–66.

[18] H.B. Stewart, Generation of an analytic semigroup by strongly elliptic operators, *Trans. Amer. Math. Soc.* **199** (1974), 141–62

[19] H. Triebel, *Interpolation Theory, Function Spaces, Differential Operators,* North-Holland, Amsterdam, 1986.

Scuola Normale Superiore di Pisa
Piazza dei Cavalieri 7, 56126 Pisa, Italy

Strongly Elliptic Problems Near Cuspidal Points and Edges

Monique Dauge

Abstract. *After an overview of the various geometrical situations occurring for two-dimensional piecewise smooth domains, we concentrate on the case of outgoing cusp points. We recall results by P. Grisvard [4] and V.G. Mazya and B.A. Plamenevskii [10]. Then, relying on a work by J.-L. Steux [14], we state a result of regularity in the space of infinitely smooth functions: if the data are C^∞, the solution is also C^∞. We extend this result to the situation of cuspidal edges (for example the domain exterior to a cylinder lying on a plane, or two tangent tori).*

Problèmes fortement elliptiques près de points ou arêtes cuspides

Résumé. *Après avoir passé en revue les différentes situations géométriques pouvant se produire pour un domaine à bord régulier par morceaux, nous nous concentrons sur le cas de point cuspides saillants. Nous rappelons des résultats de P. Grisvard [4] et V.G. Mazya and B.A. Plamenevskii [10]. Ensuite, nous basant sur un travail dû à J.-L. Steux [14], nous établissons un résultats de régularité dans l'espace des fonctions infiniment différentiables : si les données sont C^∞, la solution est aussi C^∞. Enfin nous étendons ce résultat à la situation d'une arête cuspide (par exemple le domaine extérieur à un cylindre reposant sur un plan, ou encore à deux tores tangents).*

1. Piecewise - smooth plane domains

Let Ω be a piecewise-smooth plane domain. This means that the boundary $\partial\Omega$ of Ω is the union of finitely many arcs of $\mathcal{C}^\infty$ curves, which may be straight lines, of course. We call them the *sides* of Ω. A point belonging to the intersection of two sides is called a *vertex* of Ω.

The properties that we intend to investigate being *local*, we assume for simplicity that Ω has only *one* vertex, located at the origin $\mathcal{O}$ of the coordinate axes. Let $\vec{\tau}_1$ and $\vec{\tau}_2$ be the two tangents to $\partial\Omega$ at $\mathcal{O}$ and let ω be the measure of the angle between them. Five generic situations may occur:

$\omega = 2\pi$: Ω has a crack (if the two arcs joining in $\mathcal{O}$ coincide), or a reentrant cusp point if not.

$\pi < \omega < 2\pi$: Ω has an ordinary non convex polygonal vertex.

$\pi = \omega$: Ω has a weak geometrical singularity (or is smooth if $\mathcal{O}$ is a dummy vertex!)

$0 < \omega < \pi$: Ω has an ordinary convex polygonal vertex.

$\omega = 0$: Ω has an outgoing cusp point.

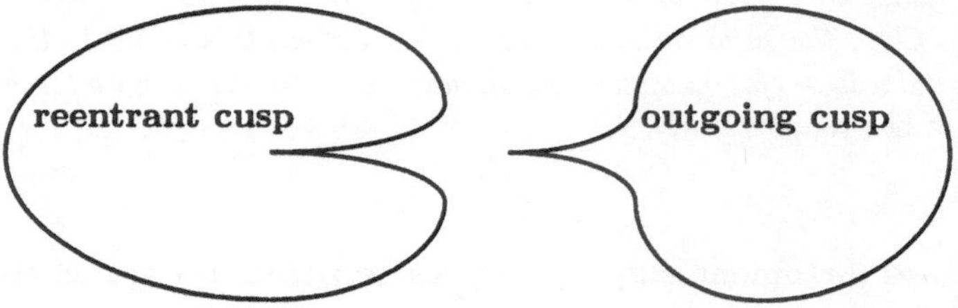

With the help of a C^∞ diffeomorphism, it is always possible to flatten *one* of the sides of Ω in the neighborhood of $\mathcal{O}$, say the side tangent to $\vec{\tau}_1$. So, from now on, we assume that one side of Ω coincides with the horizontal axis in a neighborhood of $\mathcal{O}$. When $\pi < \omega < 2\pi$ or when $0 < \omega < \pi$, by a better choice of the diffeomorphism it is also possible to flatten the other side. But when Ω has a cusp, it is of course impossible.

Our object of consideration is the behavior of solutions of elliptic boundary value problems in Ω. Let L by a properly elliptic operator of order $2m$ with C^∞ coefficients in $\mathbb{R}^2$. Let us consider the Dirichlet boundary value problem for L on Ω:

$$\begin{cases} Lu = f \text{ in } \Omega, \\ u \in \overset{\circ}{H}{}^m(\Omega). \end{cases} \tag{1.1}$$

If f is more regular than $H^{-m}(\Omega)$, say $H^{s-m}(\Omega)$ with $s > 0$, due to the presence of the corner in $\mathcal{O}$, we cannot expect that u belongs to $H^{s+m}(\Omega)$ for any s and any f.

When $\omega \neq 0$, i.e. in the four first situations, the structure of the solution u has similar properties: the function u has an asymptotic in $\mathcal{O}$, which, instead of being reduced to polynomials as in the case of the Taylor expansion of a smooth function, is made of special model functions which only depend on the geometry of Ω and the operator L. These model functions w are better described in *polar coordinates* (r,θ) centered in $\mathcal{O}$:

$$w(r,\theta) = r^{\mu} \sum_{q=0}^{Q} \psi_q(\theta) \log^q r \tag{1.2}$$

where μ is a complex (possibly real!) number, and the ψ_q are smooth functions of θ, "belonging" to w. In general, u admits a splitting:

$$u = u_{\text{sing}} + u_{\text{pol}} + u_{\text{flat}} \quad \text{with} \quad \begin{cases} u_{\text{sing}} = \sum_{k=1}^{K_s} c_k\, w_k, \\ u_{\text{pol}} \text{ a polynomial function,} \\ u_{\text{flat}} \in V^{s+m}(\Omega), \end{cases} \tag{1.3}$$

where the space $V^{s+m}(\Omega)$ is a space of flat functions contained in $H^{s+m}(\Omega)$. Note that for each fixed s, the number K_s of independent singular model functions is finite. Moreover the exponent μ_k belonging to w_k satisfies

$$m-1 < \operatorname{Re}\mu_1 \leq \ldots \leq \operatorname{Re}\mu_k \leq \ldots \leq \operatorname{Re}\mu_{K_s} < s+m-1. \tag{1.4}$$

In the case of the Laplacian Δ, the exponents μ_k are the $\frac{k\pi}{\omega}$ and $Q = Q(k)$ is equal to 1 if $\frac{k\pi}{\omega} \in \mathbb{N}$ and 0 if not. In general, the exponents μ_k are the eigenvalues of generalized Sturm-Liouville operators on the angular interval $]0,\omega[$. They are piecewise-smooth continuous functions of ω. For the opening π, all the μ_k are integers — if Ω is not smooth in $\mathcal{O}$, logarithmic terms occur in the asymptotics. For the opening 2π, all the μ_k are half-integers, i.e. belong to $\mathbb{N}/2$.

Under the form (1.2), the functions w do not depend smoothly on ω, even for $L = \Delta$. By mixing together the functions w and the polynomials, it is possible to construct *stable* linear combinations w_{stab}. Their radial behavior can be nicely described by contour integrals. The ordinary simple asymptotics can be written:

$$r^{\mu} \log^q r = \frac{q!}{2i\pi} \int_{\gamma} \frac{r^{\lambda}}{(\lambda-\mu)^{q+1}}\, d\lambda, \quad q = 0,\ldots,Q, \tag{1.5}$$

where the contour γ surrounds μ. When the exponents μ depend smoothly on ω, stable behaviors are given by divided differences of the function $r \to r^{\lambda}$:

$$S[\mu_{(0)},\ldots,\mu_{(q)};r] = \frac{1}{2i\pi} \int_{\gamma} \frac{r^{\lambda}}{(\lambda-\mu_{(0)})\cdots(\lambda-\mu_{(q)})}\, d\lambda, \quad q = 0,\ldots,Q, \tag{1.6}$$

where the $\mu_{(q)}$ occur in the exponents of the w_k or are integers — exponents of polynomials! In the general situation where the multiplicity of μ may change (for instance for $L = \Delta^2$ in the neighborhood of the angle $\omega_1 \simeq 0.813\pi$) stable behaviors are given by generalized divided differences of the function $r \to r^\lambda$:

$$S[\mu_{(0)}, \dots, \mu_{(Q)} | p_q; r] = \frac{1}{2i\pi} \int_\gamma \frac{r^\lambda p_q(\lambda)}{(\lambda - \mu_{(0)}) \cdots (\lambda - \mu_{(Q)})} \, d\lambda, \quad q = 0, \dots, Q, \tag{1.7}$$

where the p_q for $q = 0, \dots, Q$ are a basis of $\mathbb{P}_Q$ the space of polynomials of 1 variable with degree $\leq Q$.

From a very abundant literature, we quote

- G.M. Verzbinskii and V.G. Maz'ya [15, 16, 17] concerning the Dirichlet problem for the Laplace operator in all the geometrical situations quoted above,
- P. Grisvard [5], V.A. Kondrat'ev [8] and V.G. Maz'ya and B.A. Plamenevskii [10] concerning the ordinary "conical" situation where the opening is neither 0, nor π nor 2π,
- [5] again and [1] for the cracks, V.G. Maz'ya, S.A. Nazarov and B.A. Plamenevskii [9] and A.B. Movchan and S.A. Nazarov [12] for reentrant cusps,
- V.G. Maz'ya and J. Rossmann [11] and our [2] for stable asymptotics in the full range $0 < \omega \leq 2\pi$ for the opening.

As a conclusion to this section, we can say that in all the situations where the opening is > 0, the asymptotics $u_{\text{asy}} := u_{\text{sing}} + u_{\text{pol}}$ of u in the neighborhood of $\mathcal{O}$ can be described in a unified and stable way, including even the case when the opening is equal to π and the domain smooth in $\mathcal{O}$ — the function u_{asy} is then the Taylor expansion of u.

Have we still a sort of stability when the opening tends to 0?

2. When the opening tends to zero

We see that for $L = \Delta$, the first exponent occurring in the singular part u_{sing} of u is $\frac{\pi}{\omega}$ and it tends to infinity when $\omega \to 0$. The same phenomenon occurs for $L = \Delta^2$: the real part of the first exponent μ_1 tends to infinity when $\omega \to 0$. We have

Proposition 2.1 *Let L be a strongly elliptic operator. Let $\mu_1^{(\omega)}$ be the exponent with the least real part occurring in* (1.4). *Then*

$$\operatorname{Re} \mu_1^{(\omega)} \longrightarrow +\infty \quad \textit{when} \quad \omega \longrightarrow 0.$$

Proof. Let $\mathscr{L}$ be the principal part of L frozen in $\mathcal{O}$, written in the coordinates (t, θ) with $t = \log r$:

$$\mathscr{L}(\theta; \partial_t, \partial_\theta) = e^{2mt} L_{\mathrm{princ}}(\mathcal{O}; \partial_x, \partial_y).$$

For any $\eta \in \mathbb{R}$, let $\mathscr{B}_\eta^{(\omega)}$ be the operator

$$\begin{array}{rcl} \mathscr{B}_\eta^{(\omega)} : \{v \mid e^{-\eta t} v \in \mathring{H}^m(\mathbb{R}\times]0,\omega[)\} & \longrightarrow & \{g \mid e^{-\eta t} g \in H^{-m}(\mathbb{R}\times]0,\omega[)\} \\ v & \longmapsto & \mathscr{L}(\theta; \partial_t, \partial_\theta)\, v. \end{array}$$

From the general theory [8], we have for any $\eta > m-1$

$$\mathscr{B}_\eta^{(\omega)} \text{ isomorphism} \quad \Longleftrightarrow \quad \forall k \geq 1,\ \operatorname{Re} \mu_k^{(\omega)} \neq \eta.$$

Thus, we are going to prove that $\forall \eta > m-1$, $\mathscr{B}_\eta^{(\omega)}$ is always an isomorphism if ω is small enough. Setting $\mathscr{A}_\eta^{(\omega)} = e^{-\eta t} \mathscr{B}_\eta^{(\omega)} e^{\eta t}$, acting from $\mathring{H}^m(\mathbb{R}\times]0,\omega[)$ into $H^{-m}(\mathbb{R}\times]0,\omega[)$ we have

$$\mathscr{B}_\eta^{(\omega)} \text{ isomorphism} \quad \Longleftrightarrow \quad \mathscr{A}_\eta^{(\omega)} \text{ isomorphism}. \tag{2.1}$$

Let $\mathscr{A}^{(\omega)}$ be the principal part of $\mathscr{A}_\eta^{(\omega)}$. The operator $\mathscr{A}^{(\omega)}$ does not depend on η and we have the estimate

$$\begin{array}{l} \exists c > 0, \quad \forall \omega \in]0, 2\pi], \quad \forall \eta \in \mathbb{R}, \quad \forall v \in \mathring{H}^m(\mathbb{R}\times]0,\omega[), \\ \qquad \|(\mathscr{A}^{(\omega)} - \mathscr{A}_\eta^{(\omega)})\, v\|_{H^{-m}(\mathbb{R}\times]0,\omega[)} \leq c\,(1+|\eta|)^{2m} \|v\|_{H^{m-1}(\mathbb{R}\times]0,\omega[)}. \end{array} \tag{2.2}$$

Let $\mathscr{A}^{(\omega)}(0)$ be the operator $\mathscr{A}^{(\omega)}$ with its coefficients frozen in $\theta = 0$. Since

$$\partial_t = e^t\Big(\cos\theta\,\partial_x + \sin\theta\,\partial_y\Big) \quad \text{and} \quad \partial_\theta = e^t\Big(-\sin\theta\,\partial_x + \cos\theta\,\partial_y\Big)$$

we check that

$$\mathscr{A}(0)(\partial_t, \partial_\theta) = L_{\mathrm{princ}}(\mathcal{O}; \partial_t, \partial_\theta). \tag{2.3}$$

Due to the strong ellipticity of L, $\mathscr{A}^{(\omega)}(0)$ is an isomorphism for all $\omega > 0$:

$$\begin{array}{l} \exists c > 0, \ \forall \omega > 0, \quad \forall v \in \mathring{H}^m(\mathbb{R}\times]0,\omega[), \\ \qquad |v|_{H^m(\mathbb{R}\times]0,\omega[)} \leq c\,|\mathscr{A}^{(\omega)}(0)\, v|_{H^{-m}(\mathbb{R}\times]0,\omega[)}. \end{array} \tag{2.4}$$

The Poincaré inequality on the strip reads

$$\begin{array}{l} \exists c > 0, \ \forall \omega > 0, \quad \forall v \in \mathring{H}^m(\mathbb{R}\times]0,\omega[), \\ \qquad \|v\|_{H^{m-1}(\mathbb{R}\times]0,\omega[)} \leq c\,\omega\, |v|_{H^m(\mathbb{R}\times]0,\omega[)}. \end{array} \tag{2.5}$$

The regularity of the coefficients of L yields

$$\exists c > 0, \quad \forall \omega \in]0, 2\pi], \ \forall v \in \overset{\circ}{H}{}^m(\mathbb{R}\times]0,\omega[), \qquad \|(\mathscr{A}^{(\omega)}(0) - \mathscr{A}^{(\omega)})\, v\|_{H^{-m}(\mathbb{R}\times]0,\omega[)} \le c\omega\, \|v\|_{H^m(\mathbb{R}\times]0,\omega[)}. \tag{2.6}$$

From (2.4)-(2.6), we deduce that for ω small enough, $\mathscr{A}^{(\omega)}$ is an isomorphism satisfying

$$\exists c > 0, \ \forall \omega, \ 0 < \omega \le \omega_0, \quad \forall v \in \overset{\circ}{H}{}^m(\mathbb{R}\times]0,\omega[), \qquad \|v\|_{H^m(\mathbb{R}\times]0,\omega[)} \le c\, \|\mathscr{A}^{(\omega)} v\|_{H^{-m}(\mathbb{R}\times]0,\omega[)}. \tag{2.7}$$

With (2.2) and (2.5), (2.7) yields that $\mathscr{A}_\eta^{(\omega)}$ is an isomorphism if $\omega\,(1+|\eta|)^{2m}$ is small enough. With (2.1), this gives the existence of a constant $c_0 > 0$ such that

$$\forall \eta \in \mathbb{R}, \quad \forall \omega \le c_0(1+|\eta|)^{-2m}, \quad \mathscr{B}_\eta^{(\omega)} \text{ isomorphism.}$$

Therefore $\mu_1^{(\omega)}$ satisfies

$$\omega > c_0(1 + \operatorname{Re}\mu_1^{(\omega)})^{-2m} \quad \text{i.e.} \quad \operatorname{Re}\mu_1^{(\omega)} > \left(\frac{\omega}{c_0}\right)^{-\frac{1}{2m}} - 1.$$

■

The strong ellipticity has served in only one place, to insure that $\mathscr{A}^{(\omega)}(0)$ is an isomorphism for all $\omega > 0$ and satisfies the estimates (2.4). If we only assume that

$$L_{\text{princ}}(\mathcal{O}; \partial_t, \partial_\theta) : \overset{\circ}{H}{}^m(\mathbb{R}\times]0,1[) \longrightarrow H^{-m}(\mathbb{R}\times]0,1[) \text{ isomorphism,} \tag{2.8}$$

by a simple scaling argument we still obtain the estimates (2.4). By partial Fourier transform in the variable t, we obtain that (2.8) holds if

$$\forall \xi \in \mathbb{R}, \quad L_{\text{princ}}(\mathcal{O}; i\xi, \partial_\theta) : \overset{\circ}{H}{}^m(]0,1[) \longrightarrow H^{-m}(]0,1[) \text{ isomorphism.} \tag{2.9}$$

Therefore,

Proposition 2.2 *Let L be a properly elliptic operator satisfying* (2.9). *Then*

$$\operatorname{Re}\mu_1^{(\omega)} \longrightarrow +\infty \quad \textit{when} \quad \omega \longrightarrow 0.$$

So, we can expect good regularity properties for the Dirichlet problem associated with operators L such as above in the neighborhood of outgoing cusp points.

3. Outgoing cusp points: case of flat functions

We assume that in a neighborhood $[-a,a]\times[-a,a]$ of $\mathcal{O}$, Ω is determined by the inequalities

$$(x,y)\in\Omega\cap[-a,a]\times[-a,a] \iff 0<x<a \quad\text{and}\quad 0<y<\varphi(x), \tag{3.1}$$

where φ is a function $\mathcal{C}^\infty([-a,a])$, such that

$$\varphi(0)=0,\ \varphi'(0)=0 \quad\text{and}\quad \varphi>0 \text{ on }]0,a]. \tag{3.2}$$

We assume moreover that φ is not infinitely flat in 0 and let $p\in\mathbb{N}$ be the smallest integer such that

$$\varphi^{(p)}(0)\neq 0. \tag{3.3}$$

An example is given by the equation of a circle tangent to the x axis at $\mathcal{O}$: if the radius is equal to R

$$\varphi(x)=R\left(1+\sqrt{1-\frac{x^2}{R^2}}\right)=\frac{x^2}{2R^2}+\mathcal{O}(x^4).$$

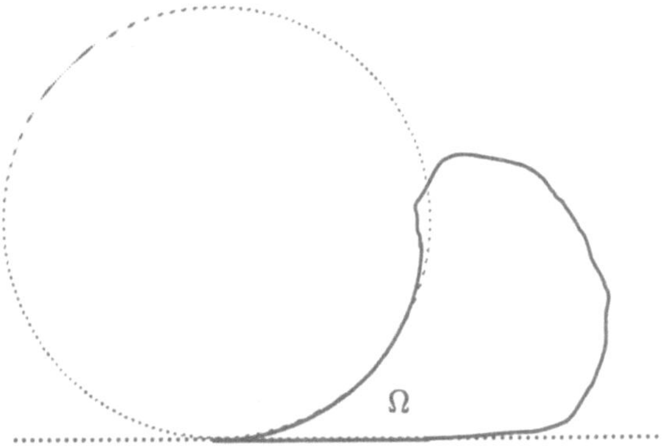

We will see later (*cf* Remark 4.4) that our results can be applied to any domain $\Omega=\mathbb{R}^2\backslash\overline{\mathcal{U}}$ exterior to the domain $\mathcal{U}$ formed by two tangent domains with analytic boundaries (for instance, $\mathcal{U}$ is the union of two tangent disks, or a disk tangent to a half plane).

As it has been proved in various frameworks by K. Ibuki [6], A. Khelif [7], V.G. Mazya and B.A. Plamenevskii [10], P. Grisvard [4] and J.-L. Steux [14], the operator of the Dirichlet problem (1.1) acts smoothly between spaces of flat functions: for any $j\in\mathbb{N}$, let

$$V^j(\Omega)=\{u\in L^2(\Omega)\mid \forall\alpha\in\mathbb{N}^2,\ |\alpha|\le j,\quad \varphi^{|\alpha|-j}\,\partial^\alpha u\in L^2(\Omega)\}.$$

Moreover, the space $V^{-j}(\Omega)$ is defined as the dual space of $\overset{\circ}{V}{}^j(\Omega)$, where $\overset{\circ}{V}{}^j(\Omega)$ is the closure of $\mathcal{D}(\Omega)$ in $V^j(\Omega)$.

Theorem 3.1 *Let L be a properly elliptic operator satisfying* (2.9). *In particular, L can be any strongly elliptic operator. Let $j \in \mathbb{Z}$, $j > -m$. Then any solution u of the Dirichlet problem* (1.1) *with right hand side $f \in V^j(\Omega)$ satisfies the optimal regularity property*

$$u \in V^{2m+j}(\Omega).$$

The proof of this theorem relies on the change of variables

$$(x, y) \longrightarrow (t, \theta) \quad \text{where} \quad \theta = \frac{y}{\varphi(x)} \quad \text{and} \quad t = -\int_x^a \frac{d\sigma}{\varphi(\sigma)}, \tag{3.4}$$

which transforms

$$\Omega_a := \Omega \cap \{(x, y) \mid 0 < x < a\} \quad \text{onto} \quad \Sigma := \{(t, \theta) \mid t < 0,\ \theta \in]0, 1[\}.$$

The spaces $V^j(\Omega)$ are transformed in a simple way: we set

$$\tilde{\varphi}(t) = \varphi(x) \quad \text{and} \quad \text{for } \eta \in \mathbb{R},\ H^j_\eta(\Sigma) = \{v \mid \tilde{\varphi}^{-\eta} v \in H^j(\Sigma)\}.$$

Then the change of variables transforms

$$V^j(\Omega_a) \quad \text{onto} \quad H^j_{j-1}(\Sigma). \tag{3.5}$$

Note that

$$\tilde{\varphi}(t) = (p-1)^p\, |t|^{-1} \Big(|t|^{-\frac{1}{p-1}} + \mathcal{O}(|t|^{-\frac{2}{p-1}}) \Big) \quad \text{when } t \to -\infty. \tag{3.6}$$

The transformation law of the operator L is

$$\varphi^{2m}(x) L(x, y; \partial_x, \partial_y) =: \widetilde{\mathscr{L}}(t, \theta; \partial_t, \partial_\theta) = L_{\text{princ}}(\mathcal{O}; \partial_t, \partial_\theta) + M(t, \theta; \partial_t, \partial_\theta)$$

where the coefficients of M are smooth functions behaving like $\mathcal{O}(|t|^{-\frac{1}{p-1}})$ when $t \to -\infty$. For any $\eta \in \mathbb{R}$, let $\mathscr{B}_\eta$ be the operator

$$\begin{aligned} \mathscr{B}_\eta : \{v \mid \tilde{\varphi}^{-\eta} v \in \overset{\circ}{H}{}^m(\Sigma)\} &\longrightarrow \{g \mid \tilde{\varphi}^{-\eta} g \in H^{-m}(\Sigma)\} \\ v &\longmapsto \widetilde{\mathscr{L}}(t, \theta; \partial_t, \partial_\theta) v. \end{aligned}$$

Setting $\mathscr{A}_\eta = \tilde{\varphi}^{-\eta}\, \mathscr{B}_\eta\, \tilde{\varphi}^{\eta}$, acting from $\overset{\circ}{H}{}^m(\Sigma)$ into $H^{-m}(\Sigma)$ we have:

$$\begin{aligned} &\mathscr{A}_\eta = L_{\text{princ}}(\mathcal{O}; \partial_t, \partial_\theta) + M_\eta(t, \theta; \partial_t, \partial_\theta) \\ &\quad \text{with} \quad \|M_\eta\|_{\overset{\circ}{H}{}^m(]-\infty,-T[\times]0,1[) \to H^{-m}(]-\infty,-T[\times]0,1[)} = \mathcal{O}(T^{-\frac{1}{p-1}}). \end{aligned} \tag{3.7}$$

Since (2.9) allows for proving that $L_{\text{princ}}(\mathcal{O}; \partial_t, \partial_\theta)$ induces an isomorphism from $\overset{\circ}{H}{}^m(\Sigma)$ onto $H^{-m}(\Sigma)$, the proof of the theorem is a consequence of (3.5) and (3.7).

The fundamental difference between the present case of a cusp point and an acute plane sector where $\varphi(x)$ is equal to γx and $\tilde{\varphi}(t)$ behaves like e^t — compare with (3.6), is the decay property of the splitting (3.7) which does not hold for a plane sector.

4. Outgoing cusp points: case of smooth functions

Following [14], we now intend to study the regularity of the u solution of problem (1.1) when $f \in \mathcal{C}^\infty(\overline{\Omega})$. We can easily prove:

Lemma 4.1 *For $f \in \mathcal{C}^\infty(\overline{\Omega})$, we set for any $\ell \geq 1$:*

$$f_\ell(x,y) = \sum_{|\alpha|<\ell} \frac{x^{\alpha_1} y^{\alpha_2}}{\alpha_1!\, \alpha_2!}\, \partial^\alpha f(\mathcal{O}).$$

Then

$$\forall \ell \geq pj, \quad f - f_\ell \in V^j(\Omega).$$

In view of Theorem 3.1, it remains to investigate the *polynomial resolution.* For $\ell \geq 0$, let $\mathcal{C}^\infty_\ell(\overline{\Omega}_a)$ denote the space of functions:

$$\mathcal{C}^\infty_\ell(\overline{\Omega}_a) = \{f \in \mathcal{C}^\infty(\overline{\Omega}_a) \mid \forall \alpha,\ |\alpha| \leq \ell, \quad \partial^\alpha f(\mathcal{O}) = 0\}.$$

We note that

$$\forall \ell \geq pj, \quad \mathcal{C}^\infty_\ell(\overline{\Omega}_a) \subset V^j(\Omega_a).$$

Lemma 4.2 *We assume that L is elliptic. Let $\alpha \in \mathbb{N}^2$. There exists a function $U_\alpha \in \mathcal{C}^\infty(\overline{\Omega}_a)$ and constants $d_{\alpha,1}, \dots, d_{\alpha,\alpha_1}$ such that*

$$LU_\alpha - x^{\alpha_1} y^{\alpha_2} - \sum_{k=1}^{\alpha_1} d_{\alpha,k}\, x^{\alpha_1 - k} y^{\alpha_2 + k} \in \mathcal{C}^\infty_{|\alpha|}(\overline{\Omega}_a) \quad \text{and} \quad \chi U_\alpha \in \mathring{H}^m(\Omega), \tag{4.1}$$

where χ is a smooth cut-off function $\equiv 1$ if $x \leq \frac{a}{2}$ and $\equiv 0$ if $x \geq a$.

In particular, if $\alpha_1 = 0$, the $d_{\alpha,k}$ are not there and the function U_α satisfies

$$LU_\alpha - x^{\alpha_1} y^{\alpha_2} \in \mathcal{C}^\infty_{|\alpha|}(\overline{\Omega}_a) \quad \text{and} \quad \chi U_\alpha \in \mathring{H}^m(\Omega), \tag{4.2}$$

Proof. The method consists of solving the boundary value problem with respect to the variable y and considering x as a parameter.

The operator ∂_θ^{2m} is continuous between the spaces of polynomials:

$$\mathbb{P}_{2m+j} \cap \overset{\circ}{H}{}^m(]0,1[) \longrightarrow \mathbb{P}_j.$$

It is one-to-one and since the dimensions of the two spaces $\mathbb{P}_{2m+j} \cap \overset{\circ}{H}{}^m(]0,1[)$ and $\mathbb{P}_j$ are equal (to $j+1$), ∂_θ^{2m} is onto. Thus, there exists a unique polynomial $P_{\alpha_2}(\theta)$ such that

$$\partial_\theta^{2m} P_{\alpha_2} = \theta^{\alpha_2} \quad \text{and} \quad P_{\alpha_2} \in \overset{\circ}{H}{}^m(]0,1[).$$

With $a_{0,2m}(x,y)$ the coefficient of ∂_y^{2m} in L and $b_0 := 1/a_{0,2m}(\mathcal{O})$, we set

$$U_\alpha(x,y) = b_0\, \varphi(x)^{2m}\, x^{\alpha_1}\, \varphi(x)^{\alpha_2}\, P_{\alpha_2}\Big(\frac{y}{\varphi(x)}\Big).$$

Since P_{α_2} is a polynomial of degree $\leq \alpha_2 + 2m$, we check that U_α has the form

$$U_\alpha(x,y) = x^{\alpha_1} \sum_{\substack{\alpha_2' \in \mathbb{N},\ \alpha_2'' \in \mathbb{N} \\ \alpha_2' + \alpha_2'' = \alpha_2 + 2m}} c_{\alpha_2', \alpha_2''}\, \varphi(x)^{\alpha_2'}\, y^{\alpha_2''}.$$

Thus U_α is smooth in a neighborhood of $\mathcal{O}$. By construction U_α satisfies the boundary conditions and

$$a_{0,2m}(\mathcal{O})\, \partial_y^{2m}\, U_\alpha(x,y) = x^{\alpha_1} y^{\alpha_2}.$$

A simple calculation proves (4.2) and (4.1). ∎

The main result of this section is the regularity result [14]:

Theorem 4.3 *Let L be a properly elliptic operator satisfying* (2.9). *In particular, L can be any strongly elliptic operator. Then any solution u of the Dirichlet problem* (1.1) *with the right hand side $f \in C^\infty(\overline{\Omega})$ satisfies the optimal regularity property*

$$u \in C^\infty(\overline{\Omega}).$$

Proof. It suffices to prove that for any $j \in \mathbb{N}$, u can be written as the sum of a function u_j belonging to $C^\infty(\overline{\Omega})$ and of a flat function $v_j \in V^{2m+j}(\Omega)$.

Let $j \in \mathbb{N}$. We begin with the following algorithm of polynomial resolution for the polynomial part f_{pj} of f given by Lemma 4.1. We start with $\alpha = (0,0)$, use Lemma 4.2 with $\alpha_1 = 0$ and put the remainder into the right hand side. Then we apply Lemma 4.2 for $(\alpha_1, \alpha_2) = (1,0)$, put the remainder into the right hand side and apply Lemma 4.2 for $(\alpha_1, \alpha_2) = (0,1)$, etc... The order in which the multi-indices α have to be treated is $|\alpha|$ increasing and α_2

increasing. In this way, we construct u_j in $\mathring{H}^m(\Omega) \cap C^\infty(\overline{\Omega})$ such that

$$Lu_j = f_{pj} + g_j \quad \text{with} \quad g_j \in C^\infty_{pj}.$$

We conclude with Theorem 3.1 since $L(u - u_j) = (f - f_{pj}) - g_j \in V^j(\Omega)$. ■

Remark 4.4 If in a neighborhood $[-a, a] \times [-a, a]$ of $\mathcal{O}$, Ω is determined by the inequalities

$$(x, y) \in \Omega \cap [-a, a] \times [-a, a] \iff -a < x < a \text{ and } \varphi_1(x) < y < \varphi_2(x), \tag{4.3}$$

where φ_1 and φ_2 are $C^\infty([-a, a])$ such that

$$\varphi_1(0) = \varphi_2(0) = 0, \quad \varphi_1'(0) = \varphi_2'(0) = 0, \quad \varphi_2 - \varphi_1 > 0 \text{ on } [-a, 0[\cup]0, a], \tag{4.4}$$

and such that $\varphi = \varphi_2 - \varphi_1$ is not infinitely flat in 0, then Theorem 4.3 still holds: the two Taylor expansions of u in $\mathcal{O}$ in the half-planes $x > 0$ and $x < 0$ are linear functions of the Taylor expansions of f, φ_1 and φ_2; therefore, one can prove that they fit together. ■

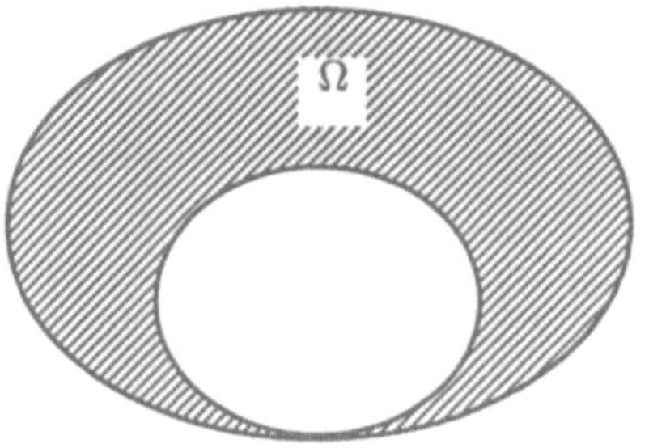

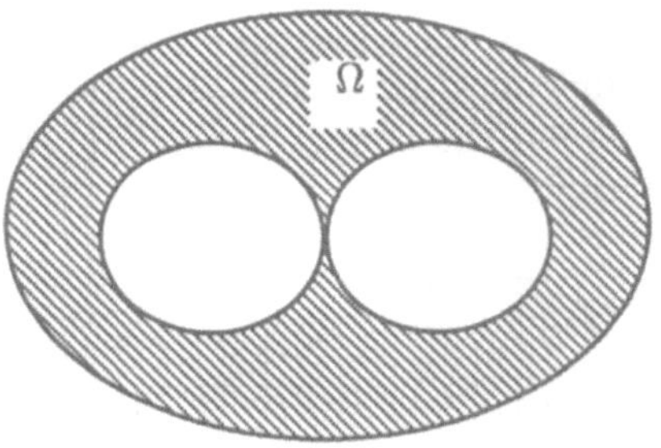

Remark 4.5 If φ has an asymptotics in non integer powers of x:

$$\varphi(x) = x^p + \gamma_1 x^{p_1} + \cdots + \gamma_N x^{p_N} + \mathcal{O}(x^{p_{N+1}}),$$

with p_n an increasing sequence tending to $+\infty$ (for example the profile of Joukowski has such a form with $p = \frac{3}{2}$ and $p_n = \frac{3}{2} + \frac{n}{2}$), Theorem 3.1 still holds [6, 7, 10, 4, 14], but the proofs of Lemma 4.2 and Theorem 4.3 yield the construction of a non smooth asymptotics for u, see [14]. ■

5. Other boundary conditions

While conditions corresponding to (2.9) are satisfied, we have the analogue of Theorems 3.1 and 4.3 for any other elliptic boundary problem. Namely, if $B_1(x,y;\partial_x,\partial_y)$ and $B_2(x,y;\partial_x,\partial_y)$ are two systems of boundary conditions on the sides $\theta=0$ and $\theta=1$ of Ω_a respectively, each of them covering L, then the behavior near $\mathcal{O}$ of the solutions of the boundary problem

$$\begin{cases} Lu = f \text{ in } \Omega, \\ B_1 u = g_1 \text{ if } y=0, \\ B_2 u = g_2 \text{ if } y=\varphi(x), \\ \partial_x^k u = 0, \ k=0,\dots,m-1 \text{ if } x=a, \end{cases} \tag{5.1}$$

depend on the problem $\big(L_{\rm princ}(\mathcal{O};\partial_x,\partial_y), B_{1,\rm princ}(\mathcal{O};\partial_x,\partial_y), B_{2,\rm princ}(\mathcal{O};\partial_x,\partial_y)\big)$ on the infinite strip $\mathbb{R}\times]0,1[$. The condition replacing (2.9) is now

$$\forall\xi\in\mathbb{R},\quad \big(L_{\rm princ}(\mathcal{O};i\xi,\partial_\theta), B_{1,\rm princ}(\mathcal{O};i\xi,\partial_\theta), B_{2,\rm princ}(\mathcal{O};i\xi,\partial_\theta)\big): \\ H^m(]0,1[) \longrightarrow H^{-m}(]0,1[)\times\mathbb{C}^{2m} \quad \text{isomorphism.} \tag{5.2}$$

Under this assumption, the real part of the exponent $\mu_1^{(\omega)}$ tends to $+\infty$ when $\omega\to 0$, the regularity results in spaces of flat functions and in spaces of smooth functions still hold — note that if B_1 and B_2 contain operators of the *same* order, the boundary data g_1 and g_2 have to satisfy a countable number of compatibility conditions at $\mathcal{O}$.

Examples are given by $B_1 = \boldsymbol{Id}$, $B_2=\partial_n$ for $L=\Delta$, or for $B_1=B_2=(\boldsymbol{Id},\Delta)$ for $L=\Delta^2$.

At the opposite, for Neumann problem, condition (5.2) is always violated in $\xi=0$: all polynomials v in $\mathbb{P}_{m-1}$ satisfy

$$\big(L_{\rm princ}(\mathcal{O};0,\partial_\theta), B_{1,\rm princ}(\mathcal{O};0,\partial_\theta), B_{2,\rm princ}(\mathcal{O};0,\partial_\theta)\big)v = (0,0,0).$$

That fact induces severe difficulties to handle flat right hand sides. Anyway for the Laplace operator for instance, it is still possible to construct an ansatz for the asymptotics of u from the Taylor expansion of the right hand side by alternating double integrations with respect to x of mean values of the type

$$x\longmapsto \frac{1}{\varphi(x)}\Big(g_1(x)-g_2(x)+\int_0^{\varphi(x)} f(x,y)\,dy\Big)$$

and double integrations in y like for Dirichlet, see S.A. Nazarov and O.R. Polyakova [13]. Such methods can be compared with what is done in elasticity for asymptotics in thin plates.

6. Cuspidal edges

Let now $\mathcal{W}$ be a three-dimensional domain with a cuspidal edge: this means that the boundary of $\mathcal{W}$ is smooth, except in the neighborhood of a smooth curve $\mathcal{E}$, the *edge* of $\mathcal{W}$, where $\mathcal{W}$ is locally diffeomorphic to $\mathbb{R} \times \Omega$, where Ω is a plane domain with an outgoing cusp in $\mathcal{O}$ as in §3-5.

Let M be a strongly elliptic operator with $\mathcal{C}^\infty$ coefficients in $\mathbb{R}^3$. We are interested in the regularity of the solutions of the Dirichlet problem:

$$\begin{cases} Mu = f \ \text{ in } \mathcal{W}, \\ u \in \overset{\circ}{H}{}^m(\mathcal{W}). \end{cases} \tag{6.1}$$

We study first the localized problem and prove that it is regular in spaces of flat and $\mathcal{C}^\infty$ functions respectively. Our method of proof is classical and relies on differential quotients.

Let (x, y) be the variables in Ω and z the variable in $\mathbb{R}$. Let φ be the function defining the boundary of Ω according to (3.1)-(3.3). The spaces $V^j(\mathbb{R} \times \Omega)$ are defined for $j \in \mathbb{N}$ by

$$V^j(\mathbb{R} \times \Omega) = \{u \in L^2(\mathbb{R} \times \Omega) \mid \forall \alpha \in \mathbb{N}^3,\ |\alpha| \leq j, \quad \varphi^{|\alpha|-j}\, \partial^\alpha u \in L^2(\mathbb{R} \times \Omega)\},$$

and by duality if $j < 0$.

We have the tensorization properties for all $j \in \mathbb{N}$:

$$V^j(\mathbb{R} \times \Omega) = H^j(\mathbb{R}, V^0(\Omega)) \cap L^2(\mathbb{R}, V^j(\Omega)) \tag{6.2}$$

and

$$V^{-j}(\mathbb{R} \times \Omega) = H^{-j}(\mathbb{R}, V^0(\Omega)) + L^2(\mathbb{R}, V^{-j}(\Omega)). \tag{6.3}$$

Proposition 6.1 *Let M be a strongly elliptic operator of order 2. Let $j \in \mathbb{N}$. Then any solution u of the Dirichlet problem* (6.1) *with compact support and the right hand side $f \in V^j(\mathbb{R} \times \Omega)$ satisfies the optimal regularity property*

$$u \in V^{2+j}(\mathbb{R} \times \Omega).$$

P. Grisvard [4] proved this result for $L = \Delta$ and $j = 0$ in L^p Sobolev spaces by a completely different technique.

Proof. Thanks to the strong ellipticity of M, we have the a priori estimate

$$\|u\|_{H^1(\mathbb{R}\times\Omega)} \leq c\,(\|Mu\|_{H^{-1}(\mathbb{R}\times\Omega)} + \|u\|_{L^2(\mathbb{R}\times\Omega)})$$

where c depends only on the support of u. This estimate can also be written as

$$\|u\|_{H^1(\mathbb{R},L^2(\Omega))} + \|u\|_{L^2(\mathbb{R},H^1(\Omega))} \leq c\Big(\|Mu\|_{H^{-1}(\mathbb{R},L^2(\Omega))+L^2(\mathbb{R},H^{-1}(\Omega))} + \|u\|_{L^2(\mathbb{R}\times\Omega)}\Big). \tag{6.4}$$

Considering for $h > 0$ small enough the function $(u(x,y,z+h) -u(x,y,z))h^{-1}$ and letting $h \to 0$, we deduce from (6.4) by recurrence over $\ell \in \mathbb{N}$ that there holds:

$$\|u\|_{H^{\ell+1}(\mathbb{R},L^2(\Omega))} + \|u\|_{H^{\ell}(\mathbb{R},H^1(\Omega))} \leq c\Big(\|Mu\|_{H^{\ell-1}(\mathbb{R},L^2(\Omega))+H^{\ell}(\mathbb{R},H^{-1}(\Omega))} + \|u\|_{L^2(\mathbb{R}\times\Omega)}\Big). \tag{6.5}$$

Integrating in y from the side $y = 0$, we easily prove that

$$\mathring{H}^1(\Omega) = \mathring{V}^1(\Omega). \tag{6.6}$$

Thus, (6.5) writes

$$\|u\|_{H^{\ell+1}(\mathbb{R},V^0(\Omega))} + \|u\|_{H^{\ell}(\mathbb{R},V^1(\Omega))} \leq c\Big(\|Mu\|_{H^{\ell-1}(\mathbb{R},V^0(\Omega))+H^{\ell}(\mathbb{R},V^{-1}(\Omega))} + \|u\|_{L^2(\mathbb{R}\times\Omega)}\Big). \tag{6.7}$$

Thus, for $f \in V^j(\mathbb{R}\times\Omega)$, the above estimate for $\ell = j+1$ yields that

$$u \in H^{2+j}(\mathbb{R}, V^0(\Omega)) \cap H^{1+j}(\mathbb{R}, V^1(\Omega)). \tag{6.8}$$

Let L be the operator

$$L(x,y,z;\partial_x,\partial_y) = M(x,y,z;\partial_x,\partial_y,0), \tag{6.9}$$

so that there exists an operator P of order ≤ 1 such that

$$M(x,y,z;\partial_x,\partial_y,\partial_z) = L(x,y,z;\partial_x,\partial_y) + P(x,y,z;\partial_x,\partial_y,\partial_z)\,\partial_z. \tag{6.10}$$

Since $Mu = f$ belongs to $H^j(\mathbb{R}, V^0(\Omega))$, we deduce from (6.8) and (6.10) that

$$Lu \in H^j(\mathbb{R}, V^0(\Omega)). \tag{6.11}$$

Theorem 3.1 applied for each z combined with an argument of differential quotients yields that (6.11) implies

$$u \in H^j(\mathbb{R}, V^2(\Omega)).$$

In that way, we prove by induction over $\ell = 0,\ldots,j$ that

$$Lu \in H^{j-\ell}(\mathbb{R}, V^{\ell}(\Omega))$$

which implies

$$u \in H^{j-\ell}(\mathbb{R}, V^{\ell+2}(\Omega)). \tag{6.12}$$

(6.12) for $\ell = j$ combined with (6.8) gives the proposition. ∎

For general operators of order $2m$, one encounters a technical difficulty in handling the norms with negative exponents. We have

Lemma 6.2 *Let M be a strongly elliptic operator of order $2m$ with $m \geq 2$. Let $j \in \mathbb{Z}$, $j > -m$. There exists an integer $k = k(m) > 0$ such that any solution u of the Dirichlet problem* (6.1) *with compact support and right hand side*

$$f \in H^{j+k}(\mathbb{R}, V^0(\Omega)) \cap H^k(\mathbb{R}, V^j(\Omega)) \tag{6.13}$$

satisfies the regularity property

$$u \in V^{2m+j}(\mathbb{R} \times \Omega).$$

Proof. Thanks to the strong ellipticity of M, just like above we obtain the a priori estimate for any $\ell \in \mathbb{N}$ — compare with (6.7)

$$\|u\|_{H^{\ell+m}(\mathbb{R},V^0(\Omega))} + \|u\|_{H^\ell(\mathbb{R},V^m(\Omega))}$$
$$\leq c\left(\|Mu\|_{H^{\ell-m}(\mathbb{R},V^0(\Omega))+H^\ell(\mathbb{R},V^{-m}(\Omega))} + \|u\|_{L^2(\mathbb{R}\times\Omega)}\right).$$

Thus, for f satisfying (6.13) the above estimate for $\ell = j+m+k$ yields that

$$u \in H^{2m+j+k}(\mathbb{R}, V^0(\Omega)) \cap H^{m+j+k}(\mathbb{R}, V^m(\Omega)).$$

The operator L is defined by (6.9) so that there exists an operator P of order $\leq 2m-1$ satisfying (6.10). Since $Mu = f$ belongs to $H^{j+k}(\mathbb{R}, V^0(\Omega))$, we deduce that

$$Lu \in H^{j+k}(\mathbb{R}, V^{1-m}(\Omega)).$$

Theorem 3.1 yields that

$$u \in H^{j+k}(\mathbb{R}, V^{1+m}(\Omega)).$$

In that way, we prove by induction over $\ell = 0, \ldots, j+m$ that

$$Lu \in H^{j+m-\ell+k_\ell}(\mathbb{R}, V^{\ell-m}(\Omega))$$

with $k_0 = k$, $k_1 = k_0 - (m-1)$, $k_2 = k_1 - (m-2)$, ... , $k_{m-1} = k_m = \ldots = k_{m+j}$, and

$$u \in H^{j+m-\ell+k_\ell}(\mathbb{R}, V^{\ell+m}(\Omega)).$$

It suffices to choose k_0 such that $k_{m-1} \geq 0$ to obtain finally for $\ell = m+j$ that u belongs to $L^2(\mathbb{R}, V^{2m+j}(\Omega))$. As u also belongs to $u \in H^{2m+j}(\mathbb{R}, V^0(\Omega))$, the lemma is proved. ∎

Theorem 6.3 *Let M be a strongly elliptic operator. Then any solution u of the Dirichlet problem* (6.1) *with compact support and right hand side $f \in \mathcal{C}^\infty(\mathbb{R} \times \overline{\Omega})$ satisfies the optimal regularity property*

$$u \in \mathcal{C}^\infty(\mathbb{R} \times \overline{\Omega}).$$

Proof. We denote by H^∞ the intersection of all spaces H^k for $k \in \mathbb{N}$. Since f belongs to $H^\infty(\mathbb{R}, V^0(\Omega))$, Proposition 6.1 and Lemma 6.2 yield that

$$u \in H^\infty(\mathbb{R}, V^{2m}(\Omega)). \tag{6.14}$$

The proof runs as above, using as spaces on Ω the spaces $V^j(\Omega)$ augmented by the spaces of polynomials:

$$\widetilde{V}^j(\Omega) = V^j(\Omega) + \mathbb{P}_{pj-1}(\Omega).$$

Indeed, Theorem 3.1 and Lemma 4.2 give that

$$\begin{cases} Lu \in \widetilde{V}^j(\Omega), \\ u \in \overset{\circ}{H}{}^m(\Omega). \end{cases} \implies u \in \widetilde{V}^{2m+j}(\Omega).$$

So, starting from (6.14), we have $Lu \in H^\infty(\mathbb{R}, \widetilde{V}^1(\Omega))$. Thus $u \in H^\infty(\mathbb{R}, \widetilde{V}^{2m+1}(\Omega))$. Going on, we prove by induction that $\forall \ell \in \mathbb{N}$,

$$u \in H^\infty(\mathbb{R}, \widetilde{V}^{2m+\ell}(\Omega)).$$

∎

Similarly to Remark 4.4, we deduce from all these statements a local regularity result in the space $\mathcal{C}^\infty(\overline{\mathcal{W}})$ for any domain $\mathcal{W} = \mathbb{R}^3 \setminus \overline{\mathcal{U}}$ where the domain $\mathcal{U}$ is the disjoint union of two (or more) domains with analytic boundaries $\mathcal{U}_1$ and $\mathcal{U}_2$ such that $\overline{\mathcal{U}}_1 \cap \overline{\mathcal{U}}_2$ is a curve $\mathcal{E}$. As examples, we can take for $\mathcal{U}_1$ and $\mathcal{U}_2$: a cylinder and a half-space, two cylinders, a torus and a cylinder, a torus and a ball, a torus and a half-space, two tori, etc...

Remark 6.4 Let us give a short description of the case where there is only one contact point: we take as $\mathcal{U}_1$ the half-space $y < 0$ and as $\mathcal{U}_2$ the ball of radius 1 and of center $(x_1, x_2, y) = (0, 0, 1)$ and we consider the Dirichlet problem for the Laplace operator. Then it is possible to prove that for a right hand side flat enough in $\mathcal{O} = (0,0,0)$, the solution is as flat as desired. The idea is to work in cylindrical coordinates (r,θ,y) with $x_1 = r\cos\theta$ and $x_2 = r\sin\theta$ and to combine arguments of differential quotients with respect to the variable θ and estimates on $\frac{u}{r}$ by considering $\Delta(\frac{u}{r})$. It is still possible to construct the asymptotics of u by the polynomial resolution, but the outcome of the construction gives not only polynomial terms but also non smooth terms. The most singular that we have found are

$$\frac{x_1 y^4}{r^2} \quad \text{and} \quad \frac{x_2 y^4}{r^2}.$$

■

Note. During the Conference, V.G. Mazya pointed out to us a work by V.I. Feigin. We only found in the literature the short note [3], where are stated results very similar to ours.

Acknowledgments. I thank Martin Costabel and Isabelle Gruais, from the Université de Rennes 1, for their valuable remarks about this paper.

References

[1] M. Costabel, M. Dauge, Singularities in presence of plane cracks. In preparation.

[2] M. Costabel, M. Dauge, Stable asymptotics for elliptic systems on plane domains with corners. *Comm. Partial Differential Equations* **9** and **10** (1994), 1677–1726.

[3] V. I. Feigin, Elliptic equations in domains with multidimensional singularities of the boundary. *Uspehi-Mat. Nauk* **27**, n°2 **(164)** (1972), 183–184.

[4] P. Grisvard, Problèmes aux limites dans des domaines avec points de rebroussement. To appear in *Ann. Fac. Sc. Toulouse.*

[5] P. Grisvard, *Boundary Value Problems in Non-Smooth Domains.* Pitman, London 1985.

[6] K. Ibuki, Dirichlet problem for elliptic equations of second order in a singular domain of $\mathbb{R}^2$. *J. Math. Kyoto Univ.* **14** n°1 (1974) 55–71.

[7] A. Khelif, Problèmes aux limites pour le laplacien dans un domaine à points cuspides. *C. R. Acad. Sci. Paris. Série A* **287** (1983) 1113–1116.

[8] V. A. Kondrat'ev, Boundary-value problems for elliptic equations in domains with conical or angular points. *Trans. Moscow Math. Soc.* **16** (1967) 227–313.

[9] V. G. Maz'ya, S. A. Nazarov, B. A. Plamenevskii, Elliptic boundary value problems in domains of the exterior-of-a-cusp type. *Problemy Matematicheskogo Analyza* **9** (1984) 105–148.

[10] V. G. Maz'ya, B. A. Plamenevskii, Estimates in L^p and in Hölder classes and the Miranda-Agmon maximum principle for solutions of elliptic boundary value problems in domains with singular points on the boundary. *Amer. Math. Soc. Transl. (2)* **123** (1984) 1–56.

[11] V. G. Maz'ya, J. Rossmann, On the asymptotics of solutions to the Dirichlet problem for second order elliptic equations in domains with critical angles on the edges. Preprint LiTH-MAT-R-91-37, Linköping University 1991.

[12] A. B. Movchan, S. A. Nazarov, Asymptotic behaviour of stress-strain state in the vicinity of sharp defects in an elastic body. *IMA J. Appl. Math.* **49** **(3)** (1992) 245–272.

[13] S. A. Nazarov, O. R. Polyakova, The asymptotic form of the stress-strain state near a spatial singularity of the boundary of the "beak tip" type. *J. Appl. Maths Mechs* **57** **(5)** (1993) 887–902.

[14] J.-L. Steux, Problème de dirichlet pour un opérateur elliptique dans un domaine à point cuspide. Technical Report 94-11/2, Université de Nantes 1994. To appear in *Ann. Fac. Sc. Toulouse.*

[15] G. M. Verzbinskii, V. G. Maz'ya, The asymptotics of solutions of the dirichlet problem near a non-regular frontier. *Dokl. Akad. Nauk SSSR* **176** (1967) 498–501.

[16] G. M. Verzbinskii, V. G. Maz'ya, Asymptotic behavior of the solutions of second order elliptic equations near the boundary. I. *Sibirsk. Mat. Z.* **12** (1971) 1217–1249.

[17] G. M. Verzbinskii, V. G. Maz'ya, Asymptotic behavior of the solutions of second order elliptic equations near the boundary. II. *Sibirsk. Mat. Z.* **13** (1972) 1239–1271.

U.R.A. 305 du C.N.R.S. – IRMAR – Université de Rennes 1
Campus de Beaulieu – 35042 Rennes Cedex 03, France
email: Monique.Dauge@univ-rennes1.fr

Star produit associé à un crochet de Poisson de rang constant

Louis Boutet de Monvel

Abstract

Nous construisons un star-produit sur une variété conique munie d'un crochet de Poisson de rang constant, en utilisant la méthode de B. V. Fedosov.

1. Description du problème, notations

On appelle cône (ou variété conique) une variété paracompacte X munie d'une action libre du groupe multiplicatif $\mathbf{R}_+$ [1]. On note $\mathcal{O}^s(X)$ (ou simplement $\mathcal{O}^s$) l'espace des fonctions f de classe C^∞ homogènes de degré s, i.e. telles que $f(\lambda x) = \lambda^s f(x)$ (on notera aussi $\mathcal{O}^s$ le faisceau correspondant). On note encore $\widehat{\mathcal{O}}^s(X)$ (ou $\widehat{\mathcal{O}}^s$) l'espace des symboles de degré s, i.e. des séries formelles:

$$a = \sum a_{s-k} \text{ avec } a_{s-k} \in \mathcal{O}^{s-k}, \ k \text{ entier}, \ k \geq 0. \tag{1.1}$$

On note $\widehat{\mathcal{O}}$ la réunion $\widehat{\mathcal{O}} = \bigcup \widehat{\mathcal{O}}^k$ ($k \in \mathbf{Z}$). L'algèbre $\mathcal{P}$ des opérateurs différentiels formels opère sur $\widehat{\mathcal{O}}$: un opérateur $P \in \mathcal{P}$ de degré m est une série formelle

$$P = \sum_{k \leq m} P_k \tag{1.2}$$

où pour chaque entier $k \leq m, P_k$ est un opérateur différentiel, homogène de degré k par rapport aux homothéties de X. On a de même un ensemble $\mathcal{P}_2$ d'opérateurs différentiels bilinéaires formels: un tel opérateur de degré $\leq m$ est une série formelle:

$$L(a,b) = \sum_{k \leq m} L_k(a,b) \tag{1.3}$$

où pour chaque $k \leq m$, L_k est un opérateur différentiel bilinéaire homogène de degré k (localement somme finie de la forme $L_k(a,b) = \sum p_{\alpha\beta}(x)\ \partial^\alpha a\ \partial^\beta b$, et homogène de degré k pour les homothéties). Un tel L définit une loi de composition (ou produit) sur $\widehat{\mathcal{O}}$.

---

[1] X est isomorphe à $Y \times \mathbf{R}_+$ où $Y = X/\mathbf{R}_+$ est la base, les homothéties étant données par: $t(y,r) = (y, tr)$; le choix d'un tel isomorphisme correspond au choix d'une fonction $r > 0$ homogène de degré 1.

et homogène de degré k pour les homothéties). Un tel L définit une loi de composition (ou produit) sur $\widehat{\mathcal{O}}$.

Un crochet de Poisson sur X est un opérateur différentiel bilinéaire $f, g \to \{f, g\}$, antisymétrique d'ordre 1, qui vérifie l'identité de Jacobi :

$$\begin{aligned} \{f, f\} &= 0 \quad &\textit{(antisymétrie)} \\ \{fg, h\} &= f\{g, h\} + \{f, h\}g \quad &(\textit{ordre } 1) \\ \{f, \{g, h\}\} + \{g, \{h, f\}\} + \{h, \{f, g\}\} &= 0 \quad &\textit{(identité de Jacobi)} \end{aligned} \tag{1.4}$$

Il est homogène de degré -1 si $\{f, g\}$ est homogène de degré $deg\,\{f, g\} = deg\,(f) + deg\,(g) - 1$ lorsque f et g sont homogènes.

Un star-produit sur X est une loi de composition $L \in \mathcal{P}_2$ associative ($L(a, L(b, c)) = L(L(a, b), c)$), et unitaire ($L(1, a) = L(a, 1) = a$). Une telle loi L définit une structure d'algèbre associative sur $\widehat{\mathcal{O}}$, unifère d'unité 1. Le terme dominant (homogène du plus haut degré) de L est alors $L_0(a, b) = ab$ (un peu plus généralement si L est seulement associative, son terme de plus haut degré L_m est nécessairement d'ordre 0, de la forme $L_m(a, b) = fab$ pour un $f \in \mathcal{O}^m$; si f est inversible, il existe une unité u de terme dominant f^{-1}, comme on voit aisément par approximations successives sur le degré): alors $u^{-1}\,L\,u$ est une loi associative et unitaire équivalente). Les relations imposées aux termes de degré -1 et -2 par l'associativité impliquent que l'application bilinéaire "partie principale du commutateur [a,b]":

$$\{a, b\} = L_{-1}(a, b) - L_{-1}(b, a) \tag{1.5}$$

est un crochet de Poisson homogène de degré -1.

Un problème naturel est de construire et de classifier, à équivalence près, les star-produits associés à un crochet de Poisson donné (deux lois L et L' sont équivalentes s'il existe un $P \in \mathcal{P}$ inversible tel que $L' = P\,L\,P^{-1}$). Ceci a été fait par M. De Wilde et P. Lecomte [DL1,2] (cf. aussi [OMY1,2]), dans le cas "semi-classique" : $X = Y \times \mathbf{R}_+$, où Y est une variété symplectique, $\{\,\}_X = h\{\,\}_Y$; la "constante de Planck" $\hbar$ est une fonction homogène de degré -1, inverse de la fonction coordonnée canonique de $\mathbf{R}_+$). Dans [BG] (cf. aussi [B1,2]) nous avons montré l'existence d'un star-produit sur une cône symplectique. Ici nous montrons le résultat suivant, en adaptant la méthode "élémentaire" de Fedosov [F2]:

Théorème. *Sur une variété conique munie d'un crochet de Poisson homogène de degré -1, de rang constant, il existe toujours un star-produit associé.*

2. Modèle local : Algèbre de Weyl fibrée

Soit X un cône, et $E \to X$ un fibré vectoriel conique à fibres symplectiques: sur chaque fibre E_x il y a un crochet de poisson $c_x = \sum c_{ij}\partial_i\partial_j$ (les ∂_i sont les dérivées verticales dans un repère convenable et la matrice c_{ij} est constante dans les fibres, i.e. ne dépend que de x, antisymétrique et inversible). Localement on peut choisir des coordonnées ξ_i linéaires dans les fibres, homogènes de degré 1/2, de sorte que les c_{ij} soient constants (le degré d'homogénéité doit être 1/2 si le crochet de Poisson est de degré -1).

Notons W l'algèbre des "sections symboles" de l'algèbre de Weyl, fibrée sur X, associée à E: les sections de degré m de W sont les séries formelles $f = \Sigma f_k(y, \xi)$, avec f_k homogène de degré $m - k$ sur E. La loi de composition est

$$\begin{aligned} f * g &= exp\,(\frac{1}{2}\; c_x(\partial_\xi), \partial_\eta)\; f(x,\xi)\; g(x,\eta)\;_{|\eta=\xi} \\ &= \sum \frac{1}{k!}(c(\partial_\xi, \partial_\eta))^k\, f(x,\xi)\; g(x,\eta)\;_{|\eta=\xi} \end{aligned} \tag{2.1}$$

qui définit bien un symbole (série formelle de fonctions homogènes) parce que c est de degré -1, donc $\frac{1}{k!}c(\partial_\xi, \partial_\eta)^k\; f(x,\xi)\, g(x,\eta)$ est de degré $\leq deg\,(f) + deg\,(g) - k \rightarrow -\infty$. ($W$ est en fait un faisceau sur les ouverts coniques de E).

Nous utiliserons aussi l'algèbre $\widehat{W}$ des jets d'ordre infini de sections de W le long de la section nulle ($\xi = 0$) de E (c'est en fait un faisceau sur les ouvertes coniques de X). Ses sections locales de degré k s'écrivent comme séries formelles (avec un choix de coordonnées ξ_j comme ci-dessus):

$$f = \sum f_\alpha(x)\, \xi^\alpha \tag{2.2}$$

où les f_α sont des symboles sur X, de degré $k - |\alpha| \deg(\xi) = k - |\alpha|/2$. Le starproduit (2.1) est également bien défini sur $\widehat{W}$.

En général nous construirons une star-algèbre comme sous algèbre d'une algèbre W de symboles de Weyl du type ci-dessus, annulée par certaines dérivations: il sera commode, comme le fait Fedosov [F2], de décrire globalement ces dérivations au moyen d'une connexion intégrable.

Rappelons d'abord la structure des dérivations de W (ou $\widehat{W}$).

Lemme 1. *Soit D une dérivation de degré $\leq k$ de W (ou $\widehat{W}$) (i.e. $D\widehat{W}^m \subset \widehat{W}^{m+k}$). Localement, dans tout bon système de coordonnées x, ξ (i.e. tel que les $\{\xi_i, \xi_j\}$ soient constants, comme ci-dessus) D s'écrit sous la forme:*

$$Df = \sum a_j(x)\; \partial f/\partial x_j \;+\; [b, f] \tag{2.3}$$

où les $a_j \in W$ sont des symboles de degré $\leq k + deg\; x_j$, et b est de degré $\leq k + 1$.

Preuve. Le centre de W est constitué des fonctions $f = f(x)$ constantes dans les fibres de E. Si f est centrale, Df l'est aussi (car $[Df, g] = D[f, g] - [f, Dg]$). Soit $D_0 = \sum a_j(x)\; \partial f/\partial x_j$ le champ de vecteurs sur X tel que $Df = D_0 f$ pour f centrale, et notons encore D_0 l'extension à W définie par la trivialisation fournie par notre choix de coordonnées. Alors $D - D_0$ est une dérivation qui annule les fonctions centrales, et une telle dérivation est intérieure (rappelons la preuve de cette deuxième assertion; choisissons des coordonnées "duales" ξ_i^* de sorte que $[\xi_i^*, \xi_j] = \delta_{ij}$; on a donc $\partial f/\partial\xi_i = [\xi_i^*, f]$. Posons $\beta_i = (D - D_0)\xi_i^*$: la forme $\beta = \Sigma\beta_i d\xi_i$ est fermée, i.e. $\partial\beta_i/\partial\xi_j = \partial\beta_j/\partial\xi_i$, comme il résulte de l'égalité

$$(D - D_0)[\xi_i^*, \xi_j^*] = [(D - D_0)\xi_i^*, \xi_j^*] + [\xi_i^*, (D - D_0)\xi_j^*] = 0 \tag{2.4}$$

Par suite β possède une primitive b homogène (le long des fibres de E) et on a $(D - D_0)f = [b, f]$. Il y a en fait une primitive globale canonique: $b = \int_0^1 \xi.\partial_\xi \,{}_L\, \beta(t\xi)dt$

3. Connexion associée à un bon système de coordonnées locales

Soit X une cône muni d'un crochet de Poisson homogène de degré -1, de rang constant. Soit F le feuilletage associé, engendré par les hamiltoniens h_f. Le fibré tangent du feuilletage TF est un fibré vectoriel symplectique conique comme ci-dessus, et il lui correspond une algèbre de Weyl W_{TF}. Comme annoncé nous construirons une star-algèbre sur X en l'identifiant à une sous-algèbre de $\widehat{W}_{TF}$, annulée par une F-connexion convenable à coefficients dans $\widehat{W}_{TF} \otimes \Omega_F$. Nous commençons par montrer comment on peut construire une telle connexion localement, dans un système de "bonnes" coordonnées locales:

Lemme 2. *Il existe au voisinage de chaque point $x_0 \in X$, un système de coordonnées $x_1, \ldots, x_n$ (homogènes de degré 1/2 tel que la matrice $\{x_i, x_j\}_{1 \le i,j \le k}$ soit constante, inversible (k= rang de $\{\ \}$).*

Preuve. On commence par choisir des coordonnées x'_j (homogènes de degré 1/2) de sorte que les champs hamiltoniens $h_{x'_j}$ forment une base de TF au point x_0, et que $h_{x'_1}, \ldots, h_{x'_{k-1}}$ soient linéairement independants du champ radial ρ, générateur infinitésimal des homothéties (remarquons que si ρ est tangent à une feuille F en un point, il l'est en tout point de F car F a des points communs avec ses homothétiques donc est invariante par homothétie). On modifie alors successivement ces coordonnées de sorte que $x_1 = x'_1$, et $\{x_i, x_j\} = constante = \{x'_i, x'_j\}(x_0)$, $x_j = x'_j$ sur une sous variétés conique initiale passant par x_0, transverse aux h_{x_i}, pour pour $1 \le i < j \le k$ (l'homogénéité de la surface initiale assure l'homògénéité de la solution; l'hypothèse de transversalité assure l'existence de sous variétés coniques initiales satisfaisant aux conditions indiquées). Nous dirons qu'un tel système de coordonnées est un "bon" système de coordonnées (seules comptent les k premières cordonnées).

Dans un bon système de coordonnées nous montrons une F-connexion canonique ∇ à coefficients dans $W_{TF} \otimes \Omega_{TF}$. Tout d'abord la F-forme tangente canonique à coefficients dans TF s'écrit $\tau = \sum dx_i \, \partial/\partial\xi_i$. On lui associe

$$\delta = \sum_{i=1}^{k} dx_i \ \xi_i^*. \tag{3.1}$$

où les ξ_i^* sont les coordonnées duales des ξ_i pour $\{\,\}$, comme ci-dessus.

Clairement τ est invariante par tout changement de coordonnées préservant les feuilles, donc δ est invariante par tout changement de coordonnées préservant $\{\,\}$, i.e. preservant les feuilles et leur structure symplectique.

La connexion modèle associée à notre bon système de coordonnées x_i est:

$$\nabla = D - \delta \tag{3.2}$$

où δ est définie ci-dessus et

$$D = \sum_{1}^{k} dx_i \ (\partial/\partial x_i)^F \tag{3.3}$$

on a noté $(\partial/\partial x_i)^F$ le champ de vecteurs tangent à F tel que $(\partial/\partial x_i)^F (x_j) = \delta_{ij}$ (symbole de Kronecker), pour $1 \leq i, j \leq k$ (c'est le champ $h_{x_i^*}$). D est le prolongement canonique de la dérivation exterieure d^F defini par notre choix de coordonnées (x_i, ξ_i) de TF. On a évidemment

$$D^2 = 0, \ [D, \delta] = 0, \ \delta^2 = 1 \otimes \omega, \tag{3.4}$$

où ω est la F-forme symplectique associée à $\{\,\}$.

Ici la forme de courbure de ∇ est centrale, et $Ad\nabla$ est intégrable. Les sections de W annulées par $Ad\nabla$ sont les symboles $f(x, \xi)$ qui, au dessus de chaque feuille, ne dépendent que de $x_1 + \xi_1, \ldots, x_k + \xi_k$. Elles forment évidemment une sous-algèbre, qui s'identifie à l'algèbre des symboles sur X muni du star-produit standard (de Moyal-Weyl) dans notre système de coordonnées.

4. Cas général: construction d'une connexion globale

La dérivation D ci-dessus n'est pas invariante par changement de coordonnées: en fait si $x' = (x_i')$ est un autre bon système de coordonnées comme ci-dessus, $\xi' = (\xi_i')$, $1 \leq i \leq k$) le système de coordonnées de TF qui s'en déduit, on a $\xi' = A\xi$ avec $A = dx'/dx$, d'où $D\xi' = dAA^{-1}\xi'$. L'opérateur linéaire $dA.A^{-1}$ est une dérivation symplectique, donc il existe un unique symbole de Weyl d'ordre 2 $\lambda = \sum \lambda_{ijk} \ \xi_i \, \xi_j \, dx_k$ symétrique (i.e. $\lambda_{ijk} = \lambda_{jik}$)) tel que $dA.A^{-1} \ \xi = [\lambda, \xi]$

Dans les nouvelles coordonnées on a alors

$$\nabla = \nabla' + \lambda$$

Plus généralement appelons connexion de Weyl une dérivation $\nabla : \widehat{W} \otimes \Omega_F \to \widehat{W} \otimes \Omega_F$ qui localement, dans tout bon système de coordonnées comme ci-dessus, s'écrive sous la forme

$$\nabla = D - \delta + \lambda$$

avec λ symbole de Weyl, de la forme

$$\lambda = \sum \lambda_j(x, \xi)\, dx_j = \sum \lambda_{i\alpha}(x)\xi^\alpha\, dx_i, \ |\alpha| \geq 2$$

forme différentielle à coefficients dans $\widehat{W}$, telle que les coefficients λ_j soient de degré $\leq 1/2$ (donc $deg\, Ad\, \lambda_i\, dx_i$ de degré ≤ 0), et $\lambda_{i,\alpha}(x)\, \xi^\alpha dx_i$ soient de degré ≤ 1, i.e. $\lambda_{i,\alpha}$ est de degré $\leq -\frac{1+\alpha}{2}$, ($|\alpha| \geq 2$). Comme nous avons vu une telle connexion existe localement (avec $deg_\xi\, (\lambda) \leq 2$), et aussi globalement parce qu'on peut les recoller au moyen d'une partition de l'unité (ceci revient à la construction d'une F-connexion symplectique).

A partir d'une telle connexion ∇_0 nous allons, par approximations successives, en construire une ∇ à courbure centrale ($Ad\nabla$ intégrable). On introduit pour cela sur $\widehat{W}$ et $\widehat{W} \otimes \Omega_F$ un nouveau poids (valuation), qui mesurera "l'ordre d'annulation" pour $\xi = 0$:

$$w(x_i) = w(dx_i) = -1, \ (w(f(x)) = -2\, deg\, f), \ w(\xi) = 0, \ w(dx) = 0.$$

(le gradué associé pour ces poids n'est pas commutatif). Avec cette convention on a

$$w\,(f * g) \leq w(f) + w(g), w([f, g]) \leq w(f) + w(g).$$

On a encore

$$w(Ad\delta) = w(\partial / \partial \xi_i) = 0, w(D) \geq 1, w(\lambda) \geq 1$$

de sorte que le terme dominant de ∇ est $-\delta$ qui s'identifie à la différentielle extérieure d_ξ. Dans les lignes qui suivent on note $\tilde{\alpha}$ la forme déduite de α par échange des dx_i , $d\xi_i$.

Notons $R = \nabla^2 = 1 \otimes \omega + r$ la courbure. On a $\nabla(r) = 0$ (parceque ∇ tue la courbure R et ω) d'où $d_\xi \tilde{r} = 0+$ des termes de poids supérieur. Soit alors α la 1-forme telle que

$$\tilde{\alpha} = \int_0^1 \xi.\partial_\xi \ {}_L\ \tilde{r}(t\xi)\, dt \ :$$

(noter que α est définie globalement). On a

$$w(\alpha) = w(R) \geq 1 \quad (deg\, \alpha = deg\, R \leq 0), \text{et} w(\tilde{R} - d_\xi \tilde{\alpha}) > w(R)$$

de sorte que la courbure de la connexion modifiée $\nabla + \alpha$ est $R_\alpha = (\nabla + \alpha)^2 = R + \nabla(\alpha) + \alpha^2 = 1 \otimes \omega +$ termes de poids $> w(r)$.

Par approximations successives on peut donc construire une connexion ∇ (i.e. λ) globale comme indiqué, de courbure centrale $\nabla^2 = 1 \otimes \omega$ et donc telle que $Ad\,\nabla$ soit integrable.

5. Fin de la construction

Soit finalement A la sous-algèbre de $\widehat{W}$ formée des sections f telles que $\nabla f = 0$. Alors $Gr\,A$ est l'algèbre des fonctions $f(x, \xi)$ constantes sur les feuilles de $gr\,\nabla$ i.e. telles que

$$df + \sum -\partial f/\partial \xi_i + \{\lambda_i, f\}_\xi dx_i = 0$$

dans n'importe quel "bon" système de coordonnées locales (x, ξ) comme ci-dessus. Le long de la section nulle $\xi = 0$ ces feuilles sont tangentes aux variétés d'équation ($x + \xi =$ *constante*), donc elles sont transverses à la section nulle $\xi = 0$. Par suite l'application de restriction $f \in A \to f \mid X$ $(f(x, \xi) \to f(x, 0))$ est bijective, et que

$$\sigma([f, g])_{\mid X} = \{\sigma f, \sigma g\}_{\xi \mid X} = \{f_{\mid X}, g_{\mid X}\}_X$$

On a donc bien construit un star-produit comme annoncé.

Remarque 1. La star-algèbre ainsi construite est "minimalement non-commutative", au sens que son star-produit s' écrit au moyen de dérivations tangentes aux feuilles de F, et qu'elle se plonge dans une algèbre de Weyl de rang k ($k =$ le rang de $\{\ \}$); son centre est également "maximal": $Z(gr\,A) = gr\,Z(A)$. Avec cette restriction on peut classifier les star-produits associés à un crochet de Poison donné comme dans Fedosov [F2]. La classification en général semble être un problème mal posé: par exemple classifier les star-produits associés au crochet de Poisson nul revient à classifier tous les star produits associés à tous les $\{\}$ de degré homogène < 1.

Remarque 2. On remarquera que la position des feuilles F par rapport aux homothéties ou au champ radial n'a pas joué de rôle dans cette analyse - en constraste avec ce qui se produit le plus souvent dans les questions concernant les E.D.P.

References

[BFFLS] Bayen F., Flato M., Fronsdal C., Lichnerowicz A., Sternheimer D., Deformation theory and quantization I , II , *Ann. Phys* **111** (1977), pp. 61–131.

[BG] Boutet de Monvel L. and Guillemin V., The spectral theory of Toeplitz operators, *Ann. of Math Studies n° 99*, Princeton University Press, 1981.

[B1] Boutet de Monvel L., Variétés de contact quantifiées, *Séminaire Goulaouic-Schwartz, 1979-80, exposé n° 3.*

[B2] Boutet de Monvel L., Toeplitz Operators, an asymptotic quantization of symplectic cones. *Research Center of Bielefeld-Bochum-Stochastics, University of Bielefeld (FDR) n° 215/86 (1986).*

[CFS] Connes A., Flato M., and Sternheimer D., Closed star-products and cyclic cohomology, *Lett. Math. Phys.* **24** (1992), pp. 1–12.

[DL1] De Wilde M. and Lecomte P., Existence of star-products and of formal deformations of the Poisson Lie algebras of arbitrary symplectic manifolds, *Lett. Math. Phys.* **7** (1983), pp. 487–496.

[DL2] De Wilde M. and Lecomte P., Formal Deformations of the Poisson Lie Algebra of a Symplectic Manifold and Star-Products: Existence, Equivalence, Derivations, in: *Deformation Theory of Algebras and Structures and Applications,* M. Hazewinkel & M. Gerstenhaber, eds., Kluwer Acad. Pub., Dordrecht (1988), pp. 897–960.

[F1] Fedosov B.V., Formal quantization, *Some topics of Modern Mathematics and their Applications to Problems of Mathematical Physics,* (in Russian), Moscow (1985), pp. 129–136.

[F2] Fedosov B.V., A simple geometrical construction of deformation quantization, *J. Diff Geom.*, (to appear).

[FS] Flato M. and Sternheimer D., Closedness of star products and cohomologies, in: *Lie Theory and Geometry in honor of B. Kostant,* J.L.Brylinski & R. Brylinski, eds., Progress in Math, Birkhäuser Boston, 1994.

[Gu] Gutt S., Equivalence of deformations of twisted products on a symplectic manifold, *Lett. Math. Phys.* **3** (1979), pp. 495–502.

[KM1] Karasev M.V. and Maslov V.P., Pseudodifferential operators and a canonical operator in general symplectic manifolds, *Math. USSR Izvestia* **23** (1984), pp. 277–305.

[KM2] Karasev M.V. and Maslov V.P., Nonlinear Poisson brackets: geometry and quantization, *Translation of mathematical monographs* vol. 119, Amer. Math .Soc., Providence, 1993.

[Mo] Moyal J., Quantum mechanics as a stastistical theory, *Proc. Camb. Phil. Soc.* **45** (1965), pp. 99–124.

[OMY1] Omori H., Maeda Y., and Yoshioka A., Weyl manifolds and deformation quantization, *Advances in Math.* **85** (1991), pp. 224–255.

[OMY2] Omori H., Maeda Y., and Yoshioka A., Existence of a closed star-product, *Lett. Math. Phys* **26** (1992), pp. 285–294.

[V] Vey J., Déformation du crochet de Poisson sur une variété symplectique, *Comment. Math. Helvet.* **50** (1975), pp. 421–454.

[W] Weinstein A, Deformation quantization, *Séminaire Bourbaki n°* **789**, Juin 1994.

Université de Paris 6, Institut de Mathématiques
UMR 9994 du CNRS, Tour 46-0, 5ième étage, boîte 172
4 place Jussieu, F-75252 Paris Cedex 05

La méthode des lâchés de tourbillons pour le calcul des efforts aérodynamiques

Philippe Destuynder et Françoise Santi

Résumé

Malgré les progrès réalisés ces dernières années dans la résolution des équations de Navier-Stokes, il est souhaitable de disposer de méthodes simplifiées et économiques permettant d'obtenir une première approximation des coefficients aérodynamiques pour une structure placée dans un écoulement.

La méthode des potentiels de vitesses a connu un grand succès dans le passé et une nouvelle approche basée sur la "création d'un sillage" par des lâchés de tourbillons, lui donne aujourd'hui un intérêt grandissant dans la communauté des ingénieurs aérodynamiciens.

Nous proposons dans cet article de donner un éclairage de cette stratégie basée sur une analyse locale en extrémité de sillage, du potentiel des vitesses.

Introduction

Considérons un profil portant bidimensionnel comme celui de la figure 1. Le problème fondamental de l'aérodynamicien est de déterminer les coefficients traduisant, pour une incidence par rapport au vent, les efforts exercés par l'écoulement sur ce profil. La résolution des équations de Navier-Stokes permet de répondre à cette préoccupation mais avec un coût de calcul important et la nécessité d'utiliser un modèle de turbulence ainsi qu'un ouvert suffisamment grand afin d'éviter les effets de confinement.

L'extension au cas tridimensionnel est par contre encore inaccessible sur le plan pratique. Seules les équations d'Euler avec un modèle de couches limites pariétales sont réellement opérationnelles, mais malheureusement conduisent à des efforts aérodynamiques nuls (c'est une version moderne du paradoxe de d'Alembert). Les modèles basés sur un écoulement dérivé d'une fonction potentielle, présentent l'avantage de la simplicité et de l'économie du coût de calcul. Mais comme pour les équations d'Euler, il est nécessaire d'avoir recours à des singularités géométriques si l'on veut éviter le paradoxe de d'Alembert dans la mesure où on se limite à un écoulement à potentiel. Rappelons que dans ce cas le champ de pression est donné à l'aide du théorème de Bernoulli par :

$$p + \frac{1}{2}\rho|\text{grad } \varphi|^2 = \text{constante}$$

p désignant la pression et φ le potentiel des vitesses alors que ρ est la masse volumique.

Les coefficients aérodynamiques caractérisant le comportement du profil placé dans un écoulement dont la vitesse à l'infini amont est V_∞ sont donnés par les expressions suivantes :

$$\begin{cases} C_D = \dfrac{\int_{\Gamma_0} p(v,e_1)}{\frac{1}{2}\rho V_\infty^2} & \text{(traînée)}, \\ C_L = \dfrac{\int_{\Gamma_0} p(v,e_2)}{\frac{1}{2}\rho V_\infty^2} & \text{(portance)}, \\ C_M = \dfrac{\int_{\Gamma_0} p(v,x,e_3)}{\frac{1}{2}\rho V_\infty^2} & \text{(coefficient de tangage à l'origine des coordonées)}, \end{cases} \qquad (1)$$

((,) désigne le produit scalaire entre vecteurs et (, ,) le produit mixte ; par ailleurs $\{e_1, e_2, e_3\}$ est la base orthonormée de la figure 1). On distinguera cette dernière de la base locale en extrémité de bord de fuite et notée $\{\underset{\sim}{e_1}, \underset{\sim}{e_2}, \underset{\sim}{e_3}\}$, sur la figure 1.

Remarque 1. Le potentiel des vitesses est proportionnel à la vitesse V_∞. La pression est donc proportionnelle au carré de cette vitesse et par conséquent les coefficients aérodynamiques C_D, C_L et C_M ne dépendent pas de V_∞. En d'autres termes, la modélisation à l'aide d'un potentiel de vitesses élimine tout effet de Reynolds. ■

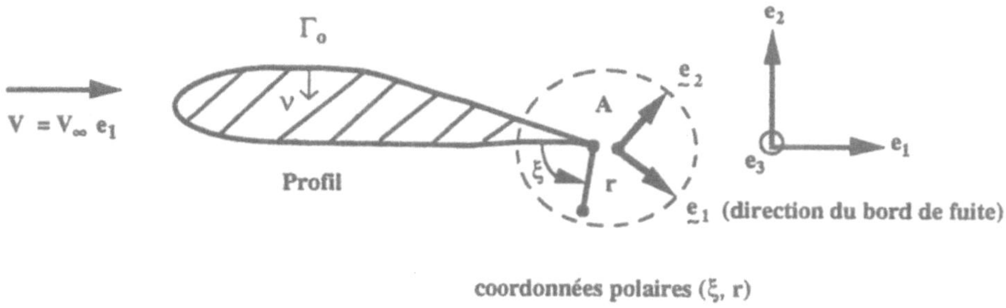

Figure 1. Le profil 2D

En fait, nous verrons que les coefficients (1) sont en général nuls lorsque l'écoulement est approché par un potentiel de vitesses et ceci sous des hypothèses très raisonnables d'invariance du comportement à l'infini sous l'effet d'un déplacement du profil (translation ou rotation). La seule façon de lever cette ambiguité est d'introduire une singularité géométrique du type fissure comme il est indiqué sur la figure 1. Mais là encore ceci ne conduit qu'à une expression non nulle pour la traînée dans la direction de la fissure, $\underset{\sim}{e}_1$. Ceci peut convenir pour un profil sous faible incidence, mais est largement insuffisant si cette dernière augmente. Il est donc nécessaire d'introduire une autre information de façon à obtenir une approximation convenable des coefficients C_D, C_L et C_M. Une idée qui connaît un succès certain est de construire une ligne de sillage issue a priori de l'extrémité du bord de fuite du profil. Elle représente la trajectoire d'un tourbillon lâché de la singularité géométrique du profil. Le long de cette courbe, le potentiel des vitesses est discontinu, ce qui permet de faire apparaître une différence de pression (via le théorème de Bernoulli) et par conséquent un effort résultant. L'équilibre global des forces permet alors de définir des forces appliquées au profil.

Le plan qui nous suivons ci-après est le suivant :

1. Rappels sur la formulation potentielle.
2. Comment calculer les coefficients aérodynamiques.
3. Comparaison entre la méthode du potentiel et les solutions des équations de Navier-Stokes.
4. Rôle du sillage dans la méthode des écoulements à potentiel.
5. Un algorithme de construction du sillage.

1. Rappels sur la formulation potentielle

Considérons un ouvert du plan noté Ω de frontière extérieure Γ_1 et donc la frontière intérieure Γ_0 représente le profil autour duquel on souhaite calculer l'écoulement. La fonction potentielle φ est alors solution du modèle suivant (la vitesse est $u = \text{grad}\,\varphi$) :

$$\begin{cases} -\Delta\varphi = 0 & \text{sur } \Omega, \\ \frac{\partial\varphi}{\partial\nu} = V_\infty(v, e_1) & \text{sur } \Gamma_1, \\ \frac{\partial\varphi}{\partial\nu} = 0 & \text{sur } \Gamma_0. \end{cases} \tag{2}$$

Compte tenu de la relation classique :

$$\int_{\Gamma_0} \nu = 0$$

le système (2) admet une solution unique, définie à une constante près dans l'espace $H^1(\Omega)$. Mais en raison de la singularité géométrique localisée au bord de fuite du profil, il n'y a pas régularité au sens $H^2(\Omega)$ de φ. En utilisant une méthode de séparation des variables, on peut trouver une expression de φ sous forme d'une série de fonctions harmoniques.

En utilisant le système de coordonnées polaires (r, ξ) indiqué sur la figure 1, on obtient (Cf. G. GEYMONAT, P. GRISVARD [3]) : (3)

$$\varphi(r, \xi) = \sum_{n \geq 1} \alpha_n r^{\frac{n}{2}} \cos\left(\frac{n\xi}{2}\right) \tag{3}$$

où α_n est un coefficient réel. On notera que la première fonction de ce développement, soit $\sqrt{r}\cos\frac{\xi}{2}$, est localement (au voisinage du point A de la figure 1), dans l'espace $H^1(\Omega)$ mais pas dans $H^2(\Omega)$. Sur le plan pratique les coefficients α_n peuvent être calculés de différentes façons. L'une d'elle consiste à remarquer que sur un cercle centré en A et de rayon R on a :

$$\alpha_n R^{\frac{n}{2}} \Pi = \int_o^{2\Pi} \varphi(R,\xi) \cos\left(\frac{n\xi}{2}\right) d\xi. \tag{4}$$

Mais il existe des stratégies plus commodes et plus générales, notamment basées sur l'utilisation des fonctions singulières duales. Pour plus de détails nous renvoyons à M. MOUSSAOUI [6] ou M. DOBROWOLSKI [2].

2. Comment calculer les coefficients aérodynamiques

L'énergie cinétique de l'écoulement contenu dans l'ouvert Ω est définie par :

$$E_c = \frac{\rho}{2} \int_\Omega |\mathrm{grad}\,\varphi|^2$$

où le potentiel φ est solution du système (1). Considérons un mouvement de corps rigide du profil décrit par un champ de vecteur θ que l'on peut représenter par un torseur sous la forme suivante :

$$\theta = P + Q \wedge x$$

où P est un vecteur translation et Q celui de la rotation. Par ailleurs x est le vecteur des coordonnées d'un point du profil. Il est alors possible de construire un prolongement régulier de θ (à l'aide par exemple d'une fonction de troncature) à l'extérieur de l'ouvert Ω et dont le support est limité à un voisinage du profil. Considérons alors la famille d'ouverts paramétrée par le scalaire η et définie par :

$$\Omega^\eta = F^\eta(\Omega)$$

où :

$$x \in \Omega,\ F^\eta(x) = x + \eta\theta(x) \tag{5}$$

Pour chaque valeur de η, il est possible de définir un problème analogue à (1) mais posé sur l'ouvert Ω^η. La solution $\varphi(\eta)$ peut alors s'exprimer en fonction des points de Ω par :

$$\varphi^\eta(x) = \varphi(\eta)\,(x + \eta\theta(x)) \tag{6}$$

(on élimine la constante additive définissant $\varphi(\eta)$ par une relation arbitraire). L'énergie cinétique de l'écoulement pour chaque position du profil paramétrée par η est alors :

$$E^\eta = \frac{\rho}{2} \int_{\Omega^\eta} |\mathrm{grad}\,\varphi(\eta)|^2.$$

Un simple calcul de géométrie différentielle permet d'obtenir l'expression de la dérivée de cette fonction de η. Pour plus de détails nous renvoyons le lecteur aux travaux de F. MURAT et J. SIMON [7]. Sans hypothèse particulière concernant la régularité de φ, nous avons ainsi :

$$\frac{\partial E}{\partial \Omega} \overset{\text{def}}{=} \lim_{\eta \to 0} \frac{E^\eta - E}{\eta} = \frac{\rho}{2} \int_\Omega |\text{grad}\,\varphi|^2 \,\text{div}\,\theta - \rho \int_\Omega \partial_i \varphi \partial_j \varphi \partial_i \theta_j. \tag{7}$$

Mais en utilisant la formule de Stokes, il est possible de transformer l'expression ci-dessus en intégrale de contour de la façon suivante. Désignons par $\mathcal{V}^\varepsilon$ un voisinage annulaire du profil et contenant le support du champ θ. La frontière interne de $\mathcal{V}^\varepsilon$ est constituée d'une portion s'appuyant sur le contour du profil et d'un cercle centré au bord de fuite (point A) et de rayon ε. Comme son nom l'indique, ce dernier est destiné à tendre vers zéro, c'est à dire que le cercle C^ε se rapprochera du point A. La surface du disque intérieur à C^ε est notée D^ε et nous avons pour tout ε :

$$\frac{\partial E}{\partial \Omega}\theta = \frac{\rho}{2} \int_{\mathcal{V}^\varepsilon \cup D^\varepsilon} |\text{grad}\,\varphi|^2 \,\text{div}\,\theta - \rho \int_{\mathcal{V}^\varepsilon \cup D^\varepsilon} \partial_i \varphi \partial_j \varphi \partial_i \theta_j \tag{8}$$

et en utilisant la formule de Stokes, compte tenu des conditions aux limites vérifiées par φ surΓ_0 :

$$\begin{aligned}
\frac{\partial E}{\partial \Omega}\theta = &\frac{\rho}{2} \int_{\Gamma_0} \left|\frac{\partial \varphi}{\partial s}\right|^2 (\theta, \nu) \\
&+ \lim_{\varepsilon \to 0} \left\{ \frac{\rho}{2} \int_{D^\varepsilon} |\text{grad}\,\varphi|^2 \,\text{div}\,\theta - \rho \int_{D^\varepsilon} \partial_i \varphi \partial_j \varphi \partial_i \theta_j \right\} \\
&+ \lim_{\varepsilon \to 0} \left\{ \frac{\rho}{2} \int_{C^\varepsilon} |\text{grad}\,\varphi|^2 (\theta, \nu) - \rho \int_{C^\varepsilon} (\text{grad}\,\varphi, \theta)\,(\text{grad}\,\varphi, \nu) \right\}.
\end{aligned}$$

Le second terme s'annule car φ a la régularité $H^1(\Omega)$. Le calcul du troisième terme peut s'effectuer analytiquement en utilisant l'expression locale de φ donnée en (3). On établit ainsi aisément la relation suivante (Ph. DESTUYNDER, M. DJAOUA [1]) :

$$\frac{\rho}{2} \int_{\Gamma_0} \left|\frac{\partial \varphi}{\partial s}\right|^2 (\theta, \nu) = \frac{\partial E}{\partial \Omega}\theta + \frac{\rho \Pi}{4} \alpha_1^2 (\theta(A), \underset{\sim}{e_1}).$$

Mais en utilisant le théorème de Bernoulli rappelé plus haut, nous obtenons le théorème de Blasius (version étendue au cas d'un champ θ pouvant inclure des rotations) :

$$((F, \theta)) = -\int_{\Gamma_0} pv = \frac{\rho}{2} \int_{\Gamma_0} \left|\frac{\partial \varphi}{\partial s}\right|^2 v = \frac{\rho \Pi \alpha_1^2}{4} \left(\theta(A), \underset{\sim}{e_1}\right) + \frac{\partial E}{\partial \Omega}\theta \tag{9}$$

où ((,)) désigne le produit scalaire entre le torseur des efforts aérodynamiques F et le champ de déplacement rigidifiant θ. Si on pose ($\mathcal{R}$ est la résultante et

$\mathcal{M}$ le moment) :

$$F = \mathcal{R} + \mathcal{M} \wedge x$$

nous avons :

$$((F, \theta)) = (\mathcal{R}, P) + (\mathcal{M}, Q).$$

En remarquant que :

$$E_c = \frac{\rho}{2} \left[\int_{\Gamma_1} \frac{\partial \varphi}{\partial \nu} \varphi - \int_{\Omega} \Delta \varphi \varphi \right] = \frac{\rho}{2} V_\infty \int_{\Gamma_1} \varphi \nu_1$$

nous pouvons conclure que si la frontière Γ_1 est suffisamment éloignée, la fonction φ, y est inchangée sous l'effet d'un déplacement du profil et par conséquent, l'énergie cinétique E_c vérifie $\frac{\partial E_c}{\partial \Omega} \theta = 0$. Ce résultat intuitif peut néanmoins se démontrer rigoureusement dans le cadre des espaces fonctionnels $H^1(\mathbb{R}^2)$, en utilisant une représentation intégrale de φ. Finalement en se plaçant dans l'hypothèse d'un milieu infini nous obtenons :

$$(\mathcal{R}, P) + (\mathcal{M}, Q) = \frac{\rho \Pi}{4} \alpha_1^2 (\theta(A), \underset{\sim}{e_1}). \tag{10}$$

Remarque 2. Sous l'hypothèse d'invariance de l'énergie cinétique par rapport à un déplacement du profil, les efforts aérodynamiques ne dépendent que du coefficient de la singularité à l'extrémité du bord de fuite. Au cas où l'angle saillant du point A (Cf. figure 1) serait inférieur à 2Π, on n'obtiendrait pas d'efforts aérodynamiques appliqués au profil. Cette conclusion surprenante n'est autre que le fameux paradoxe de d'Alembert. Notons qu'une formulation utilisant la fonction de courant n'apporterait pas de modification de cette conclusion.

Remarque 3. On notera que l'effort mécanique dû à l'écoulement est parallèle à la direction du bord de fuite (portée par le vecteur $\underset{\sim}{e_1}$).

Remarque 4. Il est possible de calculer toutes les dérivées de l'énergie cinétique par rapport à la position du profil, ceci en utilisant la même technique de dérivation par rapport à un domaine. Ces quantités sont particulièrement utiles dans le cadre d'études aéroélastiques.

3. Comparaison (sommaire) entre la méthode du potentiel et les solutions des équations de Navier-Stokes

De façon à situer la vraisemblance de la méthode du potentiel des vitesses, nous avons comparé les résultats obtenus avec ceux d'un code de calcul pour les équations de Navier-Stokes. Ce dernier utilise la méthode d'éléments finis mixtes dans laquelle vitesses et pressions sont approchées séparément. Pour plus de détails nous renvoyons le lecteur à V. GIRAULT, P.A. RAVIART [4]. Nous avons reporté sur la figure 2 le maillage et la configuration des profils utilisés. Les lignes isopotentielles obtenues pour différentes incidences sont tracées sur la figure 3 et doivent être comparées au champ de vitesses de la figure 4 qui lui est solution des équations de Navier-Stokes. On observera que la différence essentielle réside dans le sillage. Les portances résultantes des deux profils (suivant l'axe e2 de la figure2) sont indiquées pour les deux calculs sur la figure 5. On peut constater que si la coïncidence n'est pas excellente, les sens de variation vis-à-vis de l'incidence et l'apparition des phénomènes d'instabilité sont correctement reproduits par la méthode du potentiel.

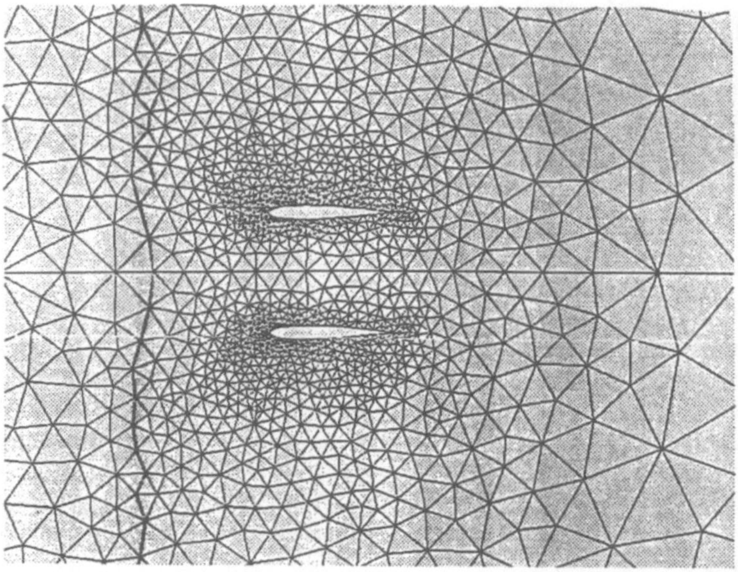

Figure 2. Maillage éléments finis pour deux profils NACA 12

4. Rôle du sillage dans la méthode des écoulements à potentiel

Considérons le système constitué d'un profil et d'une courbe AB issue de l'arrière du profil comme celui de la figure 6. La condition aux limites appliquée sur la courbe AB est du type Neumann homogène, ce qui signifie physiquement que nous avons une ligne de courant de part et d'autre de la courbe AB.

En appliquant la théorie que nous avons présentée au paragraphe 2, nous obtenons que le torseur $(\mathcal{R}, \mathcal{M})$ des efforts appliqués au système : profil + courbe AB , est tel que :

$$(\mathcal{R}, P) + (\mathcal{M}, Q) = \frac{\rho \Pi \alpha_1^2}{4}(\theta(B), \underset{\sim}{e_1})$$

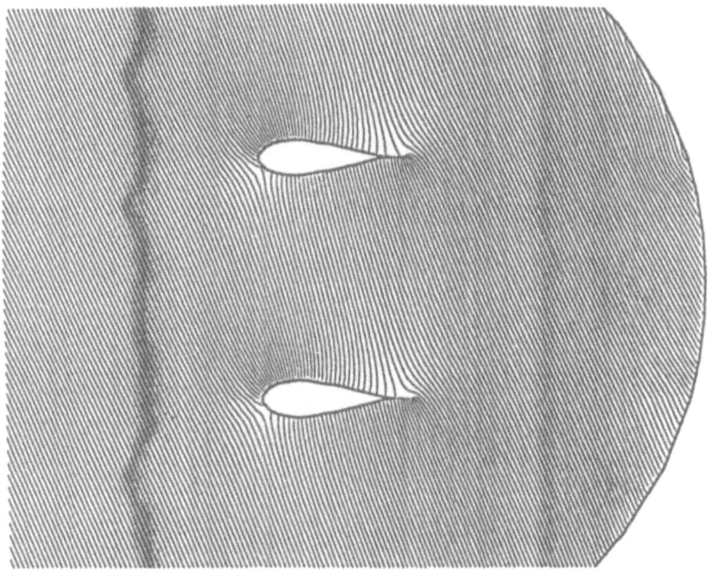

Figure 3. Lignes isopotentielles pour une incidence de 30 degrés par la méthode potentielle

où α_1 est le coefficient de la singularité de la fonction φ au point B.

Mais puisque sur le sillage AB nous avons la condition aux limites :

$$\frac{\partial \varphi}{\partial \nu} = 0,$$

le torseur des efforts uniquement appliqués au profil, et noté $(\mathcal{R}_p, \mathcal{M}_p)$, est tel que :

$$(\mathcal{R}_p, P) + (\mathcal{M}_p, Q) = \frac{\rho \Pi \alpha_1^2}{4}(\theta(B), \underset{\sim}{e_1}) - \frac{\rho}{2} \int_{\widehat{AB^\pm}} \left| \frac{\partial \varphi}{\partial s} \right|^2 (\theta, \nu) \qquad (11)$$

ν désignant la normale unitaire le long de la courbe $\widehat{AB^\pm}$ et $\underset{\sim}{e_1}$ désignant le vecteur unitaire tangent à AB au point B.

Remarque 5. Puisque la ligne de sillage $\widehat{AB}$ a une forme arbitraire, le second terme dans l'expression de gauche de l'égalité ci-dessus comprend des contributions dans les deux directions d'espace. Ceci conduit donc à une estimation simultanée de la portance et de la traînée du profil.

Remarque 6. Lorsque $\frac{\partial \varphi}{\partial s}$ est continu de part et d'autre de la ligne de sillage $\widehat{AB}$, la contribution correspondant dans le terme de gauche de (11) est nulle.

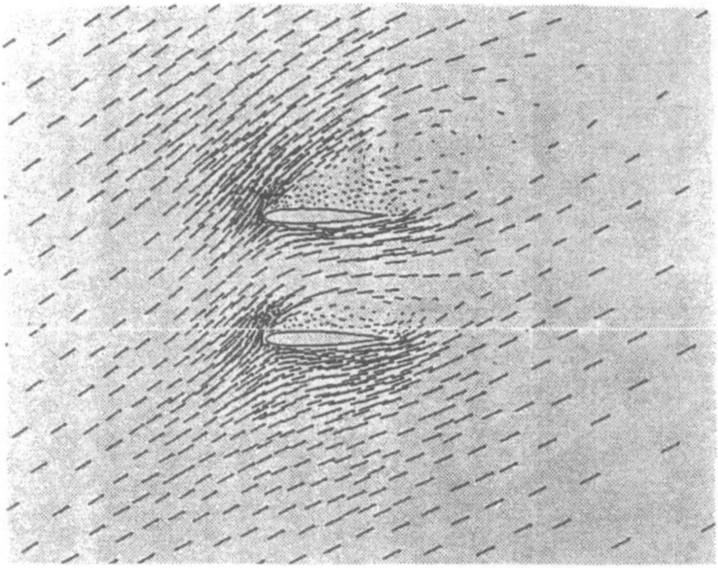

Figure 4. Champ de vitesses pour une incidence de 30 degrés par la méthode de Navier-Stokes

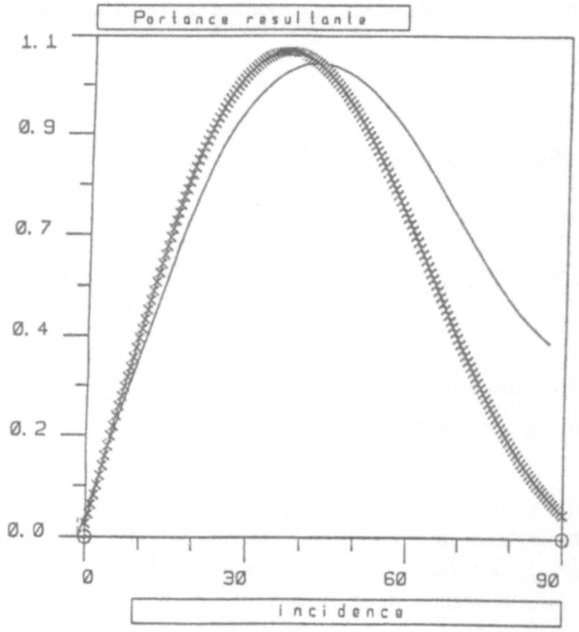

Figure 5. Portance résultante pour : ——— l'écoulement potentiel; ++++ les équations de Navier -Stokes

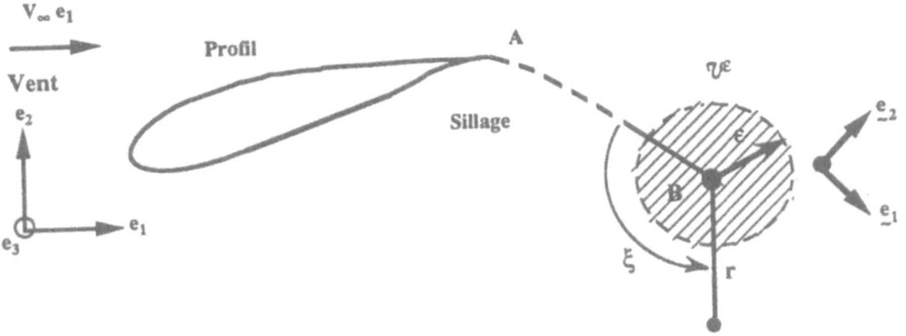

Figure 6. Voisinage circulaire servant à l'évaluation de la vitesse et de l'accélération du tourbillon

Si en outre $\frac{\partial \varphi}{\partial s}$ est continu au voisinage du point B alors $\alpha_1 = 0$ et il ne sert à rien de prolonger le sillage. On dit que ce dernier est fini ou attaché au profil.

5. Un algorithme de construction du sillage

L'idée directrice développée par plusieurs auteurs est de suivre le mouvement d'un tourbillon issu de la singularité géométrique. Cette méthode est aujourd'hui opérationnelle y compris dans le cas tridimensionnel pour lequel la ligne de sillage est une surface. Notons plus particulièrement les travaux de M. YANAGIZAWA [5] qui en couplant cette stratégie avec une méthode intégrale a proposé une méthode numérique efficace pour évaluer les efforts aérodynamiques. Le problème essentiel est alors de définir une caractérisation de la courbe du sillage. L'idée la plus naturelle est d'utiliser une méthode itérative constructive basée sur une caractérisation de la vitesse du tourbillon et éventuellement de son accélération, ceci dans le cas de schéma d'ordre plus élevé. La méthode d'évaluation que nous proposons est sans doute discutable mais présente un double avantage : d'une part elle est stable et relie vitesse et accélération du tourbillon aux coefficients α_i locaux (ceux qui interviennent dans le développement local de φ donné en (3)), d'autre part elle introduit un paramètre d'épaisseur de sillage qui physiquement est relié à la finesse du profil.

Considérons donc un voisinage circulaire $\mathcal{V}^\varepsilon$ de rayon ε du point B (Cf.figure6). La vitesse du tourbillon en ce dernier, ainsi que l'accélération sont définies par :

$$\left\{\begin{aligned} \underset{\sim}{u}(B) &= \frac{1}{|\mathcal{V}^\varepsilon|}\int_{\mathcal{V}^\varepsilon} \operatorname{grad}\varphi = \frac{1}{\Pi\varepsilon^2}\int_{\partial\mathcal{V}^\varepsilon} \varphi\nu \\ \underset{\sim}{\gamma}(B) &= \frac{1}{|\mathcal{V}^\varepsilon|}\int_{\mathcal{V}^\varepsilon} \operatorname{grad}\left[= \tfrac{1}{2}\operatorname{grad}|\varphi|^2\right]\nu \\ &= \frac{1}{2\Pi\varepsilon^2}\int_{\partial\mathcal{V}^\varepsilon} |\operatorname{grad}\varphi|^2\nu \end{aligned}\right. \tag{12}$$

où v désigne la normale unitaire sortante à l'ouvert $\mathcal{V}^\varepsilon$. Un simple calcul, utilisant l'expression (3) de φ conduit alors aux expressions suivantes :

$$\begin{cases} \underset{\sim}{u}(B) &= -\alpha_2 \underset{\sim}{e_1} - \frac{4\alpha_1}{3\Pi\sqrt{\varepsilon}} \underset{\sim}{e_2} + \dots \\ \underset{\sim}{\gamma}(B) &= -\left(\frac{3\alpha_1\alpha_3}{4\varepsilon} + 2\alpha_2\alpha_4\right) \underset{\sim}{e_1} + \left(\frac{2\alpha_1\alpha_2}{3\Pi\varepsilon^{\frac{3}{2}}} + \frac{44\alpha_1}{15\Pi}\frac{\alpha_4}{\sqrt{\varepsilon}} - \frac{2\alpha_2\alpha_3}{\Pi\sqrt{\varepsilon}}\right) \underset{\sim}{e_2} + \dots \end{cases} \tag{13}$$

Un schéma naturel d'intégration de la trajectoire du point B et qui permet de construire la ligne de sillage, consiste à poser :

$$B^{n+1} = B^n + \Delta t \, \underset{\sim}{u}(B^n) + \frac{\Delta t^2}{2} \underset{\sim}{\gamma}(B^n) \tag{14}$$

où l'indice supérieur n désigne le pas de temps (qui est fictif car nous avons envisagé un système statique avec écoulement permanent).

Remarque 7. L'utilisation d'une méthode intégrale pour évaluer φ^n à chaque itération et par conséquent $\underset{\sim}{u}(B^n)$ ainsi que $\underset{\sim}{\gamma}(B^n)$ permet d'éviter le problème du remaillage. En effet, la courbe de sillage ne passe pas nécessairement dans le réseau d'arêtes d'un maillage initial. Par contre, une méthode intégrale conduit à une matrice limitée au contour du profil ainsi qu'à la courbe de sillage. A chaque itération (permettant de passer de B^n à B^{n+1}) on ajoute un degré de liberté (en 2D) et la matrice du système linéaire associée au calcul de φ^n est bordée par une ligne (et une colonne) supplémentaire. La factorisation est inductive, c'est à dire que l'on utilise à chaque pas la factorisée de l'étape précédente. Ceci limite considérablement les calculs.

Références

[1] Ph. DESTUYNDER et M. DJAOUA, [1981]. Une interprétation mathématique de l'intégrale de Rice en mécanique de la rupture fragile, *Math. Meth. in the appl. Sci.*, **3**, p. 70–87.

[2] M. DOBROWOLSKI, [1985]. On finite element methods for non linear elliptic problems on domains with corners, *Lecture Notes in Math.*, no. 1121, Springer-Verlag, p. 85–103.

[3] G. GEYMONAT et P. GRISVARD, [1985]. Eigenfunction expansions for non self adjoint operators and separation of variables, *Lecture Notes in Math.*, no. 1121, Springer-Verlag, p. 123–165.

[4] V. GIRAULT et P.A. RAVIART, [1986]. Mixed finite element methods for Navier–Stokes equations, in: *Series in Computational Mathematics* n 5, Springer Verlag.

[5] K. KIKUCHI, A. KIKUCHI et M. YANAGIZAWA, [1994]. A study of characteristic for separation flow around body using vortex tracking method and boundary element method, ICCM3, Tokyo.

[6] M. MOUSSAOUI, [1985]. Sur l'approximation des solutions du problème de Dirichlet dans un ouvert avec coins, *Lecture Notes in Math.*, no. 1121, p. 199–206.

[7] F. MURAT et J. SIMON, [1975]. Sur le contrôle optimal par un ouvert géométrique. Publication du laboratoire d'Analyse Numérique de l'Université P.M. Curie (Paris).

Philippe Destuynder et Françoise Santi
IAT/CNAM, 15 rue Marat, 78210 St-Cyr-l'École
email: destuynder@iat.cnam.fr
santi@iat.cnam.fr

Sum of Operators' Method in Abstract Equations

Angelo Favini

Summary

The sum of operators' method of P. Grisvard is described, together with its links to real interpolation spaces. We give a generalization of this which allows us to handle degenerate differential equations of parabolic type in abstract spaces. Various examples illustrate its range of application to partial differential equations.

Introduction

In a number of very important papers, P. Grisvard, first alone, and then in collaboration with G. Da Prato, developed an abstract method (the sum of operators' method) that allows us to give a unified treatment to problems to all appearances completely different in nature, like Cauchy and Dirichlet problems. See [6], [13], [14]. In the first part of this paper we recall the mean features of the $A+B$ method, as well its links to real interpolation spaces. We also see that it can be extended (in a manner completely different from the original approach of Da Prato and Grisvard in [6]) to handle some problems of hyperbolic type. The second part of the work is an abstract generalization of some recent results by A. Yagi and A. Favini [10], [11] on degenerate differential equations. In this way the $A+B$ method becomes an $A+BC$ method, where the operator C has no inverse, in general. All abstract statements are illustrated by concrete applications to differential equations.

1. The sum of operators' method

Let us consider the very general equation in the complex Banach space E (with norm $||\cdot||_E$)

$$Au + Bu = f, \tag{1.1}$$

where f is a given element of E, A and B are two closed linear (unbounded) operators from E into itself and $u \in D(A) \cap D(B)$ is the unknown. Here and in what follows, $D(T)$ shall denote the domain of the operator T acting from E to E. In applications, B may represent a.e. the first or the second order derivative with respect to time t with homogeneous initial or boundary conditions, as well other more complicated operators, which we shall see later on. The original idea of Grisvard [13] refers to parabolic and elliptic operators and can be described shortly as follows.

Let B be an invertible operator (for simplicity) from $D(B)(\subseteq E)$ to E, with resolvent set $\rho(B)$ containing $\operatorname{Re} z \leq a_0$,

$$\|(B-z)^{-1}\|_{\mathcal{L}(\mathcal{E})} \leq C(1+|\operatorname{Re} z|)^{-1}, \tag{1.2}$$

for all complex numbers z with $\operatorname{Re} z \leq a_0$, $a_0 > 0$, so that in any sector $\pi_\phi : |\pi - \arg z| \leq \phi < \pi/2$ we have

$$\|(B-z)^{-1}\|_{\mathcal{L}(E)} \leq C(1+|z|)^{-1}, \; z \in \pi_\phi. \tag{1.3}$$

Of course, $\mathcal{L}(E)$ denotes the space of all bounded linear operators from E into itself, endowed with the uniform norm.

Let A be another closed linear operator in E such that $0 \in \rho(A)$ and $z+A$ has a bounded inverse for all $z \in \mathbb{C}, |\arg z| \leq \pi - \phi + \varepsilon$, where ϕ is given in (1.3), and

$$\|(A+z)^{-1}\|_{\mathcal{L}(E)} \leq C(1+|z|)^{-1}, |\arg z| \leq \pi - \phi + \varepsilon. \tag{1.4}$$

Now, let us assume the commutativity relation

$$B^{-1}A^{-1} = A^{-1}B^{-1}. \tag{1.5}$$

Let us introduce the path Γ in the complex plane parametrized by $z = re^{\pm i\Psi}$, $r \geq a_0$, $\Psi = \pi - \phi + \varepsilon/2$, and $z = a_0 e^{i\omega}$, $|\omega| \leq \Psi$, oriented from $(+\infty)e^{-i\Psi}$ to $(+\infty)e^{i\Psi}$. This being fixed, we notice that the candidate for the solution u of (1.1) is

$$u = (2\pi i)^{-1} \int_\Gamma (z+A)^{-1}(B-z)^{-1} f \, dz,$$

where the integral is an improper Riemann integral. It is well defined by (1.3), (1.4), but in general it does not follow that $u \in D(A) \cap D(B)$. Here it appears in all their relevance the real interpolation spaces of J.L. Lions and J. Peetre ([19], [22]).

For the sake of brevity, we confine ourselves to the space $(E, D(B))_{\theta,\infty}$, $0 < \theta < 1$. We recall that $u \in E$ belongs to $(E, D(B))_{\theta,\infty}$ if and only if $\sup_{t>0} t^\theta \|B(B+t)^{-1}u\|_E < \infty$, or, equivalently, $\sup_{z\in\Gamma} |z|^\theta \|B(B-z)^{-1}u\|_E < \infty$. These spaces will give the regularity needed to assure that u is a "true" solution of (1.1). Moreover, one obtains something much better, that is, the maximal regularity property of the solution u, according to the following basic result ([6, Theorem 3.11, p. 328):

Proposition 1.1. *Let us suppose (1.2), (1.4), (1.5). Then for all* $f \in (E, D(B))_{\theta,\infty} = D_B(\theta,\infty), 0 < \theta < 1$, *(respectively,* $D_A(\theta,\infty)$*), equation (1.1) has a unique solution* u *and* Au, $Bu \in D_B(\theta,\infty)$, *(resp.* $D_A(\theta,\infty)$*).*

In order to illustrate the large range of applications of Proposition 1.1 we give three examples. They are simple but clear up both the technique and the elegance of the method.

Example 1.1. Let τ be a fixed positive number. Given a complex Banach space X with norm $\| \ \|_X$, let

$$E = \{u \in C([0,\tau];X);\ \ u(0) = 0\} = C_0([0,\tau];X),$$

endowed with the sup-norm and define B by means of

$$D(B) = \{u \in E; u'(= du/dt) \in E\},\ \ Bu = u',\ \ \ u \in D(B).$$

It is easily seen that (1.3) holds in any sector π_ϕ, $0 < \phi < \pi/2$. We also notice that $-B$ generates the strongly continuous uniformly bounded semigroup in E

$$(e^{-tB}u)(x) = \begin{cases} u(x-t), & x \geq t \\ 0, & x < t \end{cases}$$

and hence, in view of the characterization in Lions [17], (see also [14, Theorem 4, p. 667), $D_B(\theta,\infty)$ coincides with the subspace of all $u \in E$ such that

$$t^{-\theta}\{e^{-tB} - I\}u \in L^\infty((0,\infty);E),$$

that is, $D_B(\theta,\infty) = C_0^\theta([0,\tau];X)$, the space of all Hölder-continuous X-valued functions u on $[0,\tau]$, with exponent θ, such that $u(0) = 0$.

Let Λ be a closed linear operator in X such that Λ^{-1}, $(z+\Lambda)^{-1} \in \mathcal{L}(X)$ for all $z \in \Pi_{\phi,\varepsilon} : |\arg z| \leq \pi - \phi + \varepsilon$,

$$\|(z+\Lambda)^{-1}\|_{\mathcal{L}(X)} \leq C(1+|z|)^{-1},\ \ z \in \Pi_\phi, \varepsilon.$$

If A is defined by

$$\begin{aligned} &D(A) = \{u \in C([0,\tau];D(\Lambda)),\ \ u(0) = 0\} = C_0([0,\tau];D(\Lambda)), \\ &(Au)(t) = \Lambda u(t),\ \ 0 \leq t \leq \tau, \end{aligned}$$

then (1.4) is satisfied. Therefore Proposition 1.1 allows us to state that if $f \in C_0^\theta([0,\tau];X)$, $0 < \theta < 1$, then there is a unique $u \in C_0([0,\tau];D(\Lambda)) \cap C^1([0,\tau];X)$ such that

$$u'(t) = -\Lambda u(t) + f(t),\ \ 0 \leq t \leq \tau. \tag{1.6}$$

Moreover, $u' \in C_0^\theta([0,\tau];X)$.

In order to get rid of the restriction $u(0) = 0$ and handle (1.6) with the nonhomogeneous initial condition

$$u(0) = u_0 \in D(\Lambda), \tag{1.7}$$

real interpolation spaces are used again, but this time exploiting their equivalence to trace spaces. According to Grisvard [14, pp. 679–680], given $a_0 \in (D(\Lambda^2), D(\Lambda))_{\theta,\infty}$, $a_1 \in (D(\Lambda), X)_{\theta,\infty}$, $0 < \theta < 1$, there exists w such that

$$t^\theta w \in L^\infty((0,\infty); D(\Lambda^2)), t^\theta w' \in L^\infty((0,\infty); D(\Lambda)), t^\theta w'' \in L^\infty((0,\infty); X)$$

and $w(0) = a_0$, $w'(0) = a_1$. This implies that w' and Λw belong to $C^\theta([0,\tau]; X)$ by

$$||w'(t) - w'(s)||_X \leq \int_s^t w''(\sigma)||_X d\sigma \leq C \int_s^t \sigma^{\theta-1} d\sigma = C'(t-s)^\theta,$$

$0 \leq s < t \leq \tau$, and analogously to Λw.

Let us come back to (1.6), (1.7). First of all, setting $u(t) - u_0 = v(t)$ transforms (1.6),(1.7) into

$$v'(t) = -\Lambda v(t) + f(t) - \Lambda u_0, v(0) = 0.$$

If $f(0) - \Lambda u_0 \in (X, D(\Lambda))_{\theta,\infty}$, what we recalled above says that there exists w satisfying w', $\Lambda w \in C^\theta([0,\tau]; X), w(0) = 0, w'(0) = f(0) - \Lambda u_0$.

Let us set $v - w = z$. Then the new unknown z satisfies

$$z'(t) = -\Lambda z(t) + h(t),\ 0 \leq t \leq \tau,\ z(0) = 0,$$

where $h(t) = f(t) - \Lambda u_0 - w'(t) - \Lambda w(t)$, and thus $h \in C_0^\theta([0,\tau]; X)$. We have proved

Proposition 1.2. *Under the preceding assumptions, if $f \in C^\theta([0,\tau]; X)$ and $f(0) - \Lambda u_0 \in D_\Lambda(\theta,\infty)$, then (1.6),(1.7) has a unique* strict *solution u (that is, $u \in C([0,\tau]; D(\Lambda)) \cap C^1([0,\tau]; X)$), such that $u' \in C^\theta([0,\tau]; X)$.*

The more natural extension of (1.6),(1.7) concerns the time dependent operators $\Lambda(t)$, that is, the Cauchy problem

$$u'(t) = -\Lambda(t)u(t) + f(t),\ 0 \leq t \leq \tau, u(0) = u_0.$$

This corresponds to developing a sum of operators' method for non commutative operators A and B (that is, (1.5) fails) and this has been accomplished in Grisvard [13], [14] and Da Prato and Grisvard [6].

Methods inspired by the theory of sums were used by P. Acquistapace and B. Terreni in several papers (we only quote [2] and refer to the survey paper [3] by P. Acquistapace), and by R. Labbas and B. Terreni [15].

Example 1.2. Given $f \in C([0,1]; X)$, X a complex Banach space, Λ a closed linear operator in X and λ a positive number, let us consider the "elliptic" problem to find $u \in C^2([0,1]; X) \cap C([0,1]; D(\Lambda))$ such that

$$u''(t) + \Lambda u(t) - \lambda u(t) = f(t), 0 \leq t \leq 1, \tag{1.8}$$

$$u(0) = u(1) = 0. \tag{1.9}$$

If $E = C([0,1]; X)$, we introduce the operator P in E by

$$D(P) = \{u \in E; u', u'' \in E, u(0) = u(1) = 0\}, Pu = u'', u \in D(P).$$

One can verify (see Da Prato and Grisvard [6, p. 372]) that all complex numbers z with $z \notin (-\infty, 0]$ belong to $\rho(P)$ and, if $\theta = \arg z$, then

$$||(P-z)^{-1}||_{\mathcal{L}(E)} \leq \frac{1}{|z|\cos(\theta/2)}.$$

Let $A = -P + \lambda$, λ positive and large. Then (1.4) holds in any sector $|\arg z| \leq \pi - \phi + \varepsilon$, $0 < \phi < \pi/2$ arbitrary.

Let B be defined by

$$D(B) = C([0,1]; D(\Lambda)), (Bu)(t) = -\Lambda u(t), u \in D(B),$$

and suppose that $\rho(\Lambda) \supseteq [0, \infty)$, with

$$||(\Lambda - t)^{-1}||_{\mathcal{L}(X)} \leq C(1+t)^{-1}, \ t \geq 0. \tag{1.10}$$

It is easily seen that (1.3) is satisfied for a suitable ϕ. Therefore problem (1.8),(1.9) is written in the abstract form (1.1), provided that $-f$ takes the place of f. Since $(E, D(A))_{\theta,\infty}$, $0 < \theta < 1$, is characterized (we refer to the monograph [19] by A. Lunardi, Theorem 3.1.29 for scalar-valued functions; the proof for X-valued functions is the same) by

$$(E, D(A))_{\theta,\infty} = Y_\theta = \begin{cases} \{u \in C^{2\theta}([0,1]; X); u(0) = u(1) = 0\}, & 2\theta \neq 1 \\ \{u \in C^1([0,1]; X); u(0) = u(1) = 0\}, & \theta = 1/2, \end{cases}$$

where $C^1([0,1]; X)$ denotes the Zygmund space of order 1, problem (1.8),(1.9) have a unique strict solution u such that $u'' \in Y_\theta$ provided that $f \in Y_\theta$ and (1.10) hold.

We recall that a very recent paper by R. Labbas [16] makes use of the sum method to handle the complete equation of "elliptic" type $u'' + A_1 u' + Au = f$, for suitable operators A, A_1.

Example 1.3. Let B_1 be the operator in $C([0,1]; X) = E$, where X is a complex Banach space, given by

$$D(B_1) = \{u \in E; u \in C^1((0,1]; X), t \to tu'(t) \in E\}, B_1 u = tu', \ u \in D(B_1).$$

One verifies (see Da Prato and Grisvard [7]) that if $k > 0$ is fixed and $B = B_1 + k$, then $-B$ generates a strongly continuous semigroup of negative type in E, so that (3.3) holds. Moreover, $D_B(\theta, \infty)$, $0 < \theta < 1$, is explicitly determined, because

$$(e^{-tB}u)(s) = e^{-kt}u(e^{-t}s), \ t \geq 0, 0 \leq s \leq 1, \ u \in E.$$

If Λ satisfies the same assumptions as in Example 1.1, the constant k can be suitably chosen so that the problem

$$tu'(t) = -\Lambda u(t) + f(t), \ 0 < t \leq 1,$$

has a unique strict solution provided that, for example, $f \in C^\theta([0,1]; X)$, $0 < \theta < 1$.

The method of P. Grisvard has been extended by A. Favini in [9] to treat problems of hyperbolic type; for example, when the closed linear operator A satisfies

$$||(z + A)^{-1}||_{\mathcal{L}(E)} \leq C(1 + |z|)^{m-1}, \tag{1.11}$$

for all z's in the logarithmic region

$$\Lambda : \operatorname{Re} z \geq a_0 + b \ln(1 + |z|), a_0, \ b \geq 0,$$

where m is a nonnegative integer, and

$$||(B - z)^{-1}||_{\mathcal{L}(E)} \leq C(1 + |z|)p, \tag{1.12}$$

for all z with $\operatorname{Re} z \leq a + b \ \ln(1 + |z|), 0 \leq a_0 < a, p \geq 0$, is verified by the closed linear operator B in the Banach space E.

Notice that (1.11) is precisely the assumption of J. Chazarain [4] to treat the Cauchy problem in the sense of distributions. Under hypotheses (1.11), (1.12), "regular" solutions of (1.1) exist only for sufficiently smooth data f. Precisely, we have

Proposition 1.3. *Let us assume (1.5), (1.11),(1.12). Then (1.1) has a unique solution u for all $f \in D(B^k)$, (resp., $f \in D(A^k)$), where k is the smallest integer number $> m + p + 1$.*

If $f \in D(B^k)$, the solution u to (1.1) is expressed by

$$u = (2\pi i)^{-1} \int_\gamma z^{-k} P(z)^{-1} (B - z)^{-1} B^k f \, dz,$$

where γ is the boundary of Λ oriented by increasing imaginary parts and $P(z) = z + A$. Obviously, such a result is weaker than the one by Da Prato and Grisvard relative to the true hyperbolic case, where $\rho(A), \rho(B) \supset (0, \infty)$ and there exists $M_A, M_B \geq 1$ such that

$$||(A - s)^{-k}||_{\mathcal{L}(E)} \leq M_A s^{-k}, \ ||(B - s)^{-k}||_{\mathcal{L}(E)} \leq M_B s^{-k}, \tag{1.13}$$

for all $s > 0$, ([6, Theorem 3.3, p. 320]), but it allows us to handle problems for which (1.13) is difficult to verify or even does not hold.

Example 1.4. Consider the dissipative wave equation in the Hilbert space H (with inner product $\langle\, ,\rangle$)

$$u''(t) + A_1 u'(t) + A_0 u(t) = f(t),\ 0 \le t \le \tau, \tag{1.14}$$

with the initial conditions

$$u(0) = u_0, u'(0) = u_1, \tag{1.15}$$

where $f \in C([0,\tau]; H)$. The Cauchy problem (1.14), (15) has a very large literature, but here we confine ourselves to hypotheses similar to the ones in the papers by Clément and Prüss [5], where the theory of cosine operators is used for and by Engel [8], where the *closure* of the operator $\mathcal{A}$, as defined below, acting in certain interpolation or extrapolation spaces, is investigated, generalizing commutativity of A_0 and A_1.

Precisely, we suppose that A_0 is selfadjoint and positive definite, A_1 satisfies

$\operatorname{Re}\langle A_1 x, x\rangle \ge 0$, $\operatorname{Re}\langle A_1^* f, f\rangle \ge 0, \forall x \in D(A_1), \forall f \in D(A_1^*)$, A_1, A_1^* are A_0-bounded with respective A_0-bound equal to 0, $\operatorname{Re}\langle A_1 x, A_0 x\rangle \ge 0$, $\operatorname{Re}\langle A_1^{*x}, A_0 x\rangle \ge 0$, $\forall x \in D(A_0)$.

Fix $\operatorname{Re} z \ge 0$ and let us consider

$$P(z)u = z^2 u + zA_1 u + A_0 u = f,\ u \in D(A_0). \tag{1.16}$$

By setting $zu = v$, (1.16) becomes

$$zv + A_1 v + \frac{\bar{z}}{|z|^2} A_0 v = f. \tag{1.17}$$

Computing the scalar product of (1.17) with $A_0 v$ and taking the real part gives

$$\operatorname{Re} z||A_0^{1/2} v||^2 + \operatorname{Re}\langle A_1 v, A_0 v\rangle + \frac{\operatorname{Re} z}{|z|^2}||A_0 v||^2 = \operatorname{Re}\langle f, A_0 v\rangle,$$

and hence

$$||A_0 u|| \le \langle \frac{|z|}{\operatorname{Re} z}||f||,$$

while computing the scalar product of (1.17) with v yields analogously

$$|z| \operatorname{Re} z||u|| \le ||f||.$$

Taking into account that under the above assumptions

$$P(z)^* = \bar{z}^2 + \bar{z}A_1^* + A_0,$$

we easily deduce that $P(z)$ has a bounded inverse for $\operatorname{Re} z > 0$ and

$$\max\{\|A_0P(z)^{-1}\|_{\mathcal{L}(H)}, \|zA_1P(z)^{-1}\|_{\mathcal{L}(H)}, \|z^2P(z)^{-1}\|_{\mathcal{L}(H)}\} \le C\frac{|z|}{\operatorname{Re} z},$$

for $\operatorname{Re} z > 0$. On the other hand, the operator matrix $\mathcal{A}$ in $\mathcal{H} = D(A_1) \times H$ given by

$$D(\mathcal{A}) = D(A_0) \times D(A_1), \mathcal{A}(u,v) = (-v, A_0u + A_1v), (u,v) \in D(\mathcal{A}),$$

has the resolvent $(\mathcal{A}+z)^{-1}$ for all z with $\operatorname{Re} z > 0$ and

$$(\mathcal{A}+z)^{-1}(f,g) = (P(z)^{-1}(z+A_1)f + P(z)^{-1}g,\ zP(z)^{-1}(z+A_1)f \\ - f + zP(z)^{-1}g), (f,g) \in \mathcal{H}.$$

This easily implies that in any half plane $\operatorname{Re} z \ge a > 0$ the estimate

$$\|(\mathcal{A}+z)^{-1}\|_{\mathcal{L}(\mathcal{H})} \le C(1+|z|)$$

holds. Hence a bound like (1.11) with $m = 2$ is true. Now the change of depending variables $u(t) = w(t) + \sum_{i=0}^{4} \frac{t^i}{i!}u_i$, $v(t) = z(t) + \sum_{i=0}^{4} \frac{t^i}{i!}v_i$, where u_i, $v_i \in H, i = 0, \ldots, 4$, transforms the system (1.14), (1.15) into an equivalent one

$$\mathcal{B}W(t) + \mathcal{A}W(t) = F(t), 0 \le t \le \tau, \tag{1.18}$$

$$W(0) = 0, \tag{1.19}$$

where $D(\mathcal{B}) = C_0^1([0,\tau]; D(A_1)) \times C_0^1([0,\tau]; D(H)), (C_0^1([0,\tau]; X))$ being the space of all X-valued C^1 functions on $[0,\tau]$ vanishing at $t = 0$), $\mathcal{B}(u,v) = (u', v'), F(t) = (0, f(t)), W(t) = (w(t), z(t))$.

According to Proposition 1.3, (1.18), (1.19) has a unique strict solution provided that $F \in D(\mathcal{B}^4)$. Elementary calculations show that is achieved if $f \in C^4([0,\tau]; H)$ and $u_0, u_1 \in D(A_0), v_j = u_{j+1}, j = 0, \ldots, 3$, $u_s = f^{(s-2)}(0) - A_1u_{s-1} - A_0u_{s-2}, s = 2, 3, 4, v_4 = f^{(3)}(0) - A_1u_4 - A_0u_3 \in D(A_1)$.

To finish with, we remark that Proposition 1.3 states, in particular, that if $k > m + p + 1$, then (1.1) has a unique k-integrated solution v for *any* $f \in E$, (a.e., with respect to B) in the sense that $Bv + Av = B^{-k}f$.

This definition is clearly related to the one concerning the Cauchy problem. For degenerate Cauchy problems, see Arendt and Favini [3].

2. The operator equation $\mathbf{BMu + Lu = f}$

In a recent paper [11] by A. Favini and A. Yagi, the notion and properties of multivalued linear operators were used for transforming the parabolic degenerate Cauchy problem in a complex Banach space X

$$\frac{d}{dt}(Mu(t)) = -Lu(t) + f(t), 0 < t \leq \tau, \tag{2.1}$$

$$Mu(0) = v0 \in X, \tag{2.2}$$

into the nondegenerate form

$$\frac{d}{dt}v(t) + Av(t) \ni f(t), 0 < t \leq \tau,$$
$$v(0) = v_0,$$

where $A = LM^{-1}$ is multivalued. It has been shown that if L, M are closed linear operators from X into itself, with $0 \in \rho(L), D(L) \subseteq D(M)$, such that

$$zM + L \quad \text{has a (univoque) inverse} \quad \in \mathcal{L}(X) \quad \text{for all}$$
$$z \in \Sigma_\alpha : \operatorname{Re} z \geq -c(1 + |\operatorname{Im} z|)^\alpha, c > 0, \tag{2.3}$$

$$|M(zM + L)^{-1}||_{\mathcal{L}(X)} \leq C(1 + |z|)^{-\beta}, \quad \text{for all} \quad z \in \Sigma_\alpha, \tag{2.4}$$

where $0 < \beta \leq \alpha \leq 1$, then (2.1), (2.2) has a unique *classical* solution u, i.e., $u \in C((0, \tau]; D(L)), Mu \in C^1((0, \tau]; X)$, (2.1) holds and (2.2) is verified as convergence in the seminorm $||ML^{-1} \cdot ||_X$ as $t \to 0+$, *provided that* $f \in C^\theta([0, \tau]; X), 2\alpha + \beta > 2$ and $\frac{2-\alpha-\beta}{\alpha} < \theta < 1$. Here we are interested in finding strict solutions to (2.1), (2.2), that is, $u \in C([0, \tau]; D(L)), Mu \in C^1([0, \tau]; X)$, and

$$\frac{d}{dt}(Mu(t)) = -Lu(t) + f(t), 0 \leq t \leq \tau, \tag{2.5}$$

$$Mu(0) = Mu_0, \tag{2.6}$$

where $f \in C([0, \tau]; X)$ and $u_0 \in D(L)$ are given.

Moreover, we should like to establish an analogous result of maximal regularity similar to Propositions 1.1 and 1.2. To accomplish this program we shall use a refinement of the Grisvard's technique and a previous result in Favini and Yagi' paper [10]. For sake of brevity, detailed proofs and further applications and examples will be given in monograph [12] in preparation. We also note that the theory of semigroups with weak singularity was investigated previously by Taira [21] for univalent operators. In the sequel B will denote a closed linear operator in E, (E is a complex Banach space), satisfying (1.2). M and L are two closed linear operators in E such that $D(L) \subseteq D(M)$. There exists the inverse $L^{-1} \in \mathcal{L}(E)$ and

$$||M(zM + L)^{-1}||_{\mathcal{L}(E)} \leq C(1 + |z|)^{-\beta},$$

for all $z \in \Sigma_\alpha$, where $0 < \beta \leq \alpha \leq 1$. We shall consider existence, uniqueness and regularity of the solution u of the equation

$$BMu + Lu = f, \tag{2.8}$$

where f is given in E and u is the unknown. We denote ML^{-1} by T and we know that $T \in \mathcal{L}(E)$. Our last assumption extends (1.5) and reads

$$B^{-1}T = TB^{-1}. \tag{2.9}$$

Then, setting $Lu = v$ shows that (2.8) is equivalent to find an element v in E satisfying

$$BTv + v = f. \tag{2.10}$$

We have

Theorem 2.1. *Let us assume (1.2),(2.7),(2.9), with* $0 < \beta \leq \alpha \leq 1, 2\alpha + \beta > 2$. *If* $\frac{2-\alpha-\beta}{\alpha} < \theta < 1$. *Then for all* $f \in D_B(\theta, \infty)$, *equation (2.8) has a unique solution* u. *Moreover,* u *has the (maximal) regularity* $Lu, BMu \in D_B(\omega, \infty)$, *where*

$$\omega = \alpha\theta + \alpha + \beta - 2.$$

Sketch of the proof.
Uniqueness. If $BTv + v = 0$, set $w = Tv$ so that

$$TBw + w = 0.$$

This yields

$$0 = (2\pi i)^{-1} \int_\Gamma z^{-1}(zT + I)^{-1}(B - z)^{-1}\{TBw + w\}dz = (i) + (ii),$$

where

$$(\text{i}) = (2\pi i)^{-1} \int_\Gamma z^{-1}(zT + I)^{-1}(B - z)^{-1}TBw\, dz,$$

$$(\text{ii}) = (2\pi i)^{-1} \int_\Gamma z^{-1}(zT + I)^{-1}(B - z)^{-1}w\, dz,$$

and Γ is the oriented contour

$$z = a - c(1 + |y|)^\alpha + iy, -\infty < y < \infty, a > c, a - c \text{ sufficiently small.}$$

One sees that $0 = (i) + (ii) = B^{-1}w$ and thus $w = Tv = v = Lu = 0$.

Existence. We solve (2.10) by exhibiting the explicit solution in the form

$$v = (2\pi i)^{-1} \int_\Gamma z^{-1}(zT + I)^{-1} B(B - z)^{-1} f \, dz.$$

The convergence of the integral defining v follows by the assumption $f \in D_B(\theta, \infty)$. The key point for this consists in the remark that if

$$z = a - c(1 + |y|)^\alpha + iy \in \Gamma,$$

then

$$|z|^\theta ||B(B - z)^{-1} f||_E \le C(1 + |y|)^{(1-\alpha)(1+\theta)} ||f||_{\theta,\infty},$$

where $||f||_{\theta,\infty}$ denotes the norm of f in $D_B(\theta, \infty), 0 < \theta < 1$. Hence

$$\begin{aligned} ||v||_E &\le C\{\int_\Gamma |z|^{-(1+\theta)}(1 + |z|)^{1-\beta}(1 + |y|)^{(1-\alpha)(1+\theta)} |dz|\} ||f||_{\theta,\infty} \\ &\le C'\{\int_{-\infty}^{\infty} (1 + |y|)^{-(\beta-1+\alpha+\alpha\theta)} dy\} ||f||_{\theta,\infty}, \end{aligned}$$

and the last integral converges by virtue of the assumptions on α, β and θ. Furthermore, it is easily seen that

$$\begin{aligned} Tv &= B^{-1} f - (2\pi i)^{-1} \int_\Gamma z^{-1}(zT + I)^{-1}(B - z)^{-1} f \, dz, \\ BTv &= f - (2\pi i)^{-1} \int_\Gamma z^{-1}(zT + I)^{-1} B(B - z)^{-1} f \, dz = f - v. \end{aligned}$$

So, the solvability of (2.8) has been established.

Regularity. We must verify that

$$\sup_{t>0} t^\omega ||B(B + t)^{-1} v||_E < \infty,$$

where ω is given by $\alpha\theta + \alpha + \beta - 2$. One recognizes that

$$B(B + t)^{-1}\{(2\pi i)^{-1} \int_\Gamma z^{-1}(zT + I)^{-1} B(B - z)^{-1} f \, dz\} = [\text{i}] + [\text{ii}] + [\text{iii}],$$

with $[\text{i}] = v$ and

$$[\text{ii}] = (2\pi i)^{-1} \int_\Gamma z^{-1} t(z + t)^{-1}(zT + I)^{-1} B(B + t)^{-1} f \, dz,$$

$$[\text{iii}] = -(2\pi i)^{-1} \int_\Gamma t z^{-1}(z + t)^{-1}(zT + I)^{-1} B(B + t)^{-1} f \, dz;$$

further computations lead to

$$t^{\omega}B(B+t)^{-1}v = (2\pi i)^{-1}\int_{\Gamma} t^{-\omega}(z+t)^{-1}(zT+I)^{-1}B(B-z)^{-1}f\,dz,$$

so that $t^{\omega}||B(B+t)^{-1}v||_E$ is estimated by

$$t^{\omega}\{\int_{\Gamma}|z|^{-\theta}|z+t|^{-1}(1+|y|)^{1-\beta+(1-\alpha)(1+\theta)}|dz|\}||f||_{\theta,\infty}.$$

On the other hand, the last integral can be estimated by

$$\int_0^{\infty}(1+y+t)^{-1}(1+y)^{-\theta}(1+y)^{1-\beta+(1-\alpha)(1+\theta)}dy.$$

A suitable change of variable assures our assertion.

If we repeat the argument used for Example 1.1, this time, for the sake of brevity, without introducing the interpolation spaces $(X, D(A))_{\theta,\infty}, 0 < \theta < 1$, as in the monograph [12], (notice that this time $A = LM^{-1}$ should be a multivalued operator), we have the following statement.

Theorem 2.2. *Let us assume (2.3), (2.4), with* $0 < \beta \leq \alpha \leq 1, 2\alpha + \beta > 2$ *and suppose* $(2-\alpha-\beta)/\alpha < \theta < 1$. *Then for all* $f \in C^{\theta}([0,\tau]; X)$ *and any* $u_0 \in D(L)$, *with*

$$f(0) - Lu_0 \in D(LM^{-1}) = M(D(L)),$$

the problem (2.5), (2.6) has a unique strict solution u *and* $Lu \in C^{\omega}([0,\tau]; X)$, *where* $\omega = \alpha\theta + \alpha + \beta - 2$.

As we have already observed, a refinement of Theorem 2.2 can be obtained introducing interpolation spaces. When $M = I$ and the estimate (2.7) holds, the spaces $D_L(\theta, p)$ were investigated by C. Wild in [23].

Example 2.1. Let Ω be a bounded domain in $\mathbb{R}^n$ with smooth boundary $\partial\Omega, n \geq 1$, and let $A(x, D)$ be the symmetric second order differential operator

$$A(x,D)u = -\sum_{i,j=1}^{n}\frac{\partial}{\partial x_i}\left(a_{ij}(x)\frac{\partial u}{\partial x_j}\right) + a_0 u,$$

where $a_{ij} \in C(\bar{\Omega})$ and a_0 is a positive number; these assumptions could be weakened. We add the hypotheses

$$\sum_{i,j=1}^{n} aij(x)\xi_i\xi_j \geq c_0|\xi|^2, c_0 > 0, \xi = (\xi_1,\ldots,\xi_n) \in \mathbb{R}^n, x \in \Omega, \tag{2.11}$$

$$|aij(x)| \leq M, x \in \Omega, i, j = 1, \ldots, n. \tag{2.12}$$

Let $1 < p < \infty$ and define $L(= L_p)$ in the space $X = L_p(\Omega)$ by

$$D(L_p) = W^{2,p}(\Omega) \cap W_0^{1,p}(\Omega), L_p u = A(\cdot, D)u, \ u \in D(L_p).$$

Let us define the operator M_p of multiplication by $m(x)$, where $m \in C(\bar{\Omega})$ and $m(x) \geq 0$ for all $x \in (\bar{\Omega})$. The estimate

$$\frac{|\mathrm{Im}\,\langle L_p u, u^* \rangle|}{\mathrm{Re}\langle L_p u, u^* \rangle} \leq \frac{M|p-2|}{2c_0\sqrt{p-1}},$$

where $u^* = |u|p^{-1}\bar{u}$ and $\langle u, v \rangle = \int_\Omega u(x)\bar{v}(x)\, dx$, which can be found, for example, in Pazy's monograph [20, p. 216], allows us to verify that all assumptions (2.3), (2.4) hold with $\alpha = 1, \beta = 1/p$.

Notice that if M_p is the identity we have $\alpha = \beta = 1$, as is well known. One can also handle the limit cases $p = \infty$ (in the space of continuous functions on $\bar{\Omega}$) and $p = 1$, but then we must apply an extension of Proposition 1.3.

Example 2.2. Let L, M be two closed linear operators acting in the complex Banach space X and satisfyng (2.4) in the sector Σ_1.

In view of Example 1.2, we can treat by means of Theorem 2.1 the "elliptic" problem in the space $E = C([0, 1]; X)$

$$(-Mu)''(t) + Lu(t) = f(t), 0 \leq t \leq 1,$$
$$Mu(0) = Mu(1) = 0.$$

Example 2.3. Suppose that the operators L, M satisfy (2.3), (2.4) and consider the singular equation

$$t(Mu)'(t) = -Lu(t) + f(t), 0 < t \leq \tau, \tag{2.13}$$

where $f \in C([0, \tau]; X)$. In view of the methods as in Example 1.3, one can verify that if $f \in C^\theta([0, \tau]; X), 0 < \beta \leq \alpha \leq 1, 2\alpha+\beta > 2, (2-\alpha-\beta)/\alpha < \theta < 1$; then equation (2.13) has a unique strict solution.

References

[1] P. Acquistapace, Abstract linear non-autonomous parabolic equations a survey, *Quaderni Dip. Mat. Univ. Pisa,* Serie di Analisi Matem. e Probabilità, 1992.

[2] P. Acquistapace and B. Terreni, A unified approach to abstract linear nonautonomous parabolic equations, *Rend. Sem. Mat. Univ. Padova* **78** (1987), 47–107.

[3] W. Arendt and A. Favini, Integrated solutions to implicit differential equations, *Rend. Sem. Mat. Univ. Polit. Torino* **51** (1993), 315–329.

[4] J. Chazarain, Problèmes de Cauchy abstraites et applications à quelques problèmes mixtes, *J. Funct. Anal.* **7** (1971), 386–446.

[5] Ph. Clément and J. Prüss, On second-order differential equations in Hilbert space, *Boll. U.M.I. B* **3**(7) (1989), 623–638.

[6] G. Da Prato and P. Grisvard, Sommes d'opérateurs linéaires et équations différentielles opérationnelles, *J. Math. Pures Appl.* **54** (1975), 305–387.

[7] G. Da Prato and P. Grisvard, On an abstract singular Cauchy problem, *Comm. Partial Diff. Eqs.* **3**(11) (1978), 1077–1082.

[8] K. Engel, On dissipative wave equations in Hilbert space, *J. Math. Anal. Appl.* **184** (1994), 302–316.

[9] A. Favini, An operational method for abstract degenerate evolution equations of hyperbolic type, *J. Funct. Anal.* **76** (1988), 432–456.

[10] A. Favini and A. Yagi, Space and time regularity for degenerate evolution equations, *J. Math. Soc. Japan* **44** (1992), 331–350.

[11] A. Favini and A.Yagi, Multivalued linear operators and degenerate evolution equations, *Ann. Mat. Pura Appl.* **163** (1993), 353–384.

[12] A. Favini and A.Yagi, "Degenerate differential equations in Banach spaces," in preparation.

[13] P. Grisvard, Equations différentielles abstraites, *Ann. Scien. Ec. Norm. Sup.* 4e serie, t.2, fasc. 3 (1969), 311–395.

[14] P. Grisvard, Spazi di tracce ed applicazioni, *Rend. Matem.* **5**(4) (1972), 657–729.

[15] R. Labbas and B.Terreni, Somme d'opérateurs linéaires de type parabolique; 1ère partie: *Boll. U.M.I.* **1** B(7) (1987), 545–569, 2ème partie; Applications, *Boll. U.M.I.* **2** B(7) (1988), 141–162.

[16] R. Labbas, Conditions d'ellipticité et résolution d'une équation différentielle abstraite complète du second ordre, *Boll. U.M.I.* **8**-A(7) (1984), 413–424.

[17] J.L. Lions, Théorèmes de trace et d'interpolation, I et II, *Ann. S.N.S. Pisa* **13** (1959), 389–403, **14** (1960), 317–331.

[18] J.L. Lions and J. Peetre, Sur une classe d'espaces d'interpolation, *Publications de l'I.S.H.E.*, **19** (1964), 5–68.

[19] A. Lunardi, *Analytic Semigroups and Optimal Regularity in Parabolic Problems*, Birkhäuser Basel, 1995.

[20] A. Pazy, *Semigroups of Linear Operators and Applications to Partial Differential Equations,* Springer-Verlag, New York, 1983.

[21] K. Taira, The theory of semigroups with weak singularity and its applications to partial differential equations, *Tsukuba J. Math.* **13** (1989), 513–562.

[22] H. Triebel, *Interpolation Theory, Function Spaces, Differential Operators,* North Holland, 1978.

[23] A. Wild, Semi-groupes de croissance $\alpha < 1$ holomorphes, *C.R.A.S.Paris* **285** (1977), 437–440.

Dipartimento di Matematica
Università di Bologna
40127 Bologna, Italy
email:favini@dm.unibo.it

Constructive Methods for Abstract Differential Equations and Applications

Giuseppe Geymonat and Ouro Tcha-Kondor

Abstract

The Dirichlet problem in a conical domain for some elliptic equations can be reduced by separation of variables to a linear abstract differential equation of first or second order in a Hilbert space H. We review some results giving the solution u as a superposition of exponential H-valued polynomials $e^{-\lambda_k t}p_k(t)$.

Introduction

Let Ω be a bounded conical domain of $\mathbf{R}^n$: $\Omega = \{(r,\theta);\ 0 < r < \rho \text{ and } \theta \in \Sigma\}$ where Σ is a smooth (of class C^∞) open subset of the unit sphere S^{n-1}, $n \geq 2$, with boundary $\partial\Sigma$. Let us consider the following Dirichlet boundary value problem:

$$\begin{cases} \Delta w = 0 & \text{for } (r,\theta) \in \Omega \\ w(r,\theta) = 0 & \text{for } 0 < r < \rho \text{ and } \theta \in \partial\Sigma \\ w(\rho,\theta) = g(\theta) & \text{for } \theta \in \Sigma. \end{cases} \tag{1}$$

When $g(\theta) \in H_{00}^{1/2}(\Sigma)$, i.e., its extension by zero is in $H^{1/2}(S^{n-1})$ (see J.L. Lions and E. Magenes [LM68]), then problem (1) has a *unique* solution $w \in H^1(\Omega)$. This follows, e.g., from well-known results of P. Grisvard [G66]. Moreover, it is possible to give a series expansion of w which is *convergent* in the Hilbert space $H^1(\Omega)$. Indeed, in the spherical coordinates system:

$$\Delta w = r^2 \frac{\partial^2 w}{\partial r^2} + (n-1)r\frac{\partial w}{\partial r} + \Delta_s w$$

where Δ_s denotes the Beltrami-Laplace operator on S^{n-1}. Let us perform the change of variable $\frac{r}{\rho} = e^{-t}$ and let us define: $u(t,\theta) = e^{(1-(n/2))t} w(\rho e^{-t},\theta)$ for every $t > 0$ and every $\theta \in \Sigma$. Then u verifies:

$$\begin{cases} D_u^2 u = \left[-\Delta_s + \left(1 - \frac{n}{2}\right)^2\right] u & \text{for } 0 < t < +\infty,\ \theta \in \Sigma \\ u(t,\theta) = 0 & \text{for } 0 < t < +\infty,\ \theta \in \partial\Sigma \\ u(0,\theta) = g(\theta) & \text{for } \theta \in \Sigma. \end{cases} \tag{2}$$

Let us now define in $H = L^2(\Sigma)$ the closed, positive, self-adjoint operator $A = -\Delta_s$ with dense domain $D(A) = H^2(\Sigma) \cap H_0^1(\Sigma)$. The imbedding of $D(A)$

in H being compact, A has an infinite sequence of positive eigenvalues $\{\lambda_k\}_{k\geq 1}$ (repeated according to their multiplicity) and a corresponding sequence of orthonormal (in H) eigenvectors $\{u_k\}_{k\geq 1}$. Since the sequence $\{u_k\}_{k\geq 1}$ is complete in H, $g(\theta)$ can be expanded in Fourier series: $g(\theta) = \sum_{k\geq 1} c_k u_k(\theta)$ where $c_k = (g, u_k)_H$ and, moreover, $g(\theta) \in H_{00}^{1/2}(\Sigma)$ implies $\sum_{k\geq 1} k c_k^2 < +\infty$. Therefore the solution of (2) has the following series representation, strongly convergent in H for every $t \geq 0$:

$$u(t,\theta) = \sum_{k\geq 1} c_k \exp\left(-t\sqrt{\lambda_k + \left(1 - \frac{n}{2}\right)^2}\right) u_k(\theta) \tag{3}$$

and hence the solution $w \in H^1(\Omega)$ of (1) has the following series representation:

$$w(r,\theta) = \sum_{k\geq 1} c_k \left(\frac{r}{\rho}\right)^{-\frac{n}{2}1+\sqrt{\lambda_k+(1-\frac{n}{2})^2}} u_k(\theta). \tag{4}$$

The computation of the coefficients c_k and of the powers of $\frac{r}{\rho}$ is well-known since it has been widely used to study the regularity of the solution of the problem (1) (see, e.g., V.A. Kondratiev [K67], V.G. Mazya and B.A. Plamenevsky [MP75], P. Grisvard [G85],...). However, in that context, only a *finite* number of the coefficients c_k and of the powers of $\frac{r}{\rho}$ are used. Here we stress that the series representation (4) is indeed *strongly convergent* in the space $H^1(\Omega)$, as pointed out by G. Geymonat and P. Grisvard [GG83] (see also M. Dobrowolski [D83] and M. Moussaoui [M83]).

The situation is completely different in the case of higher order operators as the iterated laplacian Δ^2. Indeed in that case the operator A is *not* self-adjoint. For simplicity let us consider directly the Dirichlet problem for the iterated laplacian in $\mathbf{R}_+ \times \Sigma$, where, with obvious notations, $\Delta^2 = \left[D_u^2 + (\Delta_s - (\frac{n}{2} - 1)^2)\right]^2$ and ν denotes the unit outer normal vector to $\partial\Sigma$:

$$\begin{cases} \left[D_u^2 + \left(\Delta_s - (\frac{n}{2} - 1)^2\right)\right]^2 w = 0 & \text{for } 0 < t < +\infty,\ \theta \in \Sigma \\ w(t,\theta) = \frac{\partial w}{\partial \nu}(t,\theta) = 0 & \text{for } 0 < t < +\infty,\ \theta \in \partial\Sigma \\ w(0,\theta) = g(\theta),\ \frac{\partial^2 w}{\partial t^2}(0,\theta) = h(\theta) & \text{for } \theta \in \Sigma. \end{cases} \tag{5}$$

If one defines $u = (w, D_u^2 w)$, then (5) gives rise in $H = H_0^2(\Sigma) \times L^2(\Sigma)$ to the second order abstract differential equation:

$$D_u^2 u = Au \tag{6}$$

where the operator A, with domain $D(A) = (H^4(\Sigma) \cap H_0^2(\Sigma)) \times H_0^2(\Sigma)$, is defined by

$$Au = \begin{pmatrix} 0 & 1 \\ \left(-\Delta_s + (1 - \frac{n}{2})^2\right)^2 & -2\left(-\Delta_s + (1 - \frac{n}{2})^2\right) \end{pmatrix} u. \tag{7}$$

The situation is essentially the same if one makes a reduction to a first order system, as can easily done also for the Lamé system. Indeed, if one defines $u = (w, \Delta w, D_t w, D_t \Delta w)$, then (5) gives rise in $H = H_0^1(\Sigma) \times L^2(\Sigma) \times L^2(\Sigma) \times H^{-1}(\Sigma)$ to the first order abstract differential equation:

$$D_t u = Au \tag{8}$$

where the operator A, with domain $D(A) = H_0^2(\Sigma) \times H^1(\Sigma) \times H^1(\Sigma) \times L^2(\Sigma)$, is defined by

$$Au = \begin{pmatrix} 0 & 0 & 1 & 0 \\ 0 & 0 & 0 & 1 \\ -\Delta_s + \left(1 - \frac{n}{2}\right)^2 & 1 & 0 & 0 \\ 0 & -\Delta_s + \left(1 - \frac{n}{2}\right)^2 & 0 & 0 \end{pmatrix} u. \tag{9}$$

The operator A, defined by (7) or by (9), is closed with a dense domain $D(A)$ having a compact imbedding in H and in both cases is *not* self-adjoint. In order to obtain a series expansion analogous to (3) many questions must be treated and at first the following:

(i) Is the spectrum of A, denoted by $\sigma(A)$, discrete, i.e., formed by a sequence $\{\lambda_k\}_{k\geq 1}$ of eigenvalues with no finite accumulation points?

When the answer to this question is positive, for every λ_k we can define the linear subspace of $D(A)$ of the generalized eigenvectors:

$$H_k \equiv H(\lambda_k) = \bigcup_{p\geq 1} \ker(A - \lambda_k I)^p.$$

Since the imbedding of $D(A)$ in H is compact $m_k \equiv m(\lambda_k) = \dim H_k$ is finite (m_k is the algebraic multiplicity of λ_k) and the H_k are linearly independent for different λ_k. The following question then naturally arises:

(ii) Is $\operatorname{span}(A) = \bigcup_{k\geq 1} H(\lambda_k)$ dense in H and in $D(A)$?

When also the answer to this question is positive, then one can ask the following:

(iii) Can every element of H, or of $D(A)$, be expanded in a convergent series of generalized eigenvectors?

Would the answer to the previous question be positive, could we obtain an expansion similar to (3) (and (4)) for the iterated laplacian, the Lamé system.

In Sections 1 and 2 we will review some abstract results concerning the questions (i)–(iii). Unfortunately the abstract results obtained for (iii) cannot be applied to (7) and (9). Therefore one can only try to set the following question:

(iv) Find for every solution of the abstract linear differential equation (8), or (6), an expansion as a series of exponential H-valued polynomials $e^{-\lambda_k t} p_k(t)$ uniformly convergent for every $t \geq t_0 > 0$.

Some answers to this question have been given in abstract framework by [GG83] and [GG91]. In some concrete situations these results can be improved; see, e.g., R.C.T. Smith [S52], D.D. Joseph [J77], [JS78], [J79], and O. Tcha-Kondor [TK94],....

1. Completeness of generalized eigenfunctions

1.1. Let H be a Hilbert space and A a closed operator with dense domain $D(A)$. We assume:

(I) *The imbedding of $D(A)$, equipped with the graph norm, into H is compact.*

In the applications indicated before this assumption is always satisfied. Let $\rho(A)$ be the resolvent set of A and $\sigma(A) = \mathbf{C} \setminus \rho(A)$ the spectrum of A. Thanks to assumption (I) either $\sigma(A) = \mathbf{C}$ or $\sigma(A)$ is formed by at most a countable sequence of isolated eigenvalues $\{\lambda_k\}_{k \geq 1}$. R.T. Seeley has given the following example where $\sigma(A) = \mathbf{C}$: $H = L^2(0, 2\pi)$, $A = e^{i\theta} \frac{\partial}{\partial\theta}$ and $D(A) = H^1_{\#} = \{u \in H^1(0, 2\pi);\ u(0) = u(2\pi)\}$. We shall therefore assume in the sequel:

(II) $0 \in \rho(A)$.

This assumption implies that on $D(A)$ one has: $\|f\|_{D(A)} \approx \|Af\|_H$.

Under the assumptions (I) and (II), the resolvent operator $R(\lambda, A) = (A - \lambda I)^{-1}$ is a meromorphic function whose only singularities except at infinity are the isolated eigenvalues $\{\lambda_k\}_{k \geq 1}$ who are poles of finite order $r_k = r(\lambda_k)$. Moreover, at each pole λ_k, $R(\lambda, A)$ has the Laurent expansion (see, e.g., A.E. Taylor [Ta58], §5.8):

$$R(\lambda, A) = \sum_{n=1}^{r_k - 1} (\lambda - \lambda_k)^{-n-1} Q_k^n + (\lambda - \lambda_k)^{-1} P_k + \sum_{n \geq 0} (-1)^n (\lambda - \lambda_k)^n S_k^{n+1} \quad (10)$$

where

$$P_k = P(\lambda_k) = \frac{1}{2\pi i} \int_{\gamma_k} R(z, A) dz, \quad Q_k = (A - \lambda_k I) P_k,$$

$$S_k = \frac{1}{2\pi i} \int_{\gamma_k} R(z, A)(z - \lambda_k)^{-1} dz$$

and γ_k is any simple, closed, Jordan contour, oriented counterclockwise and separating $\{\lambda_k\}$ from $\sigma(A) \setminus \{\lambda_k\}$. From these definitions it follows that $\operatorname{Im}(P_k) = H_k$ and that if $Q_k^n f \neq 0$, then $Q_k^n f \in H_k$.

The assumptions (I) and (II) also imply that $\lambda_k \in \sigma(A)$ if and only if $z_k = \lambda_k^{-1} \in \sigma(A^{-1})$ and that a formula analogous to (10) holds true for A^{-1}. Moreover, $f \in H$ is a generalized eigenvector of A if and only if it is a generalized eigenvector of A^{-1}, i.e., $\operatorname{span}(A) = \operatorname{span}(A^{-1})$.

Thanks to the assumptions (I) and (II), if span(A) is dense in H, i.e., the generalized eigenvectors are dense in H, then span(A) is also dense in $D(A)$ for the graph norm. However, J.P. Labrousse has communicated to us an example of an operator A satisfying (I) and (II) with $\sigma(A) = \emptyset$ and so span$(A) = \{0\}$.

In order to prove that span(A) is dense in H, the following result (see Agmon [A62], and Geymonat-Grisvard [GG67]) is very useful.

Proposition 1.1. *Let $f^* \in \text{span}(A^{-1})^{\perp}$. Then for every $f \in H$ the function:*

$$\lambda \to F(\lambda) = \begin{cases} 0 & \text{for } \lambda = 0 \\ ((A^{-1} - \lambda^{-1}I)^{-1}f, f^*) & \text{for } \lambda \neq 0 \end{cases} \tag{11}$$

is analytic in the whole complex plane and has the development:

$$F(\lambda) = \sum_{n \geq 1} \lambda^n ((A^{-1})^{n-1}, f, f^*). \tag{12}$$

Since $D(A) = \text{Im}(A^{-1})$ is dense in H, to prove that $\text{span}(A^{-1}) = \text{span}(A)$ is dense in H, it is enough to prove that for some $N \geq 1$, one has $\text{Im}((A^{-1})^N)^{\perp} \supseteq \text{span}(A^{-1})^{\perp}$. This in turn will follow if $F(\lambda)$ has a polynomial growth for $|\lambda| \to +\infty$. Such a growth is obtained using a Phragmèn-Lindelöf theorem (as in N. Dunford and J.T. Schwartz [DS63], Agmon,...) or a result on the expansion of meromorphic functions (essentially as in Titchmarsh [Ti46]).

1.2. In order to develop the argument of [Ti46], we recall at first the following result on the expansion of meromorphic functions ([Ti39], p. 110); its proof is an application of the residue theorem.

Proposition 1.2. *Let $g(z)$ be a meromorphic function whose only singularities except at infinity are poles, $a_1, a_2, \dots$ with finite orders $r_1, r_2, \dots$. Let us assume that $0 < |a_1| \leq |a_2| \leq \dots$. Suppose that there exists a sequence of closed contours C_n such that:*

(i) *the domain D_n bounded by the contour C_n includes $a_1, a_2, \dots, a_m$, but no other poles and $m = m(n) \to +\infty$ as $n \to +\infty$;*

(ii) *if $R_n = \min_{z \in C_n} |z|$, then $\lim_{n\to\infty} R_n = \lim_{n\to\infty} \min_{z\in C_n} |z| = +\infty$;*

(iii) *the length L_n of C_n is $O(R_n)$ for $n \to +\infty$;*

If, moreover, for some integer $p \geq -1$ the following holds true:

(iv; p) $\lim_{n\to\infty} \left\{ \frac{1}{R_n^{p+1}} \sup_{z\in C_n} |g(z)| \right\} = 0$, *then*

$$g(z) = \left[g(0) + zg'(0) + \cdots + z^p \frac{g^{(p)}(0)}{p!} \right] + z^{p+1} \sum_{k \geq 1} \frac{\hat{Q}_k(z)}{a_k^{p+1}(z - a_k)^{r_k}} \tag{13}$$

where $\hat{Q}_k(z)$ is a uniquely determined polynomial of degree $\leq r_k - 1$ and the term in brackets disappear for $p = -1$.

It is now easy to prove the completeness of the generalized eigenvectors if for some $p \geq -1$, the following assumption holds true:

(III; p) *The operator* A *is such that* $g(\lambda) = R(\lambda, A)$ *satisfies the conditions* (i)–(iv; p) *of Proposition* 1.2.

Indeed, under that assumption we can apply (13) and a simple computation proves that $\hat{Q}_k(\lambda)$ is polynomial in Q_k and so $\hat{Q}_k(\lambda)f \in H_k$. Since $R(\lambda^{-1}, A^{-1}) = -\lambda I - \lambda^2 R(\lambda, A)$, we find from (11) that $|F(\lambda)| \leq |\lambda|^{p+2}$.

Let us remark that one needs an accurate knowledge of the spectrum of A and of the behavior of $R(\lambda, A)$ in order to verify the assumption (III; p).

1.3. The application of the Phragmèn-Lindelöf theorem is widely developed in N. Dunford and J.T. Schwartz [DS63]. At first we assume:

(IV; p) A^{-1} *belongs to the Carleman class* C_p *with* $0 < p < \infty$.

A compact operator T in H is said to belong to C_p if $\sum_{k\geq 1} \mu_k^p < +\infty$ where $\{\mu_k\}_{k\geq 1}$ are the eigenvalues of $(T^*T)^{1/2}$, repeated according to their multiplicities. C_2 is the class of Hilbert-Schmidt operators and C_1 the class of nuclear operators (or of finite trace). The Carleman classes are described, e.g., in [DS63]. The main theorems of V.B. Lidskii [L62] and [DS63] on the operators of class C_p applied to A^{-1} give the following:

Theorem 1.3. *Let* A *verify the assumptions* (I), (II), *and* (IV; p).

(i) *Let* m_k *be the algebraic multiplicity of the eigenvalue* λ_k. *Then*

$$\sum_{k\geq 1} m_k |\lambda_k|^{-p} < +\infty. \tag{14}$$

(ii) *Let us define the Fredholm determinant of* A *with the formula:*

$$d_A(\lambda) = \prod_{k\geq 1} \left(1 - \frac{\lambda}{\lambda_k}\right)^{m_k} \exp\left\{ m_k \sum_{s=1}^{[p]} \frac{1}{s} \left(\frac{\lambda}{\lambda_k}\right)^s \right\}$$

where $\lceil p \rceil$ *is the smallest integer satisfying* $p \leq \lceil p \rceil + 1$. *By* (14), $d_A(\lambda)$ *is a well-defined entire function of* λ *of order* p *and minimum type vanishing at the points* $\lambda_k \in \sigma(A)$. *Then the following estimate is valid for all* $\lambda \in \mathbf{C}$:

$$\|d_A(\lambda)(I + \lambda R(\lambda, A))\| \leq \exp(\Gamma |\lambda|^p) \tag{15}$$

where Γ *is some constant depending only on* A *and* ρ.

For the definition and properties of entire functions, see, e.g., [B54] and [Ti39]. Since $d_A(\lambda)$ is an entire function of order $\leq p$, it follows from the "minimum modulus theorem" [Ti39] that there exists a sequence of positive numbers $r_j \to +\infty$ such that for every $\varepsilon > 0$:

$$\min_{|\lambda|=r_j} |d_A(\lambda)| > \exp(-r_j^{p+\varepsilon}) \quad j = 1, 2, \ldots . \tag{16}$$

From (11), (15), and (16) it then follows for every $\varepsilon > 0$:

$$|F(\lambda)| \leq \exp(r_j^{p+\varepsilon}) \quad \text{for } |\lambda| = r_j, \ j = 1, 2, \ldots . \tag{17}$$

This estimate is not enough in order to conclude that $F(\lambda)$ has polynomial growth. This will follow from the Phragmèn-Lindelöf theorem under the additional assumption:

(V) *There exist rays* $\arg \lambda = \theta_j$, $j = 1, \ldots, N$ *dividing the complex plane in angles all less than* π/p *and such that for all* $|\lambda|$ *sufficient large on the ray* $R(\lambda, A)$ *exists and verifies:* $\|R(\lambda, A)\| \leq C(1 + |\lambda|)^m$ *for some integer* $m \geq -1$.

Since $R(\lambda^{-1}, A^{-1}) = -\lambda I - \lambda^2 R(\lambda, A)$, we then find $|F(\lambda)| \leq C|\lambda|(1 + |\lambda|)^{m+1}$ and the Phragmèn-Lindelöf theorem applied in every angle implies that $F(\lambda)$ has polynomial growth in all the complex plane.

2. Expansion on generalized eigenfunctions

2.1. When A is a self-adjoint operator and assumptions (I) and (II) are satisfied, then every $f \in D(A)$ admits the series expansion:

$$f = \sum_{k \geq 1} (f, x_k) x_k = \sum_{k \geq 1} P_k f$$

where $\{x_k\}_{k \geq 1}$ is the sequence of the orthonormalized (in H) eigenvectors associated to the eigenvalues $\{\lambda_k\}_{k \geq 1}$ and $\{P_k\}_{k \geq 1}$ the corresponding spectral projectors. Moreover, if $\lambda \in \rho(A)$ then for every $f \in H$ also the following series expansion holds true:

$$R(\lambda, A) f = \sum_{k \geq 1} \frac{(f, x_k)}{\lambda_k - \lambda} x_k = \sum_{k \geq 1} \frac{1}{\lambda_k - \lambda} P_k f.$$

The extension of these representations to a *non* self-adjoint operator satisfying (I), (II) and one of the assumptions that imply span(A) dense in H, is in general impossible as follows from an example of Lidskii [L62]. However a first result can be obtained under the assumption (III; p):

Proposition 2.1. *Let* A *satisfy the assumptions* (I), (II) *and* (III; p) *with* $p \geq -1$. *Then every* $f \in D(A^{p+2})$ *admits the conditionally convergent series expansion (where* $D_0 = \emptyset$*)*

$$f = \sum_{n \geq 1} \left\{ \sum_{\lambda_k \in D_n \setminus D_{n-1}} P(\lambda_k) f \right\}. \tag{18}$$

If $p = -1$, then for $\lambda \in \rho(A)$ every $f \in H$ admits the conditionally convergent series expansion:

$$R(\lambda, A)f = \sum_{n\geq 1}\left\{ \sum_{\lambda_k \in D_n \setminus D_{n-1}} \frac{1}{\lambda_k - \lambda} P(\lambda_k) f \right\}. \tag{19}$$

Proof. Let D_n be the domain surrounded by C_n; then

$$\frac{1}{2\pi i}\int_{C_n} \frac{1}{z^{p+2}} R(z, A) dz = \sum_{\lambda_k \in D_n} \frac{1}{\lambda_k^{p+2}} P(\lambda_k) - A^{-(p+2)}.$$

Applying this formula to $A^{p+2}f$ with $f \in D(A^{p+2})$, one finds

$$\frac{1}{2\pi i}\int_{C_n} \frac{1}{z^{p+2}} R(z, A) A^{p+2} f\, dz = \sum_{\lambda_k \in D_n} P(\lambda_k) f - f.$$

From assumption (III; p) it follows that

$$\left\| \frac{1}{2\pi i}\int_{C_n} \frac{1}{z^{p+2}} R(z, A) A^{p+2} f dz \right\|_H \leq \frac{1}{2\pi} \frac{L_n}{R_n} \left(\frac{1}{R_n^{p+1}} \sup_{z \in C_n} \| R(z, A) A^{p+2} f \| \right) \to 0 \text{ for } n \to \infty$$

and so (18). In the same way one can prove (19). ∎

Let us remark explicitly that, in general, the *conditionally convergence*, of the series cannot be improved as it follows from the quoted example of Lidskii.

2.2. Since the assumption (III;p) can be difficult to verify it is also useful to have some results under assumptions of Carleman type. This has been done by Lidskii, whose results can be a little improved. For simplicity let us assume:

(VI) *A is the infinitesimal generator of an analytic semigroup: $t \to e^{tA}$.*

Under this assumption the spectrum of A is contained in the sector:

$$\frac{\pi}{2} + \delta \leq \arg \lambda \leq \frac{3\pi}{2} - \delta \quad \text{for some } \delta > 0 \tag{20}$$

and one has the estimate: $\|R(\lambda, A)\| \leq C|\lambda|^{-1}$ on the rays $|\arg \lambda| < \frac{\pi}{2} + \delta$. We can use the Phragmèn-Lindelöf theorem to obtain immediately the following result.

Proposition 2.2. *Let A satisfy the assumptions* (I), (II), (VI) *and* (IV;p) *with* $p < \frac{\pi}{\pi - \delta}$ (> 1). *Then* span(A) *is dense in H.*

We can now prove the following result essentially due to Lidskii [L62].

Theorem 2.3. *Let A satisfy the assumptions* (I), (II), (VI), *and* (IV;p) *with* $p < 1$. *Then for every $x \in H$ and every $t > 0$ the series*

$$\sum_{k\geq 1} P_k e^{tA} x = \sum_{k\geq 1} e^{tA} P_k x$$

is conditionally convergent to $e^{tA}x$.

Before we give the proof let us recall that under the assumptions of the theorem $\lim_{t\to 0} e^{tA}x = x$ for every $x \in H$ and so, following Lidskii, the result obtained can be considered a development in generalized eigenfunctions.

Proof. Let γ be the positively oriented path formed by the two rays arg $\lambda = (\pi + \delta)/2$ and arg $\lambda = (3\pi - \delta)/2$ where $\delta > 0$ is defined in (20). Then

$$e^{tA}x = \frac{1}{2i\pi} \int_\gamma e^{t\lambda}(\lambda I - A)^{-1} x \, d\lambda. \tag{21}$$

Let $\{r_j\}_{j\geq 1}$ be the sequence of positive numbers obtained from the "minimum modulus theorem" such that (16) holds true and let us define, for $j = 1, 2, \ldots,$ the bounded closed contours C_j by

$$\begin{aligned} C_j = &\left\{\lambda; 0 \leq |\lambda| \leq r_j \text{ and arg } \lambda = \frac{\pi}{2} + \frac{\delta}{2}\right\} \\ &\cup \left\{\lambda; 0 \leq |\lambda| \leq r_j \text{ and arg } \lambda = \frac{3\pi}{2} - \frac{\delta}{2}\right\} \\ &\cup \left\{\lambda; |\lambda| = r_j \text{ and } \frac{\pi}{2} + \frac{\delta}{2} \leq \text{arg } \lambda \leq \frac{3\pi}{2} - \frac{\delta}{2}\right\}. \end{aligned}$$

From (16) and (20) it follows that $R(\lambda; A)$ is analytic on every contour C_j and moreover on the arc

$$\left\{\lambda; |\lambda| = r_j \text{ and } \frac{\pi}{2} + \frac{\delta}{2} \leq \text{arg } \lambda \leq \frac{3\pi}{2} - \frac{\delta}{2}\right\}$$

the following estimate holds:

$$\|e^{t\lambda}(\lambda I - A)^{-1}\| \leq \frac{2}{r_j} \exp\left(-tr_j \sin\frac{\delta}{2} + \Gamma r_j^p + r_j^{p+\varepsilon}\right).$$

Therefore if one takes $\varepsilon > 0$ such that $p + \varepsilon < 1$ (21) gives for every $t > 0$:

$$e^{tA}x = \lim_{j\to\infty} \frac{1}{2i\pi} \int_{C_j} e^{t\lambda}(\lambda I - A)^{-1} x \, d\lambda$$

and so from the residue theorem,

$$e^{tA}x = \lim_{j\to\infty}\left(\sum_{|\lambda_k|<r_j} e^{tA}P_k x\right) \quad \text{for } t > 0.$$

This means that, setting $r_0 = 0$, we have obtained the series development:

$$e^{tA}x = \sum_{j\geq 0}\left(\sum_{r_j<|\lambda_k|<r_{j+1}} e^{tA}P_k x\right) \quad \text{for } t > 0 \tag{22}$$

where the series on the right is conditionally absolutely convergent in the sense

$$\sum_{j\geq 0}\left\| \sum_{r_j<|\lambda_k|<r_{j+1}} e^{tA}P_k x\right\| < +\infty \qquad \blacksquare$$

Lidskii [L62] proves also the analogous result with A replaced by A^α, $\alpha > 0$, under the assumption (IV;p) with $p \leq \alpha$.

3. Application to abstract differential equations and edge behavior

3.1. The result obtained in Theorem 2.3 can be considered as a *constructive* formula for the solution of the linear abstract evolution equation:

$$\begin{cases} D_t u(t) = Au(t) & \text{for } t > 0 \\ u(0) = x \end{cases} \tag{23}$$

Indeed (22) can be written as

$$e^{tA}x = \sum_{j\geq 0}\left(\sum_{r_j<|\lambda_k|<r_{j+1}} e^{tA}P_k x\right) = \sum_{j\geq 0} u_j(t)$$

and $u_j(t) = \sum_{r_j<|\lambda_k|<r_{j+1}} e^{tA}P_k x$ is the solution of the *ordinary* linear differential equation

$$\begin{cases} D_t u_j(t) = Au_j(t) & \text{for } t > 0 \\ u_j(0) = \sum_{r_j<|\lambda_k|<r_{j+1}} P_k x \end{cases}$$

in the finite dimensional space $\oplus_{r_j<|\lambda_k|<r_{j+1}} H_k$. From the classical results on ordinary differential equations it then follows that u_j is a superposition of exponential polynomials.

It is possible to interpret formula (3) in this context. Since A is a closed, positive self-adjoint operator, then for every $x \in D(A^{1/4})$ the equation,

$$\begin{cases} D_u^2 u(t) = Au(t) & \text{for } t > 0 \\ u(0) = x \end{cases} \tag{24}$$

has the unique solution $u(t) = e^{-tA^{1/2}}x$ under some suitable regularity requirements of the type $u \in H^1(]0,+\infty[; H) \cap L^\infty(]0,+\infty[; D(A))$. Moreover,

$$u(t) = e^{-tA^{1/2}}x = \sum_{k\geq 1} e^{-tA^{1/2}} P_k x$$

and this is exactly formula (3). The analogous of Theorem 2.3 for equation (24) can be found in [GG83], with A replaced with $A^{1/2}$ under the assumption (IV;1/2).

In order to verify the assumption (IV;p) it is enough to show that the natural injection of $D(A)$ in H is of the Carleman class C_p. This follows from a result of Agmon [A62] since this author shows that if $\Omega \subset \mathbf{R}^n$ is suitably smooth then the imbedding $H^m(\Omega) \subset L^2(\Omega)$ is of class C_p for every $p > n/m$. This result implies that Theorem 1.3, the completeness result under the additional assumption (V), and Theorem 2.3 can be applied to many elliptic operators and systems.

3.2. The situation is completely different as far as it concerns the expansion on generalized eigenfunctions for the operator A defined in (7), (resp. (9)), or equivalently, a constructive formula for (6) (resp. (8)). Indeed in that case we can only have $A \in C_p$ for every $p > (n-1)/2$ (resp. $p > n-1$) and there is no hope to apply Theorem 2.3.

In [GG83] and [GG91] Theorem 2.3 has been a little improved essentially under the assumption that (IV;p) *holds for every* $p > 1/2$ (resp. $p > 1$) and that *the corresponding Fredholm determinant* $d_A(\lambda)$ *has the right exponential growth:*

$$d_A(\lambda) = O\big(\exp(c|\lambda|^{1/2})\big) \quad (\text{resp. } d_A(\lambda) = O\big(\exp(c|\lambda|)\big)).$$

Under these supplementary assumptions the series development (22) holds for (6) (resp. (8)) *at least for all* $t \geq t_0$ for some $t_0 > 0$. We shall not give here the details of the assumptions and of the proofs that the interested reader can find in the quoted papers. We only remark that a planar sector Σ reduces to an interval and the operator A is an ordinary differential operator; moreover the corresponding Fredholm determinant $d_A(\lambda)$ is shown to be simply a Wronskian which can be calculated explicitly. This implies that *a conditionally convergent series expansion holds near the edge of a planar sector.* This results can indeed be improved in many concrete applications by a careful study of the series development: see, e.g., R.C.T. Smith [S52], D.D. Joseph [J77], [JS78], [J79], O. Tcha-Kondor [TK94],....

References

[A62] S. Agmon, On the eigenfunctions and on the eigenvalues of general elliptic boundary value problems, *Comm. Pure Appl. Math.* **15**, 1962, pp. 119–147.

[B54] R.P. Boas, Jr., *Entire Functions*, Academic Press, New York, 1954.

[D83] M. Dobrowolski, On finite element methods for nonlinear elliptic problems on domains with corners, in *Singularity and Constructive Methods for Their Treatment*, Lecture Notes in Math. **1121**, Springer-Verlag, Berlin, 1983, pp. 85–103.

[DS63] N. Dunford and J.T. Schwartz, *Linear Operators, Part II: Spectral Theory*, Interscience, New York, 1963.

[G66] P. Grisvard, Commutativité de deux foncteurs d'interpolation et applications, *J. Math. Pures Appl.* **45**, 1966, pp. 143–290.

[G85] P. Grisvard, *Elliptic Problems in Nonsmooth Domains*, Monographs and Studies in Mathematics **24**, Pitman, London, 1985.

[GG67] G. Geymonat and P. Grisvard, Alcuni risultati di teoria spettrale per i problemi ai limiti lineari ellittici, *Rend. Sem. Mat. Univ. Padova* **38**, 1967, pp. 121–173.

[GG83] G. Geymonat and P. Grisvard, Eigenfunction expansions for non selfadjoint operators and separation of variables, in *Singularity and Constructive Methods for Their Treatment*, Lecture Notes in Math. **1121**, Springer-Verlag, Berlin, 1983, pp. 123–136.

[GG91] G. Geymonat and P. Grisvard, Expansions on generalized eigenvectors of operators arising in the theory of elasticity, *Diff. Integral Eqns.* **4**, 1991, pp. 450–481.

[J77] D.D. Joseph, The convergence of biorthogonal series for biharmonic and Stokes flow edge problems, Part I, *SIAM J. Appl. Math.* **33**, 1977, pp. 337–347.

[J79] D.D. Joseph, A new separation of variables theory for problems of Stokes flow and elasticity, in *Proceedings of the Symposium "Trends in Applications of Pure Mathematics to Mechanics"*, held in Kozubnik, Poland in September 1977, Pitman, London, 1979, pp. 129–162.

[JS78] D.D. Joseph and L. Sturges, The convergence of biorthogonal series for biharmonic and Stokes flow edge problems: Part II, *SIAM J. Appl. Math.* **34**, 1978, pp. 7–26.

[K64] V.A. Kondratiev, Boundary value problems for elliptic equations in domains with conical or angular points, *Trans. Moscow Math. Soc.* **16**, 1968, pp. 227–313.

[L62] V.B. Lidskii, Summability of series in the principal vectors of nonselfadjoint operators, *Trans. Moscow Math. Soc.* **11**, 1962, pp. 193–228.

[LM69] J.L. Lions and E. Magenes, *Problèmes aux limites nonhomogènes et applications*, vol. 1, Dunod, Paris, 1968.

[MP75] V.G. Mazya and B.A. Plamenevsky, On the coefficients in the asymptotic form of solutions of elliptic boundary value problems in a cone, *J. Soviet Math.* **9**, 1978, pp. 750–764.

[M83] M. Moussaoui, Sur l'approximation des solutions du problème de Dirichlet dans un ouvert avec coins, in *Singularity and Constructive Methods for Their Treatment*, Lecture Notes in Math. **1121**, Springer-Verlag, Berlin, 1983, pp. 199–206.

[S52] R.C.T. Smith, The bending of a semi-infinite strip, *Aust. J. Sci. Res.* **5**, 1952, pp. 227–237.

[Ta58] A.E. Taylor, *Introduction to Functional Analysis*, John Wiley, New York, 1958.

[Ti39] E.C. Titchmarsh, *The Theory of Functions*, 2nd ed., Oxford University Press, Oxford, 1939.

[Ti46] E.C. Titchmarsh, *Eigenfunction Expansions Associated with Second Order Differential Equations*, Oxford University Press, Oxford, 1946.

[TK94] O. Tcha-Kondor, Série trigonométriques pour l'étude des problèmes biharmoniques dans un secteur d'ouverture quelconque, *C.R. Acad. Sci. Paris* **318**, s.I, 1994, pp. 735–738.

Giuseppe Geymonat
Laboratoire de Mécanique et Technologie
ENS de Cachan/CNRS/Université Paris 6
61, Avenue du Président Wilson
F-94235 CACHAN CEDEX (France)
email: geymonat@lmt.ens-cachan.fr

Ouro Tcha-Kondor
Laboratoire de Mathématiques
Université de Nice-Sophia-Antipolis
Parc Valrose
F-06034 NICE CEDEX (France)

On Asymptotics of Solutions of Nonlinear Second Order Elliptic Equations in Cylindrical Domains

V. A. Kondratiev and O. A. Oleinik

Dedicated to the memory of P. Grisvard

We consider the equation

$$L(u) - a_0|u|^{p-1}u \equiv \sum_{i,j=1}^{n} \frac{\partial}{\partial x_i}\left(a_{ij}(x')\frac{\partial u}{\partial x_j}\right) + \sum_{j=1}^{n} a_j(x')\frac{\partial u}{\partial x_j} - a_0|u|^{p-1}u = 0\,, \tag{1}$$

where $x = (x_1, \cdots, x_n)$, $x' = (x_1, \cdots, x_{n-1})$,

$$m_1|\xi|^2 \le \sum_{i,j=1}^{n} a_{ij}(x')\xi_i\xi_j \le m_2|\xi|^2, \;\; m_1, m_2 = \text{const} > 0\,,$$
$$a_{ij}(x') = a_{ji}(x'), \;\; a_{nn} \equiv 1, \;\; a_{in} \equiv 0 \;\text{ for }\; i < n, \;\; a_0 = \text{const} > 1, \;\; p = \text{const} > 1\,.$$

We set

$$S(a,b) = \{x : x' \subset \omega, a < x_n < b\}$$
$$\sigma(a,b) = \{x : x' \in \partial\omega, a < x_n < b\}\,,$$

where ω is a bounded smooth domain a; $b = \text{const}$. We study the asymptotic behaviour as $x_n \to \infty$ of solutions $u(x)$ of equation (1) in $S(0,\infty)$ with the boundary condition

$$\frac{\partial u}{\partial \gamma} \equiv \sum_{i,j=1}^{n} a_{ij}\frac{\partial u}{\partial x_j}\nu_i = 0 \;\text{ on }\; \sigma(0,\infty)\,, \tag{2}$$

where $\nu = (\nu_1, \cdots, \nu_n)$ is an exterior unit normal vector. The problem (1), (2) was considered in many papers (see [1]–[5]). Here we assume that $a_n(x')$ may change the sign. It is proved [2] that any solution $u(x)$ of problems (1) and (2) tends to zero as $x_n \to \infty$.

Consider the auxiliary problem

$$\mathcal{L}(u) \equiv \sum_{i,j=1}^{n-1} \frac{\partial}{\partial x_i}\left(a_{ij}(x')\frac{\partial u}{\partial x_j}\right) + \sum_{j=1}^{n-1} a_j(x')\frac{\partial u}{\partial x_j} = f(x') \text{ in } \omega\,, \tag{3}$$

$$\frac{\partial u}{\partial \gamma} = \varphi(x') \text{ on } \partial\omega \tag{4}$$

where $f(x') \in L^2(\omega)$, $\varphi(x') \in L^2(\partial\omega)$.

(3) and (4) have a solution if and only if

$$\int_\omega f(x')v_0(x')dx' = \int_{\partial\omega} \varphi(x')v_0(x')dx'\,, \tag{5}$$

where $v_0(x')$ is a solution of

$$\mathcal{L}^*(v) \equiv \sum_{i,j=1}^{n-1} \frac{\partial}{\partial x_i}\left(a_{ij}(x')\frac{\partial v}{\partial x_j}\right) - \sum_{i=1}^{n-1} \frac{\partial}{\partial x_i}(a_i(x')v) = 0 \text{ in } \omega\,, \tag{6}$$

$$\frac{\partial v}{\partial \gamma} - \sum_{j=1}^{n-1} a_j(x')\nu_j v = 0 \text{ on } \partial\omega\,. \tag{7}$$

(6) and (7) are adjoint problems to (3) and (4) and condition (5) is necessary and sufficient by virtue of Fredholm theory. The function $v_0(x)$ does not change the sign in ω. We shall suppose that $v_0(x) > 0$ in ω. If $a_i(x') \equiv 0$ in ω for $i = 1, \cdots, n-1$, (3), (4) is selfadjoint and $v_0(x') \equiv 1$. For simplicity we assume that coefficients $a_{ij}(x')$, $a_j(x')$ are sufficiently smooth in $\bar{\omega}$ and consider solutions $u(x)$ of problem (1), (2), which belong to the class $C^2(S(0,\infty)) \cap C^1(\overline{S(0,\infty)})$.

We shall use the following comparison theorem.

Theorem 1. *Assume that $V(x) > 0$ in $S(0,\infty)$ and such that*

$$L(V) - a_0V^p \equiv \sum_{i,j=1}^{n} \frac{\partial}{\partial x_i}\left(a_{ij}(x')\frac{\partial V}{\partial x_j}\right) + \sum_{j=1}^{n} a_j(x')\frac{\partial V}{\partial x_j} - a_0V^p \le 0 \tag{8}$$

in $S(0,\infty)$,

$$\frac{\partial V}{\partial \gamma} = 0 \;\; on \;\; \sigma(0,\infty)\,, \tag{9}$$

$$V(x) \ge u(x) \;\; for \;\; x_n = 0\,, \tag{10}$$

where $u(x)$ is a solution of (1), (2). Then

$$V(x) \ge u(x) \;\; for \;\; S(0,\infty)\,.$$

If

$$V(x) > 0 \ \ in \ \ S(0,\infty), \ \ L(V) - a_0 V^p \geq 0 \ \ in \ \ S(0,\infty)\,, \tag{11}$$

$$\frac{\partial V}{\partial \gamma} = 0 \ \ for \ \ x \in \sigma(0,\infty), \ \ V(x) \leq u(x) \ \ for \ \ x_n = 0\,, \tag{12}$$

then $u(x) \geq V(x)$ *in* $S(0,\infty)$.

The function $V(x)$, satisfying conditions (8)–(10) is called a supersolution or upper function. The function $V(x)$, satisfying conditions (11), (12) is called a subsolution or lower function. Theorem 1 can be easily proved using the maximum principle for linear equations.

Theorem 2. *Let* $u(x)$ *be a solution of (1) and (2) and let*

$$\int_\omega a_n(x')v_0(x')dx' < 0\,.$$

Then for any $\varepsilon > 0$ *there exist constants* T *and* τ *such that*

$$|u(x)| \leq (K_p + \varepsilon)(x_n - T)^{\frac{1}{1-p}} \ \ for \ \ x \in S(\tau + T, \infty)\,,$$

where

$$K_p = \left[\frac{\int_\omega a_n(x')v_0(x')dx'}{(1-p)a_0\int_\omega v_o(x')dx'}\right]^{\frac{1}{p-1}}\,.$$

If $\int_\omega a_n(x')v_0(x')dx' \geq 0$, *then for any* $\varepsilon > 0$ *there exist constants* T *and* τ *such that*

$$|u(x)| \leq \varepsilon(x_n - T)^{\frac{1}{1-p}} \ \ for \ \ x \in S(\tau + T, \infty)\,.$$

Proof. Consider the function

$$V(x) = cx_n^{\frac{1}{1-p}} + x_n^{\frac{p}{1-p}}\Phi(x')\,.$$

Let us choose the constant c and the function $\Phi(x')$ in such a way that the function $V(x)$ is a supersolution for the problem (1), (2). It is easy to see that

$$\mathcal{L}(\Phi) + \frac{c}{1-p}a_n(x') - a_0c^p = -\varepsilon_1^2, \ \varepsilon_1 = \text{const} > 0, \ x' \in \omega, \tag{13}$$

$$\frac{\partial \Phi}{\partial \gamma} = 0 \ \text{on} \ \partial\omega\,. \tag{14}$$

(13) and (14) have a solution if

$$\frac{c}{1-p}\int_\omega a_n(x')v_0(x')dx' - a_0c^p\int_\omega v_0(x')dx' = -\varepsilon_1^2\int_\omega v_0(x')dx'\,. \tag{15}$$

Assume that

$$\int_\omega a_n(x')v_0(x')dx' < 0\,, \tag{16}$$

and set $c = K_p + \varepsilon$. For such c we define from (15) the constant ε_1. If

$$\int_\omega a_n(x')v_0(x')dx' \geq 0\,, \tag{17}$$

then, taking $c = \varepsilon$, we can find from (15) the constant ε_1. The function V, constructed above, satisfies the inequality $\mathcal{L}(V) - a_0V^p < 0$ in $S(\tau,\infty)$, if τ is sufficiently large, and $\frac{\partial V}{\partial\gamma} = 0$ on $S(\tau,\infty)$. If T is sufficiently large, then $V(x',\tau) > u(x',T+\tau)$. According to the comparison theorem

$$V(x',x_n+\tau) \geq u(x',T+\tau+x_n) \text{ for } x_n > 0\,. \tag{18}$$

It follows from (18) that

$$u(x) \leq (K_p+\varepsilon)(x_n - T)^{\frac{1}{1-p}} \text{ for } x_n > \tau + T\,,$$

if (16) is satisfied, and

$$u(x) \leq \varepsilon(x_n - T)^{\frac{1}{1-p}} \text{ for } x_n > \tau + T\,,$$

if (17) is valid. In a similar way we can prove the same inequalities for $-u(x)$ and get the assertion of Theorem 2.

Theorem 3. *If $\int_\omega a_n(x')v_0(x')dx' < 0$ and $u(x) > 0$ is a solution of* (1), (2), *then in $S(0,\infty)$*

$$u(x) = K_p(1+0(1))(x_n - T)^{\frac{1}{1-p}} \text{ as } x_n \to \infty\,.$$

Proof. We construct the lower function in the form

$$V(x) = cx_n^{\frac{1}{1-p}} + x_n^{\frac{p}{1-p}}\Phi(x')$$

where $c = K_p - \varepsilon$,

$$\mathcal{L}(\Phi) + c\lambda a_n(x') - a_0c^p = -\varepsilon_1^2,\ x' \in \omega,\ \lambda = \frac{1}{1-p}\,, \tag{19}$$

$$\frac{\partial\Phi}{\partial\gamma} = 0 \text{ on } \partial\omega\,. \tag{20}$$

(19) and (20) have a solution, if

$$c\lambda \int_\omega a_n(x')v_0(x')dx' - a_0 c^p \int_\omega v_0(x')dx' = \varepsilon_1^2 \int_\omega v_0(x')dx'$$

This equality defines ε_1 uniquely for small ε. It is easy to see that the function $V(x)$ satisfies the following conditions: $L(V) - a_0 V^p \geq 0$, $V \geq 0$ in $S(T,\infty)$, $\frac{\partial V}{\partial \gamma} = 0$ on $\sigma(T(\varepsilon),\infty)$. If τ is sufficiently large, then

$$u(x',0) > V(x',T+\tau)\,.$$

According to Theorem 1

$$u(x) \geq (K_p - \varepsilon)x_n^{\frac{1}{1-p}} \text{ in } S(T+\tau,\infty)\,. \tag{21}$$

From (21) and Theorem 2 follows Theorem 3.

Theorem 4. *If* $\int_\omega a_n(x')v_0(x')dx' > 0$, *then* $|u(x)| \leq c_1 e^{-\alpha x_n}$, *where* c_1, α *are positive constants and* α *does not depend on* u.

Proof. Let us look for an upper function of the form

$$V(x) = e^{-\varepsilon x_n}(1 + \varepsilon\Phi(x'))\,.$$

For $V(x)$ to be an upper function, it is necessary that $V > 0$ and

$$\begin{aligned} L(V) - a_0 V^p = (\varepsilon^2 - \varepsilon a_n(x') + \varepsilon\mathcal{L}(\Phi))e^{-\varepsilon x_n} \\ - a_0 e^{-\varepsilon p x_n}(1 + \varepsilon\Phi(x'))^p \leq 0\,. \end{aligned}$$

Let $\Phi(x')$ be such that

$$\mathcal{L}(\Phi) = a_n(x') + \beta\,, \quad x' \in \omega\,,$$
$$\frac{\partial \Phi}{\partial \gamma} = 0 \text{ on } \partial\omega'\,.$$

This problem can be solved if

$$\int_\omega a_n(x')v_0(x')dx' + \beta \int_\omega v_0(x')dx' = 0\,.$$

Defining β by the above relation, we have $V > 0$, $L(V) - a_0 V^p \leq 0$ for $x \in S(T,\infty)$, if $\varepsilon > 0$ is sufficiently small and $T(\varepsilon)$ is sufficiently large. If τ is sufficiently large then $V(x', x_n + T) > u(x', x_n + T + \tau)$ for $x_n = 0$. By the comparison theorem it follows that

$$u(x', x_n) \leq V(x', x_n - T) \text{ for } x_n > T + \tau\,.$$

In an analogous manner it can be established that $-u(x', x_n) < V(x', x_n - T)$ for $x_n > T + \tau$. The theorem is proved.

Let us pass to the case

$$\int_\omega a_n(x')v_0(x')dx' = 0\,. \tag{22}$$

This case is more delicate.

Theorem 5. *If the condition* (22) *is satisfied, then for the solution* $u(x)$ *of problems* (1) and (2) *we have*

$$u(x) = 0\left(x_n^{\frac{2}{1-p}}\right)\,, \quad as\ x_n \to \infty\,.$$

Proof. Consider the function

$$V(x) = cx_n^\lambda + x_n^{\lambda-1}\Phi_1(x') + x_n^{\lambda-2}\Phi_2(x')\,,\ \lambda = \frac{2}{1-p}\,.$$

With the help of this function, let us construct an upper function. The conditions

$$L(V) - V^p \le 0\,,\ V > 0 \text{ in } S(T,\infty)\,,\ \frac{\partial V}{\partial\gamma} = 0 \text{ on } \sigma(T,\infty)$$

will be satisfied, if $c > 0$,

$$\lambda c a_n(x') + \mathcal{L}(\Phi_1(x')) = 0 \text{ in } \omega\,,\ \frac{\partial\Phi_1}{\partial\gamma} = 0 \text{ on } \partial\omega \tag{23}$$

and

$$-c^p a_0 + \lambda(\lambda-1)c + (\lambda-1)a_n(x')\Phi_1(x') + \mathcal{L}(\Phi_2(x')) = -\varepsilon_1^2 \text{ in } \omega, \tag{24}$$

$$\varepsilon_1 > 0\,,\ \frac{\partial\Phi_2}{\partial\gamma} = 0 \text{on } \partial\omega\,. \tag{25}$$

Problem (23) is solvable for any c by virtue of the conditions of Theorem 5. The condition for solvability of (24), (25) is given by the following relation:

$$\begin{aligned} &-c^p a_0\int_\omega v_0(x')dx' + (\lambda-1)\int_\omega a_n(x')\Phi_1(x')v_0(x')dx' + \\ &+\lambda(\lambda-1)c\int_\omega v_0(x')dx' < 0\,. \end{aligned} \tag{26}$$

Let $\Phi_{10}(x')$ be a solution of

$$\mathcal{L}(\Phi_{10}) + a_n(x') = 0 \text{ in } \omega,\ \frac{\partial\Phi_{10}}{\partial\gamma} = 0 \text{ on } \partial\omega\,.$$

Then $\Phi_1(x') = c\lambda\Phi_{10}(x')$. If c is taken to be sufficiently large, then (26) will be satisfied. In that case

$$V(x', x_n + T) > u(x', x_n + T + \tau) \text{ for } x_n = 0 ,$$

if τ is sufficiently large and by the comparison theorem we have

$$V(x', x_n + T) > u(x', x_n + T + \tau) \text{ for } x_n > 0 .$$

Therefore $u(x) \le C_1 x_n^{\frac{2}{1-p}}$. Similarly, it can be shown that $-u(x) \le C_1 x_n^{\frac{2}{1-p}}$. Thus, $u = 0\left(x_n^{\frac{2}{1-p}}\right)$.

Remark. The equation

$$\begin{aligned} - c^p a_0 \int_\omega v_0(x')dx' + c(\lambda - 1)\lambda \int_\omega a_n(x')\Phi_{10} v_0(x')dx' + \\ + \lambda(\lambda - 1)c \int_\omega v_0(x')dx' = 0 \end{aligned}$$

has one positive root $C_1(\omega)$ if

$$\int_\omega a_n(x')\Phi_{10}(x')dx' + \int_\omega v_0(x')dx' > 0 . \tag{27}$$

If we put $c = C_1(\omega) + \varepsilon$ in (26), then under conditions of Theorem 5 we obtain

$$|u(x)| \le (C_1(\omega) + \varepsilon) x_n^{\frac{2}{1-p}} (1 + 0(1))$$

as $x_n \to \infty$.

Theorem 6. *If condition* (27) *holds,*

$$\int_\omega a_n(x')v_0(x')dx' = 0 ,$$

and $u(x)$ *is solution of* (1), (2). *Then*

$$u(x) = (C_1(\omega) + \varepsilon) x_n^{\frac{2}{1-p}} (1 + 0(1)) \text{ as } x_n \to \infty . \tag{28}$$

Proof. From the remark after Theorem 5 it follows that

$$u(x) \le (C_1(\omega) + \varepsilon) x_n^{\frac{2}{1-p}} \text{ in } S(T, \infty) .$$

Let us construct a lower function $V(x)$ of the form

$$V(x) = cx_n^{\lambda} + x_n^{\lambda-1}\Phi_1(x') + x_n^{\lambda-2}\Phi_2(x')\,,\ \ \lambda = \frac{2}{1-p}\,.$$

Choose $\Phi_1(x')$ such that the relation (23) is satisfied. The function $\Phi_2(x')$ is taken to be a solution of

$$-c^p a_0 + \lambda(\lambda-1)c + (\lambda-1)a_n(x')\Phi_1(x') + \mathcal{L}(\Phi_2) = \varepsilon_2^2 \text{ in } \omega, \qquad (29)$$

$$\frac{\partial \Phi_2}{\partial \gamma} = 0 \text{ on } \partial\omega\,, \qquad (30)$$

ε_2 is a positive constant. Let $c = C_1(\omega) - \varepsilon$. Then (29), (30) can be solved because

$$-c^p a_0 \int_\omega v_0(x')dx' + c\lambda(\lambda-1)\int_\omega v_0(x')dx' + c\lambda(\lambda-1)\int_\omega a_n(x')\Phi_{10}v_0(x')dx' > 0\,.$$

Here, the function $\Phi_{10}(x')$ is a solution of

$$a_n(x') + \mathcal{L}(\Phi_{10}) = 0 \text{ in } \omega\,,\ \ \frac{\partial \Phi_{10}}{\partial\gamma} = 0 \text{ on } \partial\omega\,.$$

The function $V(x)$ constructed in this way is a lower function and consequently

$$u \geq (C_1(\omega) - \varepsilon)x_n^{\frac{2}{1-p}}\,. \qquad (31)$$

The equality (28) follows from (31) and the remark after Theorem 5. Theorem 6 is proved.

We now state several known lemmas without proofs.

Lemma 1. *Suppose that the line $Jm\lambda = h$ contains no points of the spectrum*

$$\mathcal{L}(u) - \lambda^2 u + i\lambda a_n(x')u = 0 \ \textit{in}\ \omega,\ \ \frac{\partial u}{\partial\gamma} = 0 \ \textit{on}\ \partial\omega\,. \qquad (32)$$

Let

$$J_h(f_1, f_2) = \int_{S(-\infty,+\infty)} e^{2hx_n}\left(f_1^2 + f_2^2\right)dx < \infty\,.$$

Then there exists a unique solution of

$$\mathcal{L}(u) = f_1 + \frac{\partial}{\partial x_n}f_2 \ \textit{in}\ S(-\infty,+\infty)\,, \qquad (33)$$

such that

$$\frac{\partial u}{\partial \gamma} = 0 \quad on \ \sigma(-\infty, +\infty) \tag{34}$$

and

$$I_h(u) = C \int_{S(-\infty,+\infty)} e^{2hx_n} \left(|u|^2 + |\nabla u|^2\right) dx < \infty .$$

In addition

$$I_h(u) \le c J_h(f_1, f_2) , \quad c = \text{const} . \tag{35}$$

This lemma is easily proved by using the Fourier transform with respect to x_n.

Lemma 2. *Let h be such that the line $Jm\lambda = h$ contains no points of the spectrum of* (32). *There exists $\varepsilon =$ const > 0 such that if $|a_i(x') - a_i^*(x)| \le \varepsilon$, $|a_0^*(x)| \le \varepsilon$, $J_h(f_1, f_2) < \infty$, then*

$$L_\varepsilon(u) \equiv \sum_{i,j=1}^{n} \frac{\partial}{\partial x_i}\left(a_{ij}(x')\frac{\partial u}{\partial x_j}\right) + \sum_{i=1}^{n} a_i^*(x)\frac{\partial u}{\partial x_i} + a_0^*(x)u =$$
$$= f_1 + \frac{\partial}{\partial x_n} f_2$$

has a unique solution satisfying the condition (34) and such that $I_h(u) < \infty$. Moreover, inequality (35) holds.

Lemma 2 follows from Lemma 1 and from the theorem on the invertibility of an operator, close to an invertible operator.

Lemma 3. *Let h_1, h_2 be such that there are no points of the spectrum of* (32) *on the lines $Jm\lambda = h_1$, $Jm\lambda = h_2$. Further, let*

$$J_{h_1}(f_1, f_2) + J_{h_2}(f_1, f_2) < \infty , \quad I_{h_1}(u) < \infty .$$

Then

$$u(x) = \sum_{h_1 \le Im\lambda_j \le h_2} e^{i\lambda_j x_n} \Phi_j(x') x_n^{K_j} +$$
$$+ \sum_{\substack{h_1 \le Im\lambda_j \le h_2 \\ l \le K \le K_j}} e^{i\lambda_j x_n} \Phi_{jK}(x') x_n^{K_j - K} + u_1(x) ,$$

where λ_j are eigenvalues of (32), *Φ_j are eigenfunctions and $\Phi_{jK}(x')$ ($K = 1, \ldots, K(j)$) are adjoint functions of* (32), *corresponding to $\lambda = \lambda_j$,*

$$J_{h_2}(u_1) \le c\left[J_{h_1}(f_1, f_2) + J_{h_2}(f_1, f_2)\right] , \quad c = \text{const} .$$

Theorem 7. *If $u(x)$, the solution of (1), (2), changes sign in every region $S(a,\infty)$, $a > 0$; then*

$$|u(x)| \le Ce^{-\alpha x_n} C, \ \alpha = const > 0 ,$$

where $\alpha > 0$ does not depend on u.

Proof. Let us remark that (32) always has the eigenvalue $\lambda = 0$ with the corresponding eigenfunction $\Psi \equiv 1$. If

$$\int_\omega a_n(x')v_0(x')dx' \neq 0$$

then there is no adjoint function corresponding to eigenvalue $\lambda = 0$. If

$$\int_\omega a_n(x')v_0(x')dx' = 0$$

then there exists an adjoint function which is a solution to

$$\mathcal{L}(\Psi_1) + ia_n(x') = 0 \ \text{ in } \omega, \ \ \frac{\partial \Psi_1}{\partial \gamma} = 0 \ \text{ on } \partial\omega .$$

We shall now show that the eigenvalue $\lambda = 0$ cannot have another adjoint function. If Ψ_2 is another adjoint function with $\lambda = 0$, then

$$\mathcal{L}(\Psi_2) + ia_n(x') - 1 = 0 \ \text{ in } \omega , \ \ \frac{\partial \Psi_2}{\partial \gamma} = 0 \ \text{ on } \partial\omega .$$

In this case, the function

$$V(x) = \frac{1}{2}x_n^2 - ix_n\Psi_1(x') - \Psi_2(x')$$

can be shown to be a solution to

$$L(V) = 0 \ \text{ in } S(-\infty,+\infty), \ \ \frac{\partial V}{\partial \gamma} = 0 \ \text{ on } \sigma(-\infty,+\infty) ,$$

which contradicts the minimum principle. Note that $\Psi_1(x')$ is a purely imaginary function. Thus, the eigenvalue $\lambda = 0$ of (32) has only one adjoint function corresponding to $\lambda = 0$.

To prove Theorem 7 we may confine ourselves to the case

$$\int_\omega a_n(x')v_0(x')dx' \le 0 ,$$

since in the contrary case, every solution of (1), (2) decreases at an exponential rate by Theorem 4. Consider the function $v(x) = \theta(x_n)u(x)$, where $u(x)$ is a solution to problem (1), (2), $\theta(x_n) \in C^\infty(\mathbb{R}^1)$, $\theta(x_n) = 0$ for $x_n < T$, $\theta(x_n) = 1$ for $x_n > T + 1$. The function $v(x)$ satisfies

$$L(v) - q(x)v = f_1 + \frac{\partial f_2}{\partial x_n} \text{ in } S(-\infty, +\infty), \tag{38}$$

where f_1 and f_2 are bounded functions with compact supports in x_n. The function $v(x)$ satisfies the boundary condition

$$\frac{\partial v}{\partial \gamma} = 0 \text{ on } \sigma(-\infty, +\infty). \tag{39}$$

The coefficient $q(x)$ of (38) is equal to $\theta(x)|u|^{p-1}$ and if T is sufficiently large we have $|q(x)| \leq \varepsilon$ since $u(x) \to 0$ as $x_n \to \infty$. From Lemma 2 it follows that (38), (39) have a solution $V(x)$ such that

$$\int_{S(-\infty,+\infty)} e^{2hx_n}(|V|^2 + |\nabla V|^2)dx < \infty. \tag{40}$$

From this and the De Giorgi estimate [6] it follows that

$$V(x) = 0\left(e^{-hx_n}\right) \text{ as } x_n \to \infty.$$

In fact, the number h in (40) can be taken to be any number such that the strip $0 < Jm\lambda \leq h$ does not contain any point of the spectrum of (32). Consider the difference $w(x) = V(x) - v(x)$. This function is a solution of

$$L(w) - q(x)w = 0 \text{ in } S(-\infty, +\infty), \quad \frac{\partial w}{\partial \gamma} = 0 \text{ on } \sigma(-\infty, +\infty).$$

The function $q(x)$ may be regarded as a predetermined zero for $x_n < T$. From Lemma 3 we then have

$$w(x) = A(x_n + \Psi_1(x')) + B + w_0(x) \tag{41}$$

where

$$\int_{S(-\infty,0)} e^{2h_1x_n}\left[w_0^2 + |\nabla w_0|^2\right] dx < \infty. \tag{42}$$

h_1 is any number such that in the strip $h_1 \leq Jm\lambda < 0$ has no point of the spectrum of (32). From (42) and from the De Giorgi estimates [6] for solutions of elliptic equations we obtain

$$|w_0(x)| \leq ce^{-h_1x_n}, \quad c = \text{const}.$$

In the representation (41) $\Psi_1(x')$ is the adjoint function of the eigenvalue equation (32), corresponding to the eigenvalue $\lambda = 0$. In the case when such adjoint functions do not exist, we have $A \equiv 0$ in (41). We now show that in (41) we always have $A = 0$.

Suppose that $A \neq 0$. First, let $A < 0$. In this case $w(x') > 0$ for $x_n \leq -T$, $T = \text{const} > 0$. From the maximum principle it follows that $w > 0$ in $S(-\infty, +\infty)$. To estimate $w(x)$ from below, let us construct a lower function of the form

$$W(x) = x_n^\lambda + x_n^{\lambda-1}\Phi(x')\,, \quad \lambda < 0\,.$$

Choose λ, $\Phi(x')$ such that

$$L(W) - qW \geq 0 \text{ in } S(T,\infty)\,, \quad \frac{\partial W}{\partial \gamma} = 0 \text{ on } \sigma(T,\infty)\,. \tag{43}$$

Recall that $|q(x)| \leq Gx_n^{-1}$ in view of Theorem 2 and 5. The relations (43) will be satisfied if

$$\lambda a_n(x') + \mathcal{L}(\Phi(x')) - C_1 \geq \delta_1^2 > 0 \text{ for } x' \in \omega\,. \tag{44}$$

$$\frac{\partial \Phi}{\partial \gamma} = 0 \text{ on } \partial\omega\,. \tag{45}$$

The solvability condition for (44), (45) is

$$\lambda \int_\omega a_n(x')v_0(x')dx' - C_1 \int_\omega v_0(x')dx' > 0\,. \tag{46}$$

So, if $\int_\omega a_n(x')v_0(x')dx' < 0$, then choosing λ to be negative and sufficiently large in modulus, we see that conditions (44), (45) are satisfied and thus the lower function is constructed. Let

$$\int_\omega a_n(x')v_0(x')dx' = 0\,. \tag{47}$$

In this case we seek a lower function of the form

$$W(x) = x_n^\lambda + x_n^{\lambda-1}\Phi_1(x') + x_n^{\lambda-2}\Phi_2(x')\,.$$

For the condition (43) to be satisfied, it is sufficient to choose λ, Φ_1, Φ_2 such that

$$\lambda a_n(x') + \lambda(\Phi_1(x')) = 0 \text{ in } \omega\,. \tag{48}$$

$$\Lambda(\lambda - 1) + (\lambda - 1)\Phi_1 a_n + \mathcal{L}(\Phi_2) - C_1 = \delta^2 > 0 \text{ in } \omega\,, \tag{49}$$

$$\frac{\partial \Phi_1}{\partial \gamma} = 0\,, \quad \frac{\partial \Phi_2}{\partial \gamma} = 0 \text{ on } \partial\omega\,. \tag{50}$$

The equation (47) with the Neumann boundary condition can be solved, since the assumption (47) is valid. In order that the second equation be solvable it is sufficient to take λ negative and sufficiently large in modulus.

Thus, in all cases, using the lower function $W(x)$, we obtain the inequality

$$w(x) \geq c_2 x_n^\lambda \text{ in } S(\tau, \infty),$$

where $\lambda < 0$. Hence

$$V(x) > c_2 x_n^\lambda + v(x). \tag{51}$$

The relation $|V(x)| \leq ce^{-hx_n}$, $h = \text{const}$, $c = \text{const}$ holds for $V(x)$ so that (51) is not possible at the points where $v(x) = u(x) = 0$. By assumption such an infinite sequence of points which tends to ∞ exists. If $A > 0$, a similar argument also leads to a contradiction. Thus, in (41) $A = 0$. In that case $w(x) \to 0$ as $x_n \to \infty$ and as $x_n \to -\infty$. From the maximum principle we then have $w \equiv 0$ in $S(-\infty, +\infty)$. This implies that $v(x) = V(x) = 0(e^{-hx_n})$. Since $v = u$ for sufficiently large x_n, we obtain the assertion of Theorem 7.

Theorem 7 in the case $a_n = 0$ is proved in [3].

Theorem 8. *If $u(x) > 0$ in $S(0, \infty)$, $a_n(x') \equiv 0$, $u(x)$ is a solution of* (1), (2) *then*

$$u(x) = \left[\frac{2(1+p)}{(1-p)^2 a_0}\right]^{\frac{1}{p-1}} (x_n - T)^{\frac{2}{1-p}} (1 + 0(e^{-\alpha x_n}))$$

as $x_n \to \infty$, where T is a constant depending on $u(x)$. In the case $L(u) \equiv \Delta u$, the relation (52) was proved in [4].

We use the same method (as in [4]) for the proof of Theorem 8.

Acknowledgment. This work was partly supported by the International Science Foundation, grant M|E000 and the Russian fundamental researchs grant 93-011-16035 and 93-013-1744.

References

[1] H. Berestycki and L. Nirenberg, Some qualitative properties of solutions of semilinear equations in cylindrical domain, *Analysis*, ed. by P. Rabinoviz, Acad. Press, 1990, 114–164.

[2] V. A. Kondratiev and O. A. Oleinik, On asymptotic behaviour of solutions of some nonlinear elliptic equations in unbounded domains, in *Partial Differential Equations and Related Subjects*, Pitman Research Notes in Math. Series, v. 269, Longman, 1992, 163–195.

[3] V. A. Kondratiev and O. A. Oleinik, Boundary value problems for nonlinear elliptic equations in cylindrical domains, *J. Partial Diff. Equations*, v. 6, N 1, 1993, 10–16.

[4] V. A. Kondratiev and O. A. Oleinik, Some results for nonlinear elliptic equations in cylindrical domains, in *Operator Calculus and Spectral Theory*, Basel: Birkhauser Verlag, 1992, 185–195.

[5] V. A. Kondratiev and O. A. Oleinik, On asymptotic of solutions of nonlinear elliptic equations, *Uspekhi Mat. Nauk*, v. 48, N 4, 1993, 184–185.

[6] E. De Giorgi, Sulla differeziabilita el'analiticita delle estremali degli integrali, *Mem. Accad. Sci. Torino*, 1957, 1–19.

V. A. Kondratiev
Moscow State University
MGU
119899 Moscow, Russia

O. A. Oleinik
Moscow State University
Korpus, K, ap 133
117234 Moscow, Russia

Singularities in Solutions to Mathematical Physics Problems in Non-Smooth Domains

V.A. Kozlov† and V.G. Maz'ya ‡

During the last three decades the theory of linear boundary value problems in "bad" domains was developed. However, there exists a certain gap between the requirements of applications and this theory which includes the Fredholm property, estimates of solutions in different function spaces and general structure of asymptotics of solutions near irregularities of the boundary (see [1],[2]). Probably, the most important problem is to find explicit information about singularities of solutions.

The aim of the present paper is twofold. We review some known results containing such information and obtain new theorems on singularities of solutions to the Dirichlet problem for second order matrix elliptic differential operators in angles and 3-d cones.

1. Known results on singularities of solutions near the vertex of a cone

1.1. Three-dimensional elasticity

As a typical example consider the boundary value problems of 3-d linear and isotropic elasticity. Its solutions may exhibit infinite stresses near singularities of the boundary, such as edges and vertices. Their occurrence is related to the displacement field U which, near an isolated vertex, has the form $U = r^\lambda u(\omega)$, where r denotes the distance to the vertex and ω are spherical coordinates with origin at the vertex. The quantity λ and the function u depend only on the geometry, the type of the boundary condition and the Poisson ratio, but are independent of the particular right-hand side and the boundary data. The knowledge of λ and u enables one to determine not only the asymptotics of stresses near conic points, but also the regularity of a weak solution in scales of Sobolev spaces. Moreover, λ and u are needed for the computation of coefficients in the asymptotic expansion of the stresses near the conic points, the so-called stress intensity factors. Knowledge of λ and u is crucial for

† V.A. Kozlov was supported by the Royal Swedish Academy of Science, the Swedish Council for Engineering Sciences (TFR) under "Linköping Centre for Applied and Industrial Mathematics" and by the London Mathematical Society.

‡ V.G. Maz'ya was supported by the C.N.R.S. during his stay at the Centre de Mathematiques Appliquees, Ecole Polytechnique, France.

the design of proper numerical approximation schemes (mesh grading in finite element schemes, for example).

The pairs (λ, u) can be characterized as eigenvalues and eigenvectors of a rather complicated operator pencil in a domain on the unit sphere. This pencil is Fredholm and depends nonlinearly on λ. Thus, for cones with piecewise smooth boundaries there always exists a countable sequence of eigenvalues λ, in general complex and eigenfunctions. By considering general singularities

$$r^{\lambda} \sum_{k=0}^{N} \frac{(\log r)^{N-k}}{(N-k)!} u_k(\omega) \tag{1}$$

one arrives at generalized eigenfunctions of the pencil.

We turn to precise definitions. Let K be an open cone in $\mathbf{R}^3$ with vertex at the origin and let Ω denote the intersection of K with the unit sphere S^2. Consider the Lamé system

$$\Delta U + (1-2\nu)^{-1} \nabla\, \nabla \cdot U = 0 \text{ on } K, \tag{2}$$

where U is the displacement vector subject to the Dirichlet boundary condition

$$U = 0 \text{ on } \partial K \backslash \{0\} \tag{3}$$

and ν is the Poisson ratio. We are interested in special solutions of (2), (3) having the form (1) with

$$u_k \in \left(\overset{o}{H}{}^1(\Omega) \right)^3,$$

where $\overset{o}{H}{}^1(\Omega)$ is the completion of $C_0^\infty(\Omega)$ in the norm of the Sobolev space $H^1(\Omega)$.

The importance of solutions to problems (2), (3), which have the form (1), is explained by the fact that their linear combinations describe the asymptotics of solutions of the Dirichlet problem with arbitrary data near the vertex of a cone.

The above mentioned operator pencil

$$\mathcal{L}(\lambda) : \left(\overset{o}{H}{}^1(\Omega) \right)^3 \to \left(H^{-1}(\omega) \right)^3$$

is introduced by the formula

$$\mathcal{L}(\lambda) u = r^{2-\lambda} \big(\Delta + (1-2\nu)^{-1} \nabla\, \nabla \big) (r^{\lambda} u)$$

The eigenvalues of this pencil are placed symmetrically on the complex plane with respect to the line $\operatorname{Re} \lambda = -1/2$.

The first nontrivial result on the spectrum of $\mathcal{L}$ was obtained in [3]. It was shown that the strip

$$|\operatorname{Re}\lambda + 1/2| \le \frac{(3-4\nu)\mu}{(\mu+6-4\nu)} + \frac{1}{2}, \tag{4}$$

where $\mu > 0$ and $\mu(\mu+1)$ is the first eigenvalue of the Dirichlet problem for the Laplace-Beltrami operator $-\delta$ on Ω, and is free of eigenvalues of $\mathcal{L}$. This implies, in particular, that solutions of the Dirichlet problem are either Hölder continuous near the vertex of a cone or have a singularity stronger than r^{-1}.

Detailed analysis of singularities generated by $\mathcal{L}(\lambda)$ was performed in [4]. First it was proved that in a certain strip of the complex plane which is centered about $\operatorname{Re}\lambda = -1/2$ and which contains $|\operatorname{Re}\lambda + 1/2| \le 3/2$, only real eigenvalues of $\mathcal{L}$ may occur and these eigenvalues do not admit generalized eigenfunctions. A more precise formulation of this result given in the following theorem involves the function

$$F_\nu(\Omega) = \sqrt{\Gamma_\nu(\Omega)/2 + 3/4 + 2(1-\nu)(3-4\nu)}, \tag{5}$$

where

$$\Gamma_\nu(\Omega) = \inf\left\{Q(u_\omega) + \frac{1}{2(1-\nu)}\int_\Omega |\nabla_\omega u_\omega|^2 d\omega\right\}\Big/\int_\Omega |u_\omega|^2 d\omega, \tag{6}$$

$$Q(u_\omega) = \int_\Omega \left\{|\partial_\Theta u_\Theta|^2 + |\partial_\Theta u_\varphi|^2 + \left|\frac{1}{\sin\Theta}\partial_\varphi u_\Theta - \cot\Theta u_\varphi\right|^2\right.$$

$$\left. + \left|\frac{1}{\sin\Theta}\partial_\varphi u_\varphi - \cot\Theta u_\Theta\right|^2\right\} d\omega.$$

Here $\omega \in S^2$ and Θ, φ are spherical coordinates of ω, $\Theta \in [0,\pi]$, $\varphi \in [0, 2\pi)$. By $(u_r, u_\Theta, u_\varphi)$ we denote spherical components of a vector $u = u(r,\Theta,\varphi)$ and we put $u_\omega = (u_\Theta, u_\varphi)$ and

$$\nabla_\omega u_\omega = \frac{1}{\sin\Theta}\partial_\Theta(\sin\Theta u_\Theta) + \frac{1}{\sin\Theta}\partial_\varphi u_\varphi.$$

The infimum in (6) is taken over all non-zero u_ω, which satisfy zero Dirichlet condition on $\partial\Omega$.

In what follows we write $A \Subset B$ for subsets A, B of S^2, if $A \subset B$ and $B \backslash A$ contains an open and nonempty set. By $A \subset B$ we denote the ordinary inclusion of sets, with $A = B$ admissible. The value $F_\nu(\Omega)$ monotonically depends on Ω, $F_\nu(\Omega_1) > F_\nu(\Omega_2)$ if $\Omega_1 \Subset \Omega_2$. Moreover, if $\overline{\Omega} \ne S^2$ then $F_\nu(\Omega) > 3/2$.

Theorem 1. [4] 1) *All eigenvalues of the pencil* $\mathcal{L}$ *in the strip*

$$-1/2 \le \operatorname{Re}\lambda \le F_\nu(\Omega) - 1/2$$

are real

2) *Eigenvalues in* $[-1/2, F_\nu(\Omega) - 1/2)$ *have no generalized eigenfunctions.*

By using a variational principle for the pencil $\mathcal{L}$ the next monotonicity result is proved in [4].

Theorem 2. *Eigenvalues of* $\mathcal{L}$ *in the interval*

$$\big(-1/2, \min\{3 - 4\nu, F_\nu(\Omega) - 1/2\}\big)$$

are decreasing functions of Ω.

Since the eigenvalues of $\mathcal{L}$ are written explicitly for the sphere S^2 and the half-sphere S^2_+ one can use them to derive the following facts from Theorem 2.

Corollary. a) *If* $\Omega \Subset S^2_+$ *then the interval* $[-1/2, 1]$ *is free of eigenvalues;*

b) *if* $S^2_+ \Subset \Omega \Subset S^2$ *then there are exactly three real eigenvalues* $\lambda_k \in (0, 1]$.

Hence, in a neighbourhood of the reentrant vertex, the solution of the Dirichlet problem with finite energy integral has the asymptotics

$$U(x) = \sum_{k=1}^{3} r^{\lambda_k} u_k(\omega) + O\big(r^{1+\varepsilon}\big),\ \varepsilon > 0, \tag{7}$$

where $0 < \lambda_1 \le \lambda_2 \le \lambda_3 < 1$.

The above results on the spectrum of $\mathcal{L}$ together with a technique developed in [5] enable us to draw conclusions about the regularity of solutions of the nonhomogeneous Dirichlet problem

$$\Delta U + \gamma^{-1}\nabla\,\nabla\cdot U = f \text{ in } D,$$
$$U = 0 \text{ on } \partial D$$

where D is a domain in $\mathbf{R}^3$ with corners p and edges e, in particular, a smooth cone or a polyhedron. For example,

(i) $U \in \big(C^{1,\alpha}(D)\big)^3$ for some positive α if $f \in \big(H^1(D)\big)^3$ and all tangent cones at vertices p are contained in a half-space and all interior edge angles are not greater than π.

(ii) $U \in \big(H^2(D)\big)^3$ if $f \in \big(L^2(D)\big)^3$ and the tangent cones at the vertices p are contained in a circular cone for which real parts of the eigenvalues $\lambda_k(\Omega)$ are greater than $\frac{1}{2}$ and if the interior edge angles are not greater than π.

The method used in [4] does not apply directly to other boundary conditions and, in particular, for the time being there are no similar results for the Neumann problem of elasticity. The only information on the spectrum of the corresponding operator pencil was obtained in [6], [7]. The result of [6] is stated as

Theorem 3. *Let the boundary of K admit a single-valued projection into a certain plane and let $\partial\Omega$ be a piecewise smooth curve. Then the strip $|\operatorname{Re}\lambda + 1/2| \le 1/2$ contains two eigenvalues 0 and -1 of geometric multiplicity 3. There are no generalized eigenfunctions corresponding to these eigenvalues.*

This assertion implies the Hölder continuity of solutions to the Neumann problem in a neighbourhood of a vertex of a polyhedral angle. In [7] the Hölder continuity was established also for the displacement field in a nonhomogeneous anisotropic medium with a polygonal crack.

1.2. Three-dimensional Stokes system

Consider the Dirichlet problem for the Stokes system in a cone K

$$-\Delta U + \nabla P = 0 \quad , \quad \nabla \cdot U = 0 \text{ on } K, \tag{8}$$

$$U = 0 \text{ on } \partial K \backslash \{0\}.$$

Let Ω be a Lipschitz domains. We seek special solutions (U, P) which have the form

$$U(x) = r^{\lambda_0} \sum_{0 \le k \le \kappa} \frac{(\log r)^{\kappa-k}}{(\kappa-k)!} u^{(k)}(\omega), \tag{9}$$

$$P(x) = r^{\lambda_0 - 1} \sum_{0 \le k \le \kappa} \frac{(\log r)^{\kappa-k}}{(\kappa-k)!} p^{(k)}(\omega), \tag{10}$$

for some complex λ_0. To this end we define the operator pencil

$$\mathcal{L}(\lambda) : \left(\overset{\circ}{H}{}^1(\Omega)\right)^3 \times L^2(\Omega) \to \left(H^{-1}(\Omega)\right)^3 \times L^2(\Omega)$$

by

$$\mathcal{L}(\lambda) \begin{pmatrix} u \\ p \end{pmatrix} = \begin{pmatrix} r^{2-\lambda}(\Delta(r^\lambda u) - \nabla(r^{\lambda-1}p)) \\ r^{1-\lambda}\nabla \cdot (r^\lambda u) \end{pmatrix}.$$

The operator pencil $\mathcal{L}$ is Fredholm. The vector function (9), (10) is a solution of (8) if and only if λ_0 is an eigenvalue of $\mathcal{L}$ and col $(u^{(0)}, p^{(0)})$ and col $(u^{(j)}, p^{(j)})$,

$j = 1, \ldots, \kappa$, are eigenvectors and generalized eigenvectors of $\mathcal{L}$ corresponding to λ_0. The eigenvalues of $\mathcal{L}$ are placed symmetrically on $\mathbf{C}$ with respect to the line $\operatorname{Re}\lambda = -1/2$.

It was proved in [3] that the strip

$$|\operatorname{Re}\lambda + \frac{1}{2}| \le \frac{1}{2} + \frac{\mu}{\mu+4},$$

where μ is the same as in (4), is free of the spectrum of $\mathcal{L}$. Several other estimates of the same nature were obtained in [8]. They imply, in particular, that the strip $|\operatorname{Re}\lambda + 1/2| \le 3/2$ does not contain eigenvalues of the Stokes pencil except $\lambda = 1$ and $\lambda = -2$, provided that Ω is contained in a half-sphere.

An analog of the theory developed for the Lamé system in [4] was given in [9] for the pencil $\mathcal{L}$ corresponding to (8). We formulate the principal result obtained in [9].

Theorem 4. *Let*

$$F(\Omega) = \sqrt{\Gamma(\Omega)/2 + 7/4},$$

where $\Gamma(\Omega)$ *is given by the right–hand side of (6) with* $\nu = 1/2$.

1) *All eigenvalues of the pencil* $\mathcal{L}$ *in the strip*

$$-1/2 \le \operatorname{Re}\lambda \le F(\Omega) - 1/2$$

are real.

2) *Eigenvalues of* $\mathcal{L}$ *in* $[-1/2, 1) \cup (1, F(\Omega) - 1/2)$ *have no generalized eigenvectors*

3) *Eigenvalues of* $\mathcal{L}$ *in the interval* $(-1/2, 1)$ *are decreasing functions of* Ω.

Special attention is paid in [9] to $\lambda_0 = 1$, which is an eigenvalue of $\mathcal{L}$ with the eigenvector col $(0,0,0,1)$. The information obtained is collected in

Theorem 5. 1) *The eigenvalue* $\lambda_0 = 1$ *has more than one eigenvector if and only if the scalar problem*

$$(\delta + 6)w = 0, \quad w \in \overset{o}{H}{}^1(\Omega),$$
$$\int_\Omega w \, d\omega = 0$$

has a non-trivial solution. Here δ *is the Laplace–Beltrami operator on* S^2.

2) *The eigenvalue* $\lambda_0 = 1$ *has a generalized eigenvector if and only if the problem*

$$(\delta + 6)w = 1, \quad w \in \overset{o}{H}{}^1(\Omega),$$
$$\int_\Omega w \, d\omega = 0$$

has a non-trivial solution. Further generalized eigenvectors do not exist.

We introduce the eigenvalues $N_1(\Omega) \leq N_2(\Omega) \leq \ldots$ of the quadratic form

$$\int_\Omega |\nabla_\omega v|^2 d\omega$$

defined on all functions which vanish on $\partial\Omega$ and are orthogonal to 1 in $L_2(\Omega)$. In [9] some estimates for the eigenvalues of $\mathcal{L}$ formulated in terms of $N_k(\Omega)$ were obtained. It is proved, for example, that the interval $(0,1)$ contains exactly k eigenvalues of $\mathcal{L}(\lambda)$, provided $N_k(\Omega) < 6 \leq N_{k+1}(\Omega)$. If, on the other hand, $N_1(\Omega) \geq 6$, the strip $0 < \operatorname{Re}\lambda < 1$ is free of the spectrum of the pencil $\mathcal{L}(\lambda)$. If $N_1(\Omega) > 6$, the strip $0 < \operatorname{Re}\lambda \leq 1$ contains exactly one simple eigenvalue $\lambda = 1$ of $\mathcal{L}(\lambda)$. For the case $N_1(\Omega) \leq 6$ it follows from one of the variational principles in [9] that the strip

$$|\operatorname{Re}\lambda + \frac{1}{2}| < \frac{1}{2}\Big(13 - 4(13 - 2N_1(\Omega))^{\frac{1}{2}}\Big)^{\frac{1}{2}}$$

does not contain any eigenvalues of $\mathcal{L}(\lambda)$. The bound in the right-hand side is the best possible one for all Ω with either $N_1(\Omega) = 6$ or $N_1(\Omega) = 2$.

These results imply the following assertions on the spectrum of the pencil $\mathcal{L}(\lambda)$. We denote the spherical domain Ω corresponding to the right circular cone with solid opening angle 2α by D_α, i.e.

$$D_\alpha = \{\omega = (\Theta, \varphi) | 0 \leqq \varphi < 2\pi, 0 \leqq \Theta < \alpha\}.$$

Then

(i) If $S^2_+ \Subset \Omega \Subset D_{\frac{2\pi}{3}}$, the strip $-\frac{1}{2} < \operatorname{Re}\lambda \leqq 1$ contains exactly 3 eigenvalues. One of them is $\lambda = 1$ with multiplicity one.

(ii) If $D_{\frac{2\pi}{3}} \Subset \Omega$, the strip $-\frac{1}{2} < \operatorname{Re}\lambda \leqq 1$ contains exactly 4 eigenvalues. One of them is $\lambda = 1$ with multiplicity one.

In case when $S^2_+ \Subset \Omega$ and the boundary $\partial\Omega$ is smooth this leads to an asymptotic representation near the vertex for solutions U, P of the Dirichlet problem for the Stokes and the Navier-Stokes systems with smooth data. For $\Omega \subset S^2$, we denote by $\lambda_i(\Omega)$, $i = 1, 2, \ldots$ the eigenvalues of $\mathcal{L}$ with $\operatorname{Re}\lambda_i(\Omega) \geqq 0$ ordered with respect to increasing real part and with multiplicity counted.

(i) If $S^2_+ \Subset \Omega \Subset D_{\frac{2\pi}{3}}$, then

$$U = r^{\lambda_1(\Omega)} u_1^{(0)}(\omega) + r^{\lambda_2(\Omega)} u_2^{(0)}(\omega) + 0(r),$$
$$P = r^{\lambda_1(\Omega)-1} p_1^{(0)}(\omega) + r^{\lambda_2(\Omega)-1} p_2^{(0)}(\omega) + 0(1)$$

with $0.4996 < \lambda_1(\Omega) \leqq \lambda_2(\Omega) < 1$. Here the lower bound for λ_1 results from numerical values for $\lambda_1(D_{\frac{2\pi}{3}})$.

(ii) If $D_{\frac{2\pi}{3}} \Subset \Omega$, then

$$U = \sum_{1 \leqq k \leqq 3} r^{\lambda_k(\Omega)} u_k^{(0)}(\omega) + O(r),$$

$$P = \sum_{1 \leqq k \leqq 3} r^{\lambda_k(\Omega)-1} p_k^{(0)}(\omega) + O(1)$$

where $0 \leqq \lambda_1(\Omega) \leqq \lambda_2(\Omega) \leqq \lambda_3(\Omega) < 1$.

This is in contrast to the Dirichlet problem for linear elasticity in piecewise smooth domains. For this problem, according to (7), in the case $S^2_+ \Subset \Omega$ there are exactly three terms in the asymptotics which give rise to unbounded stresses at the vertex.

Information on the spectrum of $\mathcal{L}$ obtained here imply also regularity results for the Dirichlet problem for the nonhomogeneous Stokes system

$$\begin{gathered} -\Delta U + \nabla P = f, \quad \nabla \cdot U = 0 \quad \text{on } G \\ U = 0 \quad \text{on } \partial G, \end{gathered} \tag{11}$$

as well as for the Dirichlet problem for the stationary incompressible Navier-Stokes system

$$\begin{gathered} -\Delta U + \nabla P + U \cdot \nabla U = f \quad \text{on } G, \\ \nabla \cdot U = 0 \quad \text{on } G, \\ U = 0 \quad \text{on } \partial G. \end{gathered} \tag{12}$$

We confine ourselves to two results of this kind.

Consider the case where G is a domain in $\mathbf{R}^3$ with compact closure and piecewise smooth boundary ∂G without zero angles. The boundary ∂G is then a union of finitely many smooth boundary surface pieces, edges η and vertices ϱ. To formulate regularity results, we introduce the following notation: denote by $\theta_\eta(\varsigma)$ the angle between the two tangent planes to ∂G at the edge point ς and by $\Omega_\varrho \subset S^2$ the domain cut out of the unit sphere S^2 by the tangent cone to ∂G at the vertex ϱ. Both θ_η and the tangent cone are taken on the side of G. Assuming that $f \in (L_2(G))^3$, the following regularity results hold.

1. Stokes System: Theorem 6.1 of [5] implies: *if $\Omega_\varrho \Subset D_{\alpha^*}$ and $\theta_\eta < \pi$ for all vertices ϱ and edges η then the solution $(U, P) \in [H^1(G)]^3 \times L_2(G)$ of* (11) *is in*

$$[H^2(G)]^3 \times H^1(G).$$

Here $2\alpha^*$ is the solid opening angle of a right circular cone for which $\lambda_1(D_{\alpha^*}) = 1/2$ ($\alpha^* \approx 0.6665\pi$). In the case that G is a convex polyhedron this is the H^2-regularity result obtained by M. Dauge in [8].

2. Navier-Stokes System: Theorem 10.3 in [5] along with our results implies: *Let again $\Omega_\varrho \Subset D_{\alpha^*}$ and $\theta_\eta < \pi$ for all vertices and edges. Then the solution $(U, P) \in [H^1(G)]^3 \times L_2(G)$ of the Navier Stokes system (12) satisfies*

$$(U, P) \in [H^2(G)]^3 \times H^1(G).$$

1.3. Biharmonic equation

Let K be a cone in $\mathbf{R}^n$, $n > 2$, and let $\Omega = K \cap S^{n-1}$. We consider the Dirichlet problem for the biharmonic operator Δ^2 in K. The corresponding operator pencil

$$\mathcal{L}(\lambda) : \overset{o}{H}{}^2(\Omega) \to H^{-2}(\Omega)$$

is given by

$$\mathcal{L}(\lambda)u = \delta^2 u + 2[\lambda(\lambda + n - 4) - (n-4)]\delta u \\ + \lambda(\lambda + n - 4)(\lambda - 2)(\lambda + n - 2)u$$

In [10] some estimates for the width of the strip

$$\{\lambda : |\operatorname{Re}\lambda + (n-4)/2| < T\}$$

free of eigenvalues of $\mathcal{L}$ were given. Namely, the following assertion was proved, where μ denotes a positive number such that $\mu(\mu + n - 2)$ is the first eigenvalue for the Dirichlet problem for the operator $-\delta$ in Ω.

Theorem 6. 1) *If $n \geq 4$ or $n = 3$ and $2\mu \geq 1$ then the strip $4 - n - \mu \leq \operatorname{Re}\lambda \leq \mu$ does not contain eigenvalues of $\mathcal{L}$.*

2) *If $n = 3$ and $2\mu < 1$ then the spectrum of $\mathcal{L}$ is placed outside the strip $0 \leq \operatorname{Re}\lambda \leq 1$.*

As a corollary of this theorem one can show for example that the solution of the Dirichlet problem

$$\Delta^2 u = f \quad \text{in } Q \tag{13}$$

$$u = \partial_n u = 0 \quad \text{on } \partial Q, \tag{14}$$

where $f \in L_2(Q)$ and Q is a three-dimensional cube, belongs to the Sobolev space $H^4(Q)$. The fact that for some cones in $\mathbf{R}^4$ the line $\operatorname{Re}\lambda = 1$ contains eigenvalues of $\mathcal{L}$ was noticed in [11] and later in [12]. This shows that starting with $n = 4$ the Miranda-Agmon maximum principle may fail for domains with conic points on the boundary.

More complete information in comparison with Theorem 6 on the spectrum of $\mathcal{L}$ was obtained in [13].

We shall use the notation

$$\sigma = 2 - n/2 + \sqrt{(\mu + n/2 - 1)^2 + 1}.$$

Since the spectrum of $\mathcal{L}$ is symmetric with respect to the line $\operatorname{Re}\lambda = 2 - n/2$ it suffices to describe its location in the half–plane $\operatorname{Re}\lambda \geq 2 - n/2$.

Theorem 7. [13] *All eigenvalues of the pencil $\mathcal{L}$ in the strip*

$$2 - n/2 \leq \operatorname{Re}\lambda \leq \sigma \tag{15}$$

are real and have no generalized eigenfunctions.

It is shown in [13] that the eigenvalue λ_0 of $\mathcal{L}$ in the strip (15) nearest to the line $\operatorname{Re}\lambda = 2 - n/2$ can be obtained from a variational principle.

Theorem 8. *The equality holds*

$$\lambda_0 = \inf \left\{ \Lambda(u),\ u \in \overset{o}{H}{}^2(\Omega),\ u \neq 0 \right\}$$

where $\overset{o}{H}{}^2(\Omega)$ is the completion of $C_0^\infty(\Omega)$ in the norm of $H^2(\Omega)$ and $\Lambda(u)$ is the least real root of the equation

$$\left(\mathcal{L}(\lambda)u, u\right)_{L_2(\Omega)} = 0$$

belonging to $[2 - n/2, \sigma]$. If there are no such roots for all u then the strip (15) *is free of eigenvalues of $\mathcal{L}$.*

By this variational principle λ_0 is a decreasing function of Ω, which leads to the following explicit information on the spectrum of $\mathcal{L}$.

Corollary. a) *If $\overline{\Omega} \neq S^{n-1}$ then the strip*

$$2 - n/2 \leq \operatorname{Re}\lambda \leq \max\{0, (5 - n)/2\}$$

is free of the eigenvalues of $\mathcal{L}$.

b) *If $\Omega \Subset S_+^{n-1}$ then the strip*

$$2 - n/2 \leq \operatorname{Re}\lambda \leq 2$$

does not contain eigenvalues of $\mathcal{L}$.

c) *If $\Omega \supset S_+^{n-1}$ then the strip*

$$2 - 2/n \leq \operatorname{Re}\lambda \leq 2$$

contains at least one eigenvalue of $\mathcal{L}$.

This corollary implies, in particular, that the solution of (13), (14), where $f \in L_\infty(Q)$ and Q is a convex polyhedron, belongs to $C^{2,\alpha}(\bar{Q})$, $\alpha > 0$.

1.4. Dirichlet problem for 2m-order elliptic systems in a cone

Here we review some results from [6], [13]-[18].

1.4.1. Estimates of eigenvalues for Lipschitz cones. Let K be a conical domain in $\mathbf{R}^n$, $n \geq 2$. Also let

$$\mathcal{A}(\partial_x) = \sum_{|\alpha|=2m} A_\alpha \partial_x^\alpha \tag{16}$$

be a strongly elliptic $\ell \times \ell$-system of differential operators, i.e.

$$\operatorname{Re}\big(\mathcal{A}(\xi)f, f\big) \geq c_0 |\xi|^{2m} |f|^2, \; c_0 > 0, \tag{17}$$

for any $\xi \in \mathbf{R}^n$ and $f \in \mathbf{C}^\ell$.

We are interested in solutions of the problem

$$\mathcal{A}(\partial_x) U(x) = 0 \quad \text{on} \quad K \tag{18}$$

$$\partial_\nu^j U(x) = 0 \quad \text{on} \quad \partial K \backslash \{0\}, \; j = 0, \dots, m-1, \tag{19}$$

where ν is the normal to $\partial K \backslash \{0\}$, having the form

$$U(x) = |x|^{\lambda_0} \sum_{k=0}^{N} \frac{(\log|x|)^k}{k!} u_{N-k}(x/|x|) \tag{20}$$

with $u_k \in \big(\overset{o}{H}{}^m(\Omega)\big)^\ell$, $\Omega = K \cap S^{n-1}$. Here $\overset{o}{H}{}^m(\Omega)$ is the completion of $C_0^\infty(\Omega)$ in $H^m(\Omega)$.

We introduce a polynomial operator pencil

$$\mathcal{L}(\lambda) : \big(\overset{o}{H}{}^m(\Omega)\big)^\ell \to \big(H^{-m}(\Omega)\big)^\ell \tag{21}$$

defined on functions $u \in \big(\overset{o}{H}{}^m(\Omega)\big)^\ell$ by the equality

$$\mathcal{L}(\lambda) u = |x|^{2m-\lambda} \mathcal{A}(\partial_x)(|x|^\lambda u(x/|x|)). \tag{22}$$

The function (20) is a solution of (18), (19) if and only if λ_0 is an eigenvalue of $\mathcal{L}$, u_0 is an eigenfunction and $u_1, \dots, u_N$ are generalized eigenfunctions of $\mathcal{L}$ corresponding to λ_0.

Using the strong ellipticity of $\mathcal{A}$ one can obtain that the line $\operatorname{Re}\lambda = m - n/2$ is free of eigenvalues of $\mathcal{L}$ if $\overline{K} \neq \mathbf{R}^n$. As was shown in [22] for $n = 2, m = 1$ eigenvalues of $\mathcal{L}$ can be arbitrary close to the line $\operatorname{Re}\lambda = 0$. Hence without additional assumptions on $\mathcal{A}$ or K we cannot expect that some strip centered about the line $\operatorname{Re}\lambda = m - n/2$ is free of eigenvalues of $\mathcal{L}$.

The next assertion was proved in [14].

Theorem 9. *Suppose that $\mathcal{A}(\xi)$ is a symmetric matrix for all $\xi \in \mathbf{R}^n$ and that the cone K is Lipschitz which means that*

$$K = \{x \in \mathbf{R}^n : x_n > \varphi(x'),\ x' \in \mathbf{R}^{n-1}\}, \tag{23}$$

where φ is a real valued positive homogeneous function of degree 1 belonging to $C^\infty(K \backslash \{0\})$. Then the strip

$$|\operatorname{Re}\lambda - m + n/2| \leq 1/2 \tag{24}$$

is free of eigenvalues of $\mathcal{L}$.

It was shown for $2m \geq n$ (see [14]) and for $2m = n - 1$ (see [15], Ch.10) that if $\varphi(x') = -\varepsilon^{-1}|x'|$ with a small ε, then some eigenvalues of $\mathcal{L}$ are located in sufficiently small neighbourhoods of $\lambda = m - n/2 \pm 1/2$. Thus, the estimate (24) cannot be improved under the conditions of theorem.

In [17], [18] the case of nonsmooth functions φ was treated. It was proved in [17] that the interior of the strip (24) is free of eigenvalues of $\mathcal{L}$ if the cone K is given by (23) with continuous function φ. For the second order symmetric systems the strip (24) does not contain eigenvalues of $\mathcal{L}$ if φ is continuous and

$$\nabla\varphi|_{S^{n-2}} \in L_2(S^{n-2})$$

(see [18]).

1.4.2. Special cases. There are several cases when all solutions of the problem (18), (19) having the form (20) and hence all eigenvalues and eigenfunctions and generalized eigenfunctions of $\mathcal{L}$ can be found.

Here we assume that the differential operator (16) satisfies (17)

a) $K = \mathbf{R}^n \backslash \{0\}$. The exponent λ_0 in (20) may only be equal to $0, 1, \ldots$ or $2m - n,\ 2m - n - 1, \ldots$ The function (20) is a solution of (18), (19) if and only if

$$U(x) = \sum_{|\alpha| = \lambda_0} p_\alpha x^\alpha + \sum_{|\beta| = 2m - n - \lambda_0} \partial_x^\beta E(x) q_\beta, \tag{25}$$

where E is the fundamental matrix of $\mathcal{A}(\partial_x)$, p_α, q_β are constant vectors and the first sum in the right-hand side of (25) satisfies (18). If $\lambda_0 < 0$ or $\lambda_0 > 2m - n$ then the corresponding sum in (25) is omitted.

b) $K = \mathbf{R}^n_+ = \{x = (x', x_n) : x_n > 0\}$. The exponent λ_0 in (20) may only be equal to $m, m+1, \ldots$ or $m-n,\ m-n-1,\ m-n-2, \ldots$. In order to describe special solutions (18), (19) we introduce the Poisson kernels Q_k, $k = 0, \ldots, m-1$, as homogeneous solutions of the Dirichlet problem

$$\begin{cases} \mathcal{A}(\partial_x) Q_k(x) = 0 \quad \text{on} \quad \mathbf{R}^n_+ \\ \partial^j_{x_n} Q_k(x) = I\delta(x')\delta^j_k \quad \text{for} \quad x_n = 0,\ j = 0, \ldots, m-1, \end{cases}$$

where I is the identity in $\mathbf{C}^n$. Then (20) is a solution of (18), (19) if and only if

$$\begin{aligned} U(x) = x_n^m &\sum_{|\alpha| = \lambda_0 - m} p_\alpha x^\alpha \\ + \sum_{k=0}^{m-1} &\sum_{|\beta| = k+1-n-\lambda_0} \partial^\beta_{x'} Q_k(x) q_\beta \end{aligned} \tag{26}$$

where p_α, q_β are vectors and the first term in the right-hand side of (26) satisfies (18), (19). If $\lambda_0 < m$ or $\lambda_0 > k+1-n$ for some $k = 0, \ldots, m-1$ then the corresponding terms in (26) are omitted.

We turn to another case considered in [14], when the spectrum is given explicitly. Let $\mathcal{R}$ be the ray $\{x \in \mathbf{R}^n : x' = 0,\ x_n \geq 0\}$. We put $K = \mathbf{R}^n \backslash \mathcal{R}$.

Theorem 10. [14]. *Let $2m \geq n$.*

(i) *Let n be odd. The spectrum of $\mathcal{L}(\lambda)$ is exhausted by the eigenvalues $k = 0, \pm 1, \pm 2, \ldots$. If u is an eigenfunction corresponding to k then*

$$|x|^k u(x/|x|) = \sum_{|\alpha| = k} p_\alpha x^\alpha + \sum_{|\beta| = 2m-n-k} \partial^\beta_x E(x) q_\beta, \tag{27}$$

where $p_\alpha, q_\beta \in \mathbf{C}^\ell$. If $k < 0$ then there is no the first sum on the right. If $k > 2m - n-$ then the second sum is omitted.

(ii) *Let n be even. Then the integers k, $k \neq m - n/2$, are eigenvalues of $\mathcal{L}(\lambda)$ and the corresponding eigenfunctions admit the representation* (27). *The semi-integers*

$$m - n/2 + 1/2 + k \quad \text{where} \quad k = 0, \pm 1, \pm 2, \ldots,$$

are also eigenvalues, with the geometric multiplicities $\ell \binom{m + n/2 - 1}{n-1}$. There are no other eigenvalues. In the case (i) and (ii) there are no generalized eigenfunctions.

For the operator $\mathcal{A}(\partial_x) = (-\Delta)^m$, $2m \geq n$, all eigenfunctions of the corresponding pencil are presented in [14].

1.4.3. Exterior of a thin cone. Let ω be a domain in $\mathbf{R}^{n-1}$ with compact closure and

$$k_\varepsilon = \{x = (x', x_n) \in \mathbf{R}^n : \varepsilon^{-1}x'/x_n \in \omega\}. \tag{28}$$

Denote by K_ε the complement of k_ε in $\mathbf{R}^n$.

Consider the Dirichlet problem

$$\begin{cases} \mathcal{A}(D_x)U(x) = 0, \ x \in K_\varepsilon \\ D_x^\alpha U(x) = 0, \ x \in \partial K_\varepsilon \setminus \{0\}, |\alpha| < m \end{cases} \tag{29}$$

where $\mathcal{A}$ is a scalar operator satisfying (17) and $n > 2m$. Denote by $\mathcal{L}_\varepsilon$ the corresponding operator pencil

$$\mathcal{L}_\varepsilon(\lambda) : \overset{o}{H}{}^m(\Omega_\varepsilon) \to H^{-m}(\Omega_\varepsilon),$$

where $\Omega_\varepsilon = K_\varepsilon \cap S^{n-1}$. Let also $E(x)$ be the fundamental solution of the operator $\mathcal{A}^*(D_x)$ in $\mathbf{R}^n$.

In [15], Ch.10, the following asymptotic formulae for a small eigenvalue of $\mathcal{L}_\varepsilon$ was obtained. Here we use the notation

$$\mathcal{A}(\partial_{x'}, 0) = (-1)^m \sum_{|\alpha|=|\beta|=m} A_{\alpha\beta} \partial_{x'}^{\alpha+\beta},$$

where α and β are $(n-1)$–dimensional multi–indices and

$$k = \int_{\mathbf{R}^{n-1}\setminus\omega} \sum_{|\alpha|=|\beta|=m} A_{\alpha\beta} \partial_{x'}^{\beta} w(x') \overline{\partial_{x'}^{\alpha} w(x')} dx',$$

with w being the solution of the Dirichlet problem

$$\begin{aligned} &\mathcal{A}(\partial_{x'}, 0) w(x') = 0, \ x' \in \mathbf{R}^{n-1}\setminus\omega \\ &\omega(x') = 1, \ \partial_{x'}^\alpha w(x') = 0, \ x' \in \partial\omega, \ 0 < |\alpha| < m, \end{aligned}$$

which vanishes at infinity.

Theorem 11. (i) *Let* $n - 1 = 2m$. *Then*

$$\lambda(\varepsilon) = |2 \log \varepsilon|^{-1}(1 + o(1)).$$

(ii) *Let* $n - 1 > 2m$. *Then*

$$\lambda(\varepsilon) = \varepsilon^{n-1-2m}\Big(k\overline{E(0, \ldots, 0, 1)} + o(1)\Big). \tag{30}$$

The value k is a generalization of the m-harmonic capacity $\mathrm{cap}_m(\Omega)$ of the domain $\Omega \subset \mathbf{R}^{n-1}$ (see [19]). In the special case $\mathcal{A}(D_x) = (-\Delta)^m$ the asymptotic formula (30) takes the form

$$\lambda(\varepsilon) = \varepsilon^{n-1-m}\Big(2^{m-2}\pi^{n/2}\Gamma\big((n-2m)/2\big)\Gamma(m)^{-1}\mathrm{cap}_m(\omega) + o(1)\Big)$$

For the second order elliptic operator

$$\mathcal{A}(\partial_x) = -\sum_{j,k=1}^{n} a_{jk}\partial^2_{x_j x_k}, \quad n > 3$$

the eigenvalue $\lambda(\varepsilon)$ satisfies

$$\lambda(\varepsilon) = \varepsilon^{n-3}\Big(\big((n-2)|S^{n-1}|\big)^{-1}\mathrm{cap}(\omega; \mathcal{A}(\partial_{x'}, 0))(\det\,(a_{jk})_{j,k=1}^{n-1})^{(2-n)/2} \times \big(\det\,(a_{jk})_{j,k=1}^{n}\big)^{(n-3)/2} + o(1)\Big),$$

where $|S^{n-1}|$ denotes the area of the $(n-1)$-dimensional unit sphere and

$$\mathrm{cap}\big(\omega; \mathcal{A}(\partial_{x'}, 0)\big) = \int_{\mathbf{R}^{n-1}\backslash\omega} \sum_{j,k=1}^{n-1} a_{jk}\partial_{x_k} w \partial_{x_j}\bar{w}\,dx'$$

with w being the solution of

$$\mathcal{A}(\partial_{x'}, 0)w(x') = 0, \quad x' \in \mathbf{R}^{n-1}\backslash\omega$$

which equals 1 on $\partial\omega$ and vanishes at infinity.

Similar asymptotic formulae were obtained in [15], Ch.10, for matrix operators. In particular, for the Lamé operator it was shown that there are exactly three small eigenvalues $\lambda_j(\varepsilon)$, $j = 1, 2, 3$, with the asymptotics

$$\lambda_j(\varepsilon) = |2\log\varepsilon|^{-1}\big(1 + o(1)\big).$$

The same is true for the Stokes operator.

According to [15], Ch.10, there are no eigenvalues of the pencil $\mathcal{L}$ generated by the equation (29) in the strips

$$\big\{\lambda \in \mathbf{C} : \mathrm{Re}\,\lambda \in [2m - n + \varepsilon^{n-1-2m}\mu, \varepsilon^{n-1-2m}\mu]\big\} \quad \text{for} \quad n-1 > 2m,$$
$$\big\{\lambda \in \mathbf{C} : \mathrm{Re}\,\lambda \in [-1 - |\log\varepsilon|^{-1}\mu, |\log\varepsilon|^{-1}\mu]\big\} \quad \text{for} \quad n-1 = 2m,$$

where $\mu < \mathrm{Re}\,\big(k\overline{E(0,\ldots,0,1)}\big)$ if $n-1 > 2m$ and $\mu < 1/2$ if $n-1 = 2m$.

One can deduce from this fact that in the case $n-1 = 2m$ all H^m-solutions of the Dirichlet problem for $\mathcal{A}(\partial_x)$ are continuous at the vertex of K_ε provided ε is sufficiently small. If the coefficients of $\mathcal{A}$ are real the same is true for an arbitrary Lipschitz cone.

Let us turn to the case $n-1>2m$. If

$$\operatorname{Re}\left(k\overline{E(0,\ldots,0,1)}\right)>0 \tag{31}$$

then the above statement about the strip, which is free of the spectrum, guarantees the continuity of solutions to the Dirichlet problem in K_ε with sufficiently small ε. There are examples, which show that continuity may fail. This may happen in the case $n>4$ for the second order strongly elliptic operator with complex coefficients

$$\mathcal{A}(\partial_x)=(1+i\beta)\partial_{x_1}^2+\partial_{x_2}^2+\ldots+\partial_{x_{n-1}}^2+\alpha\partial_{x_n}^2$$

where α and β are some numbers such that $\operatorname{Im}\beta=0$, $\operatorname{Re}\alpha>0$ (see [15], 10.6.1). Another example of the loss of continuity is provided by the operator

$$\mathcal{A}(\partial_x)=\Delta^2+a^2\partial_{x_n}^4,\quad n\geq 8,$$

where $(n-3)\arctan a\in(2\pi,4\pi)$ (see [16] and [15],10.6.2).

1.4.4. Cones close to half–space. Following from Sect.1.4.2 the operator pencil (21), where Ω is a half–sphere, has noneigenvalues in the strip $m-n<\operatorname{Re}\lambda<m$. One can conjecture that this strip (or its closure) is free of eigenvalues when Ω is contained in a half–sphere. This is true for the n-dimensional biharmonic equation and for the three-dimensional Lamé and Stokes systems. In the general case this conjecture fails. Here we present corresponding results from [13].

Let $\varphi=\varphi(x')$ be a real-valued function on $\mathbf{R}^{n-1}$ positive homogeneous of degree one, i.e. $\varphi(x')=|x'|\varphi(\omega')$, $\omega'=x'/|x|$. We suppose that φ is smooth outside the origin. Denote by K_ε the cone

$$\{x=(x',x_n)\in\mathbf{R}^n : x_n>\varepsilon|x'|\},$$

where $\varepsilon\in\mathbf{R}^1$. We set $\Omega_\varepsilon=K_\varepsilon\cap S^{n-1}$.

Let $\mathcal{A}(\partial_x)$ be a homogeneous elliptic operator of order $2m$ with constant real coefficients. Also let $\mathcal{L}_\varepsilon(\lambda):\overset{o}{H}{}^m(\Omega_\varepsilon)\to H^{-m}(\Omega_\varepsilon)$ be the operator pencil which is given by (22). If $|\varepsilon|$ is a small number, then the pencil has a single simple eigenvalue in some neighbourhood of $\lambda=m$, denoted by λ_ε. Denote by $E_{m-1}(x)=E_{m-1}(x',x_n)$ the homogeneous solution of the boundary value problem

$$\begin{cases}\mathcal{A}(\partial_x)E_{m-1}(x)=0 \quad \text{on} \quad \mathbf{R}^n_+,\\ \partial^j_{x_n}E_{m-1}(x)=0 \quad \text{for} \quad x_n=0,\ j=0,\ldots,m-2,\\ \partial^{m-1}_{x_n}E_{m-1}(x)=\delta(x') \quad \text{for} \quad x_n=0.\end{cases}$$

Theorem 12. [13] *The asymptotic formula is valid*

$$\lambda_\varepsilon = m + \varepsilon \int_{S^{n-2}} \partial_{x_n}^m E_{m-1}(\omega', 0)\varphi(\omega')d\omega' + o(\varepsilon^2). \tag{32}$$

This theorem implies that for sufficiently small $|\varepsilon|$ the strip

$$m - n - \varepsilon\mu \le \operatorname{Re}\lambda \le m + \varepsilon\mu \tag{33}$$

is free of eigenvalues of $\mathcal{L}_\varepsilon$, where μ is a real number satisfying

$$\mu < \int_{S^{n-2}} \partial_{x_n}^m E_{m-1}(\omega', 0)\varphi(\omega')d\omega'. \tag{34}$$

Let $\mathcal{A}(\partial_x) = (-\Delta)^m$. Then

$$\partial_{x_n}^m E_{m-1}(\omega', 0) = m\Gamma(n/2)\pi^{-n/2}$$

and the inequality (34) becomes

$$\mu < m\Gamma(n/2)\pi^{-n/2} \int_{S^{n-2}} \varphi(\omega')d\omega'.$$

Therefore, if φ is nonnegative and $\varphi(\omega') > 0$ at least for one value of ω', then the constant μ in (33) can be taken as positive. This implies, in particular, that all H^m-solutions of the Dirichlet problem for $(-\Delta)^m$ have continuous derivatives up to order m in a neighbourhood of the vertex of K_ε.

The next example shows that the coefficient of ε in (32) can be negative for positive φ.

Let $\mathcal{A}(\partial_x) = \Delta^2 + b^2\partial_{x_1}^4$, $b > 0$. Then (see [13])

$$\partial_{x_n}^2 E_1(x', 0) = \frac{2\Gamma(n/2)}{\pi^{n/2}} \operatorname{Re}\Big\{\frac{1}{\sqrt{1+ib}}\Big(\frac{x_1^2}{1+ib} + x_2^2 + \ldots + x_{n-1}^2\Big)^{-n/2}\Big\}.$$

Hence

$$\partial_{x_n}^2 E_1(x', 0)\big|_{x_1=0} = \frac{2\Gamma(n/2)}{\pi^{n/2}} \operatorname{Re}\,(1+ib)^{-1/2}(x_2^2 + \ldots + x_{n-1}^2)^{-n/2} > 0;$$

$$\partial_{x_n}^2 E_1(x', 0)\big|_{x_2=\ldots=x_{n-1}=0} = \frac{2\Gamma(n/2)}{\pi^{n/2}} \operatorname{Re}\,(1+ib)^{(n-1)/2}|x_1|^{-n/2}. \tag{35}$$

Let $\alpha = \arg(1+ib)$. If $\alpha(n-1)/2 \in (\pi/2 + 2k\pi, 3\pi/2 + 2k\pi)$, $k \in Z$, then the quantity (35) is negative. In particular, if $n = 4$ and $\alpha \in (\pi/3, \pi/2)$, then the right–hand side of (35) is negative. Therefore, the function $\partial_{x_n}^2 E_1(\omega', 0)$ changes sign. Consequently, there exists a positive function φ such that the coefficient of ε in (32) is negative.

1.5. The Neumann problem

Consider a system of differential operators

$$\mathcal{A}(\partial_x) = (-1)^m \sum_{|\alpha|=|\beta|=m} A_{\alpha\beta}\partial_x^{\alpha+\beta}$$

where $A_{\alpha\beta}$ are constant $\ell \times \ell$-matrices such that $A^*_{\alpha\beta} = A_{\beta\alpha}$. We shall suppose that

$$\sum_{|\alpha|=|\beta|=m} (A_{\alpha\beta}f_\beta, f_\alpha)_{\mathbf{C}^\ell} \geq c \sum_{|\alpha|=m} |f_\alpha|^2, \; c > 0$$

for all $f_\alpha \in \mathbf{C}^\ell$. It is clear that this property implies the ellipticity of $\mathcal{A}$.

We suppose that the cone K is determined by (23), where φ is a real valued positive homogeneous of degree 1 function on $\mathbf{R}^{n-1}$ satisfying the Lipschitz condition

$$|\varphi(x') - \varphi(y')| \leq c|x' - y'| \tag{36}$$

for all $x', y' \in \mathbf{R}^{n-1}$ with constant c independent of x', y'. Since we do not suppose that the surface $\partial K \backslash \{0\}$ is smooth we state the Neumann problem for the operator $\mathcal{A}(\partial_x)$ in a weak form.

Consider solutions U of the equation

$$\int_K \sum_{|\alpha|=|\beta|=m} \left(A_{\alpha\beta}\partial_x^\alpha U, \; \partial_x^\beta V\right) dx = 0, \tag{37}$$

which must be satisfied for all $V \in \left(H^m(K)\right)^\ell$ with support in $\overline{K}\backslash\{0\}$. If the boundary $\partial K\backslash\{0\}$ is sufficiently smooth then the equation (37) can be written as a boundary value problem in $\overline{K}\backslash\{0\}$ for the operator $\mathcal{A}(\partial_x)$. We are interested in solutions U of (37) having the form (20) with $u_k \in \left(H^m(\Omega)\right)^\ell$.

We connect an operator pencil with the equation (37) in the following way. First we introduce differential operators $Q_\alpha(\lambda)$ on Ω by

$$\left(Q_\alpha(\lambda)u\right)(x/|x|) = |x|^{-\lambda+|\alpha|}\partial_x^\alpha\left(|x|^\lambda u(x/|x|)\right),$$

for $u \in \left(H^m(\Omega)\right)^\ell$. We put

$$L(u, v; \lambda) = \sum_{|\alpha|=|\beta|=m} \int_\Omega \left(A_{\alpha\beta}Q_\beta(\lambda)u, Q_\alpha\left(-\overline{\lambda} + 2m - n\right)v\right)_{\mathbf{C}^\ell} d\omega$$

where $d\omega$ is the Lebesque mesure on S^{n-1}. For every $\lambda \in \mathbf{C}$ the form L induces a continuous operator

$$\mathcal{L}(\lambda) : \left(H^m(\Omega)\right)^\ell \to \left(H^m(\Omega)\right)^\ell.$$

The operator pencil $\mathcal{L}$ is Fredholm and its spectrum consists of isolated eigenvalues with finite algebraic multiplicity. Moreover, the number λ_0 is an eigenvalue of $\mathcal{L}$ if and only if $-\overline{\lambda}_0 + 2m - n$ is an eigenvalue of $\mathcal{L}$. The geometric, partial and algebraic multiplicities of the eigenvalues λ_0 and $-\overline{\lambda}_0 + 2m - n$ coincide. Furthermore, the function (20) with $u_k \in (H^m(\Omega))^\ell$ satisfies (37) if and only if λ_0 is an eigenvalue of $\mathcal{L}$, u_0 and $u_1, \dots, u_N$ are an eigenfunction and generalized eigenfunctions corresponding to λ_0.

Theorem 13. [6],[20] *Let $\varphi \in C^\infty(\mathbf{R}^{n-1}\backslash\{0\})$. Then the following propositions are valid:*

(i) *If $2m < n - 1$ then the strip* (24) *does not contain any eigenvalues of the operator pencil $\mathcal{L}$.*

(ii) *If $2m \geq n - 1$ and n is even then the strip* (24) *contains exactly one eigenvalue $\lambda_0 = m - n/2$. The vector function u is an eigenfunction if and only if $u = p|_\Omega$, where p is homogeneous vector polynomial in $\mathbf{R}^n$ of degree $m - n/2$. Every eigenfunction has exactly one generalized eigenfunction.*

(iii) *If $2m \geq n - 1$ and n is odd then the strip* (24) *contains exactly two eigenvalues $m - n/2 \pm 1/2$, whose geometric and algebraic multiplicities are equal to $\ell C^{n-1}_{m+(n-1)/2}$ and $\ell[C^{n-1}_{m+(n-1)/2} + C^{n-1}_{m+(n-3)/2}]$ respectively. The vector-function u is an eigenfunction of $\mathcal{P}$ corresponding to the eigenvalue $m - n/2 + 1/2$ if and only if $u = p|_\Omega$, where p is a homogeneous vector polynomial in $\mathbf{R}^n$ of degree $m - n/2 + 1/2$. Every eigenfunction has at most one generalized eigenfunction.*

The next theorem deals with the case of nonsmooth φ.

Theorem 14. [20] *Let φ satisfy* (36). *Then the following assertions are valid.*

(i) *If $2m < n - 1$ or $2m \geq n - 1$ and n is odd then the strip*

$$|\operatorname{Re}\lambda - m + n/2| < 1/2 \tag{38}$$

is free of eigenvalues of $\mathcal{L}$.

(ii) *If $2m > n - 1$ and n is even then the assertion* (ii) *of Theorem 13 is valid if we replace the inequality* (24) *by* (38) *in its formulation.*

Let G be a bounded domain in $\mathbf{R}^n$ and $\mathcal{O} \in \partial G$. For simplicity we assume that $G \cap \mathcal{D} = K \cap \mathcal{D}$, where $\mathcal{D}$ is the unit ball in $\mathbf{R}^n$ and K is the same cone as in Theorem 13.

We shall consider the Neumann problem in two formulations.

Problem I. Let f be a vector-valued function from $(L_2(G))^l$ orthogonal to all vector polynomials of degree no greater than $m-1$. By a solution of Problem I we mean a function $\mathcal{U}$ from $(H^m(G))^l$ satisfying the integral equation

$$\int_G \sum_{|\alpha|=|\beta|=m} (A_{\alpha\beta}\partial_x^\beta \mathcal{U},\, \partial_x^\alpha \mathcal{V})dx = \int_G (f, \mathcal{V})dx, \tag{39}$$

where $\mathcal{V}$ is an arbitrary element of the space $(H^m(G))^l$.

Obviously, Problem I is solvable and its solution is determined up to an additive polynomial term of degree no greater than $m-1$.

To formulate Problem II we need the function space $\overset{o}{H}{}^m(G;\mathcal{O})$, which is the completion of $C_0^\infty(\overline{G}\backslash\{0\})$ in the norm of the space $H^m(G)$. This space differs from $H^m(G)$ only for $2m > n$.

We denote by $\dot{\Pi}$ the set of polynomials of degree no greater than $m-1$ vanishing at the point $\mathcal{O}$, together with the derivatives of order less than $m - n/2$.

Problem II. Let $2m > n$ and let f be a vector-valued function from $(L_2(G))^l$, orthogonal to $\dot{\Pi}^l$. By a solution of Problem II we mean a vector-valued function $\mathcal{U}$ from $(\overset{o}{H}{}^m(G;\mathcal{O}))^l$ satisfying the integral equation (39) for $\mathcal{V} \in (\overset{o}{H}{}^m(G;\mathcal{O}))^l$. Clearly, this problem is also solvable and its solution is determined up to polynomials from $\dot{\Pi}^l$.

Let f be a smooth function. From results of the paper [1] and from Theorem 13 it follows that the solution $\mathcal{U}_{II}$ of Problem II has the asymptotics

$$\mathcal{U}_{II}(x) = \sum_{|\alpha|=m-n/2} c_\alpha x^\alpha + O(r^{m-(n-1)/2+\varepsilon}),$$

provided n is even, and the asymptotics

$$\begin{aligned}\mathcal{U}_{II}(x) = &\sum_{|\beta|=m-(n-1)/2} c_\beta x^\beta \\ &+ \sum_{|\gamma|=m-(n+1)/2} c_\gamma\big(p_\gamma(x)\log r + h_\gamma(x)\big) + O(r^{m-(n-1)/2+\varepsilon})\end{aligned}$$

if n is odd. Here c_α, c_β are constant vectors, c_γ are scalars, p_γ are vector-valued polynomials of degree $|\gamma|+1$, h_γ are vector-valued functions positive homogeneous of degree $|\gamma|+1$ and ε is a positive number.

The following asymptotic formula is valid for the solution $\mathcal{U}_I$ of Problem I:

$$\mathcal{U}_I(x) = \sum_{|\alpha|\le m-(n-1)/2} c_\alpha x^\alpha + O(r^{m-(n-1)/2+\varepsilon}), \tag{40}$$

where c_α are constant vectors (if $2m < n-1$, the sum in (40) is absent).

Comparing the asymptotics of the solutions $\mathcal{U}_I$ and $\mathcal{U}_{II}$ for $2m > n$ and odd n, we deduce that $U_I \in (H^{m+1/2+\delta}(G))^l$ for some $\delta > 0$ and $\mathcal{U}_{II} \in (H^{m+1/2-\delta}(G))^l$ for all $\delta > 0$. For other dimensions $\mathcal{U}_I$ and $\mathcal{U}_{II}$ belong to the class $(H^{m+1/2+\delta}(G))^l$.

In [25] (also, see [15], 10.7) the asymptotics of solutions to the Neumann problem for the Laplacian in the exterior of the thin $3d$–cone (28) was studied.

It was shown that

$$u(x) \sim c_0 + \sum_{j=1}^{3} c_j r^{\lambda_j(\varepsilon)} \psi_j(\varepsilon, \omega),$$

where $c_0, \dots, c_3$ are constants and

$$\lambda_j(\varepsilon) = 1 - \pi^{-1}\mu_j \varepsilon^2 + O(\varepsilon^3 |\log \varepsilon|), \; j = 1, 2, \tag{41}$$

$$\lambda_3(\varepsilon) = 1 + 4\pi^{-1}\varepsilon \operatorname{mes}_2 \omega + O(\varepsilon^3 |\log \varepsilon|).$$

Here μ_1, μ_2 denote the eigenvalues of a certain positive definite constant matrix depending on ω are denoted. The asymptotics (41) implies that, in general, the gradient of the solution u is unbounded.

1.6. The Dirichlet problem for higher-order elliptic differential equations in angles

In the two-dimensional case it is possible to write transcendental equations for eigenvalues of operator pencils generated by general boundary value problems for elliptic systems. Such equations are well known for various boundary value problems for the two-dimensional Lamé and Stokes systems, for the biharmonic equation (see for example, [21]). For arbitrary equations the above mentioned transcendental equations are rather complicated. However, in [22], [23] the following explicit information about their roots was obtained.

We shall denote by (r, Θ) the polar coordinates of $(x, y) \in \mathbf{R}^2$. Let

$$K_\varphi = \{(x, y) \in \mathbf{R}^2 : 0 < r < \infty, \; \Theta \in (0, \varphi)\}$$

where $\varphi \in (0, 2\pi]$. Consider an elliptic operator of order $2m$:

$$\mathcal{A}(\partial_x, \partial_y) = \sum_{k=0}^{2m} A_k \partial_x^{2m-k} \partial_y^k,$$

where A_k are real numbers. We introduce the polynomial operator pencil

$$\mathcal{L}_\varphi(\lambda) : H^{2m}(0, \varphi) \cap \overset{o}{H}{}^m(0, \varphi) \to L_2(0, \varphi),$$

in accordance with the rule

$$\mathcal{L}_\varphi(\lambda)\Phi(\Theta) = r^{-\lambda+2m} \mathcal{A}(\partial_x, \partial_y)(r^\lambda \Phi(\Theta)),$$

where $\Phi \in H^{2m}(0, \varphi) \cap \overset{o}{H}{}^m(0, \varphi)$. This operator pencil corresponds to the Dirichlet problem for the operator $\mathcal{A}$ in K_φ. The simplest representative of such an operator is the Laplacian. In this case we have $\mathcal{L}_\varphi(\lambda) = \lambda^2 + \partial_\varphi^2$ and the eigenvalues are $\pm k\pi/\varphi$, $k = 1, 2, \ldots$ with the eigenfunctions being equal to $\sin(k\pi\Theta/\varphi)$.

In the general case it is well known that the spectrum of $\mathcal{L}_\varphi$ consists of eigenvalues of finite algebraic multiplicity (see [1]). Moreover, if λ_0 is an eigenvalue of $\mathcal{L}_\varphi$ then $\overline{\lambda}_0$, $m - 1 - \overline{\lambda}_0$ are also eigenvalues of $\mathcal{L}_\varphi$ and the geometric, partial and algebraic multiplicities of these three eigenvalues coincide. Therefore it is sufficient to describe the location of eigenvalues and their multiplicities in the half-plane $\operatorname{Re}\lambda \geq m - 1$. When $\varphi = \pi$ or $\varphi = 2\pi$ all eigenvalues of $\mathcal{L}_\varphi$ can be evaluated explicitly.

Proposition [22]. a) *The spectrum of $\mathcal{L}_\pi$ consists of the eigenvalues $m - 1 \pm k$, $k = 1, 2, \ldots$, with the multiplicity k for $k \leq m$ and m for $k > m$.*

b) *The spectrum of $\mathcal{L}_{2\pi}$ consists of the eigenvalues $m - 1 \pm k$, $k = 1, 2, \ldots$, of the same multiplicity as in the case* a) *and of the eigenvalues $m - 1/2 \pm k$, $k = 0, 1, \ldots$, of multiplicity m.*

In both cases there are no generalized eigenfunctions.

The next theorem pertains to the case $\varphi \in (0, \pi)$.

Theorem 15. [22] *Let $\varphi \in (0, \pi)$. Then the strip $m - 1 \leq \operatorname{Re}\lambda \leq m$ contains no eigenvalues of the operator pencil $\mathcal{L}_\varphi$. Furthermore, the eigenvalues of this pencil do not coincide with $m + k$, $k = 0, 1, \ldots, m - 1$.*

As an application consider the Dirichlet problem

$$\mathcal{A}(\partial_x, \partial_y)u = f \quad \text{on} \quad Q \tag{43}$$

$$\partial_\nu^j u = 0 \quad \text{on} \quad \partial Q,\ j = 0, \ldots, m - 1, \tag{44}$$

where Q is a convex polygon and ν is the normal to ∂Q. If $f \in L_2(Q)$ then the problem (43), (44) has a solution u in $\overset{o}{H}{}^m(Q)$. Theorem 15 implies that $u \in H^{m+1}(Q)$.

Now we turn to the case $\varphi \in (\pi, 2\pi)$.

Theorem 16. [23] *Let $\varphi \in (\pi, 2\pi)$. Then*

(i) *All eigenvalues of $\mathcal{L}_\varphi$ in the strip $m - 1 \leq \operatorname{Re}\lambda \leq m$ are real and located in $(m - 1/2, m]$.*

(ii) *All eigenvalues of $\mathcal{L}_\varphi$ in $(m - 1/2, m)$ are simple and strictly decreasing with respect to $\varphi \in (\pi, 2\pi)$.*

(iii) *The total number of eigenvalues in the interval $(m - 1/2, m)$ changes from m to 1 when φ changes from 2π to π.*

The above theorem together with results of [1] implies that H^m–solution u of the Dirichlet problem for the operator $\mathcal{A}$ in the angle K_φ, $\varphi \in (\pi, 2\pi)$, with the right–hand side from L_2, admits the representation

$$u(x,y) = \sum_{j=1}^{k} c_j r^{\lambda_j} u_j(\Theta) + v(x,y)$$

where $m - 1/2 < \lambda_1 < \ldots < \lambda_k < m$, $1 \le k \le m$,u_j are smooth functions independent of f, and c_j are constants, v belongs to $H^{m+1-\varepsilon}$ in a neighbourhood of the vertex with arbitrary positive ε. When the angle φ is close to π or 2π, it is possible to give an explicit asymptotic formulae for eigenvalues of $\mathcal{L}_\varphi$ close to m and to $m - 1/2$.

Let $z_1, \ldots, z_m$, denote the roots of the equation $\mathcal{A}(1, z) = 0$ with positive imaginary parts (if there is a multiple root then we take into account its multiplicity).

Theorem 17. [23] (i) *When φ is close to $\ell\pi$, $\ell = 1, 2$, there is only one eigenvalue $\lambda_\ell(\varphi)$ of $\mathcal{L}_\varphi$ located near m and*

$$\begin{aligned}\lambda_\ell(\varphi) &= m + \operatorname{Im}(z_1 + \ldots + z_m)(1 - \varphi/\ell\pi)\\ &+ O(|1 - \varphi/\ell\pi|^2).\end{aligned}$$

(ii) *When φ is close to 2π there are exactly m eigenvalues $\mu_k(\varphi)$, $k = 1, \ldots, m$, located near $m - 1/2$ and*

$$\mu_k(\varphi) = m - 1/2 + \gamma_k (2\pi - \varphi)^{2k-1} + 0(|2\pi - \varphi|^{2k}),\ k = 1, \ldots, m,$$

where

$$\gamma_k = \frac{-1}{4\pi}\Big[\frac{(k-1)!\Gamma(k-1/2)}{(2k-2)!\Gamma(1/2)}\Big]^2 \frac{\det D_{2k}}{\det D_{2k-2}}.$$

Here D_{2k} is a $2k \times 2k$ matrix determined as follows.

Let

$$\sum_{0 \le s \le m} (\alpha_s + i\beta_s) z^{m-s} = \prod_{1 \le k \le m} (z - z_k)$$

then $D_0 = 1$ and

$$D_{2k} = \begin{pmatrix} \alpha_0 & \alpha_1 & \alpha_2 & \cdots & \alpha_{2k-1} \\ \beta_0 & \beta_1 & \beta_2 & \cdots & \beta_{2k-1} \\ 0 & \alpha_0 & \alpha_1 & \cdots & \alpha_{2k-2} \\ 0 & \beta_0 & \beta_1 & \cdots & \beta_{2k-2} \\ \vdots & \vdots & \vdots & \cdots & \vdots \\ \vdots & \vdots & \vdots & \cdots & \vdots \end{pmatrix}$$

where $\alpha_s = \beta_s = 0$ for $s > m$. (The value $\det D_{2k}$ is positive for even k and negative for odd k. Therefore, γ_k is positive.)

The following example shows that Theorems 14,15 may fail if coefficients of the operator $\mathcal{A}(\partial_x)$ are complex.

Consider the strongly elliptic operator

$$\mathcal{A}(\partial_x, \partial_y) = \partial_y^2 + (t-1)i\partial_x\partial_y + t\partial_x^2 \tag{45}$$

in the angle K_φ, where $t > 0$. The set

$$\{\lambda_k\},\ k = \mp 1,\ \mp 2, \dots,$$

where

$$\begin{aligned}\operatorname{Re}\lambda_k &= 2\pi k[\rho + (\varphi - \alpha)^2/\rho]^{-1},\\ \operatorname{Im}\lambda_k &= -\operatorname{Re}\lambda_k(\varphi - \alpha)/\rho,\end{aligned}$$

and

$$\begin{aligned}\rho &= \log(\cos^2\varphi + t\sin^2\varphi),\\ \alpha &= \arg(\cos\varphi - it\sin\varphi) \in [-2\pi, 0],\end{aligned}$$

is the spectrum of the corresponding pencil. If t is sufficiently large and $\varphi \neq \pi, 2\pi$, then $\operatorname{Re}\lambda_1$ is close to 0. If we consider the solution $u \in H^1(Q)$ of the problem (43), (44), where Q is a polygon then $u \notin H^{1+\alpha}(Q)$, $\alpha > 0$, for some $f \in L_2(Q)$ and for $\mathcal{A}$ given by (45) with sufficiently large t.

2. The Dirichlet problems for the second order differential systems in angles and in three-dimensional polyhedral cones

The second part of this paper deals with the operator pencils generated by the problems mentioned in the title. We prove that the strip $|\operatorname{Re}\lambda| \leq 1$ is free of the eigenvalues of the pencils corresponding to convex plane angles. If the angle is greater than π then each of the strips $-1 < \operatorname{Re}\lambda < -1/2$, $1/2 < \operatorname{Re}\lambda < 1$ contains exactly ℓ eigenvalues where ℓ is the size of the system.

In the three-dimensional case we consider polyhedrons whose inside angles between faces are less than π except, possibly, one, which may take an arbitrary value from $(0, 2\pi]$. We prove that the strip $|\operatorname{Re}\lambda + 1/2| \leq 1$ does not contain eigenvalues of the corresponding pencil.

2.1. The Dirichlet problem for the second order system in an angle

We shall denote by (x_1, x_2) the Cartesian coordinates and by (r, θ) the polar coordinates in $\mathbf{R}^2$. Let K_φ, $\varphi \in (0, 2\pi]$, be the same angle as in Sect. 1.6 and let

$$\mathcal{A}(\partial_{x_1}, \partial_{x_2}) = \sum_{i,j=1}^{2} A_{ij} \partial_{x_i} \partial_{x_j}$$

where A_{ij} are $\ell \times \ell$-matrices and $A_{ij}^* = A_{ji}$. We assume that

$$\sum_{i,j=1}^{2} (A_{ij} f_j, f_i)_{C^\ell} \geq c_0(\parallel f_1 \parallel^2 + \parallel f_2 \parallel^2), \; c_0 > 0, \qquad (1)$$

for all $f_1, f_2 \in C^\ell$. It is clear that (1) implies the ellipticity of $\mathcal{A}$.

We introduce the operator pencil

$$\mathcal{L}(\lambda) : \left(\overset{o}{H}{}^1(0,\varphi) \cap H^2(0,\varphi) \right)^\ell \to \left(L_2(0,\varphi) \right)^\ell$$

by the formula

$$\left(\mathcal{L}(\lambda)u\right)(\theta) = r^{2-\lambda} \mathcal{A}(\partial_{x_1}, \partial_{x_2}) \left(r^\lambda u(\theta) \right).$$

Lemma 1. (i) *If $\varphi \neq 2\pi$ then the strip $|\operatorname{Re} \lambda| \leq 1/2$ contains no eigenvalues of $\mathcal{L}$.*

(ii) *If $\varphi = 2\pi$ then the spectrum of $\mathcal{L}$ consists of the eigenvalues $\pm k/2$, $k = 1, 2, \ldots$, with the geometric multiplicity ℓ. There are no generalized eigenfunctions.*

(iii) *If $\varphi = \pi$ then the spectrum of $\mathcal{L}$ consists of the eigenvalues $\pm k$, $k = 1, 2, \ldots$, with the geometric multiplicity ℓ. There are no generalized eigenfunctions.*

Proof. Assertion (i) follows from Theorem 9. Theorem 10 (ii) implies (ii). Assertion (iii) is a consequence of 1.4.2b).

Now, we formulate the main result of this section.

Theorem 1. (i) *If $\varphi \in (0, \pi)$ then the strip $|\operatorname{Re} \lambda| \leq 1$ includes no eigenvalues of $\mathcal{L}$.*

(ii) *If $\varphi \in (\pi, 2\pi)$ then the strip $|\operatorname{Re} \lambda| \leq 1$ contains exactly 2ℓ eigenvalues of $\mathcal{L}$. Half of them lies in the strip $1/2 < \operatorname{Re} \lambda < 1$ and another half in the strip $-1 < \operatorname{Re} \lambda < -1/2$.*

The proof of this theorem is based on the following assertion.

Lemma 2. *Let $\phi \neq \pi$, $\phi \neq 2\pi$ and let the conditions of Theorem 1 be valid. Then the line $\operatorname{Re} \lambda = 1$ contains no eigenvalues of $\mathcal{L}$.*

Proof. Let

$$K_1 = \Big\{(x_1, x_2) : 1 < r < 2,\ \theta \in (0, \varphi)\Big\}$$
$$\Gamma_1 = \Big\{(x_1, x_2) : 1 < r < 2,\ \theta = 0 \text{ or } \theta = \varphi\Big\}.$$

Suppose that λ is an eigenvalue of $\mathcal{L}$ with $\operatorname{Re}\lambda = 1$. We take an eigenfunction u of $\mathcal{L}$ corresponding to λ and introduce $U(x_1, x_2) = r^\lambda u(\theta)$. Since $\mathcal{A}(\partial_{x_1}, \partial_{x_2})U = 0$ we have

$$\sum_{i,j=1}^{2} \int_{K_1} \big(\partial_{x_i} B_{ij} \partial_{x_j} U, \Delta U\big)_{\mathbb{C}^\ell} dx_1 dx_2 = 0 \tag{2}$$

where Δ is the Laplacian and

$$B_{ij} = \frac{1}{2}\big(A_{ij} + A_{ij}^*\big).$$

Integrating by parts in (2) and using the fact that the integrals over the arcs of the circles $r = 1$ and $r = 2$ are cancelled we get

$$\sum_{i,j,k=1}^{2} \int_{K_1} \Big(B_{ij}\partial_{x_j}\partial_{x_k} U,\ \partial_{x_i}\partial_{x_k} U\Big)_{\mathbb{C}^\ell} dx_1 dx_2$$
$$= \sum_{i,j=1}^{2} \Bigg\{ \int_{\Gamma_1} \Big(B_{ij}\partial_{x_j} U,\ \partial_{x_i}\partial_\nu U\Big)_{\mathbb{C}^\ell} dr \tag{3}$$
$$- \int_{\Gamma_1} \nu_i \Big(B_{ij}\partial_{x_j} U,\ \Delta U\Big)_{\mathbb{C}^\ell} dr \Bigg\}$$

where $\nu = (\nu_1, \nu_2)$ is the outward normal to Γ_1. We show that the right-hand side of the last equality vanishes. Let $\partial_\tau = \nu_2 \partial_{x_1} - \nu_1 \partial_{x_2}$. This is a differential operator with respect to the direction tangent to Γ_1. Since

$$\partial_{x_1} = \nu_1 \partial_\nu + \nu_2 \partial_\tau,\ \partial_{x_2} = \nu_2 \partial_\nu - \nu_1 \partial_\tau,$$

on Γ_1 it follows that the right-hand side of (3) is equal to

$$\int_{\Gamma_1} \big(B\partial_\nu U,\ \partial_\tau \partial_\nu U\big)_{\mathbb{C}^\ell} dr, \tag{4}$$

where

$$B = \sum_{j=1}^{2} \big(\nu_2 B_{1j} \nu_j - \nu_1 B_{2j} \nu_j\big),$$

is a symmetric matrix. Since the left-hand side of (3) is real we can write (4) as

$$\frac{1}{2}\int_{\Gamma_1} \partial_r(B\partial_\nu U,\ \partial_\nu U)_{\mathcal{C}^\ell} dr,$$

and, hence, it is zero. Thus, we get

$$\sum_{i,j,k=1}^{2} \int_{K_1} \Big(B_{ij}\partial_{x_j}\partial_{x_k}U,\ \partial_{x_i}\partial_{x_k}U\Big)_{\mathcal{C}^\ell} dx_1 dx_2 = 0.$$

Using (1) we obtain that $\partial_{x_j}\partial_{x_k}U = 0$ for $k, j = 1, 2$. So $U = x_1 f_1 + x_2 f_2$ for some $f_1, f_2 \in \mathcal{C}^\ell$. However, the equality $U(0) = U(\phi) = 0$ excludes this possibility for $\phi \neq \pi$ and for $\phi \neq 2\pi$. The lemma is proved.

Proof of Theorem 1. Since the spectrum of $\mathcal{L}$ is invariant with respect to the transformation $\lambda \longrightarrow -\overline{\lambda}$, it suffices to prove the statement of the theorem relating to the spectrum in the half-plane $\operatorname{Re}\lambda \geq 0$.

Let I be the $\ell \times \ell$ identity matrix. We put

$$\mathcal{A}_t(\partial_{x_1}, \partial_{x_2}) = (1-t)I\Delta + t\mathcal{A}(\partial_{x_1}, \partial_{x_2}),\quad t \in [0,1].$$

The operator pencil corresponding to $\mathcal{A}_t$ will be denoted by $\mathcal{L}_t$. We note that the operators $\mathcal{A}_t$ satisfy the conditions of Theorem 1.

(i) If $\phi \in (0, \pi)$ then by Lemma 1 (i) and Lemma 2 the operator pencil $\mathcal{L}_t$ has no eigenvalue on the lines $\operatorname{Re}\lambda = 1/2$ and $\operatorname{Re}\lambda = 1$. It is readily verified in a standard way that $\mathcal{L}_t$ has no eigenvalues in the strip $1/2 \leq \operatorname{Re}\lambda \leq 1$ for large $|\lambda|$. By the Rouché operator theorem [24] we conclude that the operator pencils $\mathcal{L}_t$ have the same number of eigenvalues in the strip $1/2 \leq \operatorname{Re}\lambda \leq 1$ (with account taken to their algebraic multiplicity). It is known that the operator pencil $\mathcal{L}_0$ has no eigenvalues in the strip $1/2 \leq \operatorname{Re}\lambda \leq 1$; therefore the same is true for the operator pencil $\mathcal{L}_1 = \mathcal{L}$.

(ii) Let $\phi \in (\pi, 2\pi)$. By the same argument as above we conclude that the number of eigenvalues of the operator pencil $\mathcal{L}_t$ in the strip $1/2 < \operatorname{Re}\lambda < 1$ does not depend on t. The operator pencil $\mathcal{L}_0$ has the only eigenvalue π/ϕ of multiplicity ℓ in the above mentioned strip. This completes the proof.

2.2. Dirichlet problem for the second order systems in a three-dimensional cone

Let (r, ω), $r > 0$, $\omega \in S^2$, be spherical coordinates in $\mathbf{R}^3$. By K we denote a cone and by Γ its boundary. We suppose that

$$\Gamma = \bigcup_{j=1}^{N} S_j \cup \bigcup_{j=1}^{N} \mathcal{R}_j,\ \ N > 2,$$

where $\{S_j\}$ are plane angles, $\{\mathcal{R}_j\}$ are rays, and we assume that the boundary of S_j coincides with $\mathcal{R}_j \cup \mathcal{R}_{j+1}$ when $j = 1, \dots, N-1$ and with $\mathcal{R}_N \cup \mathcal{R}_0$ when $j = N$. Let α_j denote the interior angle between two plane parts of the boundary Γ intersecting along the edge $\mathcal{R}_j$.

Consider the differential operator

$$\mathcal{A}(\partial_x) = \sum_{1\le i,j\le 3} A_{ij}\partial_{x_i}\partial_{x_j}$$

where A_{ij} are $\ell \times \ell$ matrices such that $A_{ij}^* = A_{ji}$. We assume that

$$\sum_{1\le i,j\le 3} (A_{ij}f_j, f_i) \ge c_0 \sum_{1\le j\le 3} |f_j|^2, \quad c_0 > 0 \tag{5}$$

for any $f_1, f_2, f_3 \in \mathbb{C}^l$. It is clear that $\mathcal{A}(\xi)$ is symmetric and satisfies (1.17).

We are interested in solutions of the Dirichlet problem

$$\mathcal{A}(\partial_x)U = 0 \quad \text{on} \quad K \tag{6}$$

$$U = 0 \quad \text{on} \quad \Gamma\backslash\{0\} \tag{7}$$

having the form (1.20), where λ_0 is a complex number and $u_k \in \big(\overset{o}{H}{}^1(\Omega)\big)^\ell$ with $\Omega = K \cap S^2$.

We introduce the operator pencil

$$\mathcal{L}(\lambda) : \big(\overset{o}{H}{}^1(\Omega)\big)^\ell \to \big(H^{-1}(\Omega)\big)^\ell$$

by the formula

$$\mathcal{L}(\lambda)u(\omega) = r^{2-\lambda}\mathcal{A}(\partial_x)\big(r^\lambda u(\omega)\big), \quad u \in \big(\overset{o}{H}{}^1(\Omega)\big)^\ell.$$

The main result of this section is the following.

Theorem 2. *Let $\alpha_j \in (0,\pi)$ for $j = 2, \dots, N$ and $\alpha_1 \in (0, 2\pi]$. Then the strip*

$$|\operatorname{Re}\lambda + 1/2| \le 1 \tag{8}$$

is free of eigenvalues of the operator pencil $\mathcal{L}$.

The main step in the proof of Theorem 2 is contained in

Lemma 3. *Let the conditions of Theorem 2 be valid. Then there are no eigenvalues of the operator pencil $\mathcal{L}$ on the line* $\operatorname{Re}\lambda = 1/2$.

Proof. Let λ be an eigenvalue of $\mathcal{L}$ with $\operatorname{Re}\lambda = 1/2$ and let u be a corresponding eigenfunction. This means that

$$\mathcal{L}(\lambda)u = 0 \quad \text{on} \quad \Omega \quad \text{and} \quad u \in \big(\overset{o}{H}{}^1(\Omega)\big)^\ell.$$

Since the boundary of Ω is smooth outside the points $\omega_k = \mathcal{R}_k \cap \partial\Omega$, $k = 1, \dots N$, the vector-function u is smooth outside these points.

One can readily verify that the principal part of the differential operator $\mathcal{L}(\lambda)$ at ω_k satisfies the condition of Theorem 1. Using Theorem 1 (i) and Lemma 1 (i),(ii) together with results of the paper [1], we conclude that

$$u(\omega) = 0\big(|\omega - \omega_k|^{1+\delta}\big)$$

in a neighbourhood of the point $\omega_k, k = 2, \dots, q$, for some positive δ and

$$u(\omega) = 0\big(|\omega - \omega_k|^{1+\delta}\big) \tag{10}$$

in a neighbourhood of ω_1. Moreover, both relations (9) and (10) may be differentiated.

We introduce the truncated cone

$$K_1 = \{x \in \mathbf{R}^3 : 1 < r < 2,\ \omega \in \Omega\},$$

and the set

$$\Gamma_1 = \{x \in \Gamma : 1 < r < 2\}.$$

Let $\nu = (\nu_1, \nu_2, \nu_3)$ denote the unit outward normal to Γ and let τ be the vector directed along the ray $\mathcal{R}_1$. Also, let $U(x) = r^\lambda u(\omega)$. Since $\mathcal{A}(\partial_x)U = 0$ on K, we have

$$\begin{aligned} 0 &= \int_{K_1} \Big(\mathcal{A}(\partial_x)U,\ \partial_\tau^2 U\Big)_{C^\ell} dx \\ &= \sum_{i,j=1}^{3} \int_{K_1} \Big(A_{ij}\partial_{x_j}\partial_\tau U,\ \partial_{x_i}\partial_\tau U\Big)_{C^\ell} dx \\ &+ \sum_{i,j=1}^{3} \int_{\Gamma_1} \Big\{\nu_i\big(A_{ij}\partial_{x_j}U,\ \partial_\tau^2 U\big)_{C^\ell} - (\tau,\nu)\big(A_{ij}\partial_{x_j}U,\ \partial_{x_i}\partial_\tau U\big)_{C^\ell}\Big\} d\Gamma. \end{aligned} \tag{11}$$

Here, we made use of the fact that the integrals over the boundary lying on the spheres $r = 1$ and $r = 2$ mutually cancel. According to (9), (10) all the integrals in (11) make sense.

We show that the last sum in the right-hand side of (11) vanishes. Keeping this aim in mind consider the integral

$$\operatorname{Re} \sum_{1 \le i,j \le 3} \int_{\Gamma_{n1}} \Big\{\nu_i\big(A_{ij}\partial_{x_j}U, \partial_\tau^2 U\big)_{C^\ell} - (\tau,\nu)\big(A_{ij}\partial_{x_j}U, \partial_{x_i}\partial_\tau U\big)_{C^\ell}\Big\} d\Gamma, \tag{12}$$

where $\Gamma_{n1} = \{x \in S_n : 1 < r < 2\}$, $n = 1, \dots, q$. Without loss of generality one can assume that Γ_{n1} lies in the plane $x_3 = 0$ (one can always gain this by virtue of a linear change of coordinates) and that $\nu = (0, 0, 1)$ on Γ_{n1}. Then (12) is equal to

$$\operatorname{Re} \int_{\Gamma_{n1}} \Big\{ \big(A_{33}\partial_{x_3}U, \partial_\tau^2 U\big)_{C^\ell} - \tau_3 \sum_{i=1}^{3} \big(A_{i3}\partial_{x_3}U, \partial_{x_i}\partial_\tau U\big)_{C^\ell} \Big\}$$

$$= \tau_3 \operatorname{Re} \int_{\Gamma_{n1}} \Big\{ \sum_{j=1}^{2} \tau_j \partial_{x_j} \big(A_{33}\partial_{x_3}U, \partial_{x_3}U\big)_{C^\ell}$$

$$- \sum_{i=1}^{2} \tau_3 \partial_{x_i} \big(A_{i3}\partial_{x_3}U, \partial_{x_3}U\big)_{C^\ell} \Big\} d\Gamma = 0$$

Thus, we get

$$\sum_{1 \le i,j \le 3} \int_{K_1} \big(A_{ij}\partial_{x_j}\partial_\tau U, \partial_{x_i}\partial_\tau U\big)_{C^\ell} dx = 0$$

Using the estimate (5) we obtain

$$\partial_\tau U = c \quad \text{on} \quad K,$$

where $c = const$. Since $\partial_\tau U$ is a positive homogeneous function of degree $-1/2$, we get $\partial_\tau U = 0$. If K is not a dihedral angle, this together with $U = 0$ on ∂K implies $U = 0$ on K.

Proof of Theorem. Since the number λ is an eigenvalue of $\mathcal{L}$ simultaneously with $-1 - \overline{\lambda}$, we derive from Lemma 3 that the line $\operatorname{Im} \lambda = -3/2$ contains no eigenvalues of $\mathcal{L}$.

Let I be the $\ell \times \ell$ identity matrix. We put

$$\mathcal{A}_t(\partial_x) = (1-t)I\Delta + t\mathcal{A}(\partial_x), \quad t \in [0,1].$$

By $\mathcal{L}_t$ we denote the operator pencil corresponding to $\mathcal{A}_t$. We note that the operators $\mathcal{A}_t$ satisfy the conditions of Theorem 2. By Lemma 3 the operator pencils $\mathcal{L}_t$ have no eigenvalues on the lines $\operatorname{Re} \lambda = -3/2$ and $\operatorname{Re} \lambda = 1/2$. Using the strong ellipticity of $-\mathcal{A}_t(\partial_x)$ one can check that the operator pencils $\mathcal{L}_t$ have no eigenvalues in the strip $-3/2 \le \operatorname{Re} \lambda \le 1/2$ for large $|\lambda|$. By Rouché's operator theorem (see [24], Ch.11) we conclude that the operator pencils $\mathcal{L}_t$ have the same number of eigenvalues (taking into account their multiplicity) in the strip (8). Thus, it suffices to show that the operator pencil $L_0 = I(\delta + \lambda(\lambda + 1))$ has no eigenvalues in the strip (8).

Let K_0 be the dihedral angle with the opening α_1 and with the edge directed along the ray $\mathcal{R}_1$. Also let $\Omega_0 = K_0 \cap S^2$. Then the operator pencil

$$\delta + \lambda(\lambda + 1) : \overset{o}{H}{}^1(\Omega_0) \to H^{-1}(\Omega_0)$$

does not have any eigenvalue in the strip $|\operatorname{Re}\lambda + 1/2| < 1/2 + \pi/\alpha_1$. Since $\Omega \subset \Omega_0$ the strip (8) is free of spectrum of $\mathcal{L}_0$. The proof is complete.

As an example of application of Theorems 1 and 2 we consider the Dirichlet problem

$$\mathcal{A}(\partial_x)u = f \quad \text{on} \quad Q,$$
$$u = 0 \quad \text{on} \quad \partial Q,$$

where Q is a three-dimensional convex polyhedron.

If $f \in (L_2(Q))^\ell$ then the H^1-solution of this problem belongs to $(H^2(Q))^\ell$.

This is a direct corollary of Theorems 1, 2 and of results of a general theory of boundary value problems in nonsmooth domains (see [1], [2]).

References

1. V.A. Kondrat'ev, Boundary value problems for elliptic equations in domains with conical or angular points, Trudy Moskov. Mat. Obshch. **16** (1967), 209–292; translation in: Trans. Moscow Math. Soc. 1967 (1968).

2. V.G. Maz'ya and B.A. Plamenevskii, Elliptic boundary value problems, *Amer. Math. Soc. Transl.* (2) **123** (1984).

3. V.G. Maz'ya and B.A. Plamenevskii, On properties of solutions of three dimensional problems of elasticity theory and hydrodynamics in domains with isolated singular points, *AMS Translations* **123** (2) (1984),109–123.

4. V.A. Kozlov, V.G. Maz'ya and C. Schwab, On singularities of solutions of the displacement problem of linear elasticity near the vertex of a cone, *Arch. Rat. Mech. Anal.* **119** (1992),197–227.

5. V.G. Maz'ya and B.A. Plamenevskii, The first boundary value problem for the classical equations of mathematical physics on piecewise smooth domains (in Russian), Parts I and II, *Zeitschr. f. Anal. Anw.* **2** (1983),335–359 and 523–551.

6. V.A. Kozlov and V.G. Maz'ya, Spectral properties of operator pencils (in Russian), *Functional Analysis and its Applications* **22** (1988),38–46; translation in: *Functional Anal. Appl.* **22** (1988) No.2, 114–121.

7. V.A. Kozlov and V.G. Maz'ya, On stress singularities near the boundary of a polygonal crack, *Proc. Roy. Soc. Edinburgh* **117A** (1991), 31–37.

8. M. Dauge, Stationary Stokes and Navier-Stokes systems on two- and three-dimensional domains with corners. Part I: Linearized Equations, *SIAM J. Math. Anal.* **20** (1989), 74–97.

9. V.A. Kozlov, V.G. Maz'ya and C. Schwab, On singularities of solutions to the Dirichlet problem of hydrodynamics near the vertex of 3-D cone; to appear in *Journal für die reine und angewandte Mathematik.*

10. V.G. Maz'ya and B.A. Plamenevskii, On the maximum principle for the biharmonic equation in domains with conical points (in Russian), *Mathematika/Izv. VUZ* **2** (1981), 52–59.

11. V.G. Maz'ya and J. Rossmann, On the Agmon-Miranda maximum principle for solutions of strongly elliptic equations in domains of $\mathbf{R}^n$ with conical points, *Anal. of Global Analysis and Geometry*, **10** (1992), 125–150.

12. J. Pipher and G. Verchota, A maximum principle for biharmonic functions in Lipschitz and $\mathbf{C}^1$-domains, *Comm. Math. Helvetici*, **68**, **3** (1993), 385–414.

13. V.A. Kozlov, The Dirichlet problem for elliptic equations in domains with conical points. (Russian), *Differentialnye Uravnenija*, **26**, **6** (1990),1014-1023; translation in: *Differential Equations*, **26**, **6** (1990), 739–747.

14. V.A. Kozlov and V.G. Maz'ya, On the spectrum of an operator pencil generated by the Dirichlet problem in a cone (in Russian), *Mat. Sbornik* **182** (1919), 638–660; translation in: *Math. USSR Sbornik* **73** (1) (1992), 27–48.

15. V.G. Maz'ya, S.A. Nazarov and B.A. Plamenevskii, Asymptotishe Theorie Elliptischer Randwertanfgaben in Singulär Gestörten Gebieten I, Akademie-Verlag Berlin, 1991.

16. V.G. Maz'ya and S.A. Nazarov, The apex of a cone can be irregular in Wiener's sense for a fourth-order elliptic equation, (Russian), *Mat. Zametki* **39** (1) (1986), 24–28.

17. V.A. Kozlov and J. Rossmann, On the behaviour of the spectrum of parameter-depending operator under small variation of the domain and application to operator pencils generated by elliptic boundary value problems in a cone, *Math. Nachr.* **153** (1991), 123–129.

18. V.A. Kozlov and J. Rossmann, Singularities of solutions of elliptic boundary value problems near conical points; to appear in *Math. Nachr.*

19. V.G. Maz'ya, The Dirichlet problem for elliptic equations of arbitrary order in unbounded regions, *Soviet Mathem. Dokl.* **4** (1963), 860–863.

20. V.A. Kozlov and V.G. Maz'ya, On the spectrum of an operator pencil generated by the Neumann problem in a cone (in Russian), *Algebra i Analiz* **3** (1991),111–131; translation in: *St. Petersburg Math. J.* **3** (2) (1992), 333–353.

21. P. Grisvard, Singularities in boundary value problems. Masson, Springer-Verlag, 1992.

22. V.A. Kozlov, Singularities of solutions to the Dirichlet problem for elliptic equations in neighbourhoods of angular points (in Russian), *Algebra i Analiz* **4** (1989),161–177; translation in: *Leningrad Math. J.* **1** (1990), 967–982.

23. V.A. Kozlov, On the spectrum of the pencil, generated by the Dirichlet problem for elliptic equations in an angle (in Russian), *Sib. Math. J.* **32**, (1991),74–87; translation in: *Siberian Math. J.* **32** (2) (1991), 238–251.

24. I. Gohberg, S. Goldberg and M.A. Kaashoek, Classes of linear operators. Vol.1. Operator Theory: Advances and Applications, Vol. 49. Birkhäuser-Verlag, Basel, Boston, Berlin, 1990.

25. V.G. Maz'ya and S.A. Nazarov, Singularities of solutions of the Neumann problem at a conical point, *Siberian Math. Journal* **30** (3) (1989), 52–63.

V.A. Kozlov
Department of Mathematics
University of Linköping
58183 Linköping, Sweden
email: vkozlov@math.liu.se

Vladimir G. Maz'ya
Department of Mathematics
University of Linköping
58183 Linköping, Sweden
email: vlmaz@math.liu.se

Problèmes sensitifs et coques élastiques minces

J.-L. Lions et E. Sanchez-Palencia

A la mémoire de Pierre Grisvard

Résumé

Les problèmes sensitifs sont des problèmes aux limites pour des équations ou systèmes dont l'étude de l'existence et de l'unicité des solutions fait intervenir des espaces fonctionnels inhabituels (notamment non contenus dans l'espace des distributions) ayant la propriété suivante :

La solution peut cesser d'exister si le second membre f est remplacé par $f + \delta f$ où δf peut être une fonction C^∞ à support compact arbitrairement petite ainsi que chacune de ses dérivées.

Nous considérons ici un problème de perturbation singulière $\varepsilon \searrow 0$ qui est classique mais dont la limite est sensitive. Il s'agit d'un modèle simplifié (à coefficients constants) des coques élastiques minces ayant une partie du bord libre et dont la surface moyenne a une courbure totale positive. On montre sur un exemple explicite que les solutions du problème limite ne sont pas des distributions et l'on commente les difficultés du calcul numérique correspondant.

1. Introduction et généralités

Considérons une équation (ou système) aux dérivées partielles ayant une formulation variationnelle dans un espace fonctionnel V et soit V' son dual (lorsque L^2 est identifié à son dual). Dans le cadre (par exemple) du théorème de Lax-Milgram, et en écrivant le problème sous la forme

$$Au = f, \qquad u \in V, \qquad f \in V' \tag{1.1}$$

l'opérateur A définit un isomorphisme entre l'espace V (dit "d'énergie") et l'espace des données V'.

Il peut arriver que l'espace V ne soit pas contenu dans l'espace des distributions sur Ω, noté $\mathcal{D}'(\Omega)$. Alors, l'espace $\mathcal{D}(\Omega)$ des fonctions indéfiniment dérivables à support compact dans Ω ne sera pas contenu dans V'. En vertu de ce qui précède, il y a des données $f \in \mathcal{D}(\Omega)$ (aussi petites que l'on veut, ainsi que chacune de leurs dérivées) pour lesquelles il n'y a pas de solution dans V. C'est le phénomène dit de *sensitivité* [9],[10].

Dans le cas des problèmes elliptiques, la sensivité peut se présenter lorsque les conditions aux limites ne satisfont pas à la condition dite de *recouvrement* (ou de Shapiro-Lopatinskii) (cf. [8] dans le cas des équations ou [1] pour des systèmes). L'étude du problème sort alors du cadre classique des problèmes aux limites elliptiques.

Dans [9] nous avions démontré la sensitivité d'un problème modèle concernant l'équation biharmonique. Nous présentons ici un problème sensitif pour un système elliptique, qui apparaît comme limite d'un problème de perturbation singulière faisant intervenir un petit paramètre $\varepsilon \geq 0$. Pour $\varepsilon > 0$ le problème est elliptique classique mais la limite $\varepsilon = 0$ est sensitive. Il s'agit d'un modèle simplifié (à coefficients constants) des coques élastiques minces à courbure totale positive, encastrées par une partie Γ_0 du bord et libres sur le reste Γ_1, dans le cadre du modèle de Koiter [11]. Le paramètre ε est proportionnel au carré du rapport de l'épaisseur de la coque à une longueur caractéristique de sa surface moyenne. Le problème limite $\varepsilon = 0$ constitue l'*approximation membranaire*. Il est connu que dans le cadre de l'approximation membranaire, les conditions sur bord libre ne satisfont pas à la condition de recouvrement, on peut alors construire des suites de Weyl montrant que 0 appartient au spectre de l'opérateur A, tout en n'étant pas une valeur propre (cf. [5], p. 258-261). Par contre, si tout le bord est encastré ou fixé, ce qui implique dans le cadre de l'approximation membranaire que les composantes tangentielles du déplacement s'annulent, la condition de recouvrement est satisfaite (cf. [3] par exemple). On trouvera aussi dans [12] une description des différents comportements $\varepsilon \searrow 0$ dans les coques.

Comme nous avons dit, le problème limite est sensitif, si bien qu'on ne sait pas étudier l'équation (1.1) correspondante avec un second membre f "quelconque". En fait nous étudions le problème de perturbation singulière avec des conditions de Dirichlet non homogènes sur Γ_0 (ce qui correspond à un "encastrement forcé") et forces nulles. Le problème se ramène alors un autre avec $f \in V'$. Nous montrons sur un exemple l'allure des solutions, démontrant qu'elles ne sont pas des distributions même pour certaines données régulières.

Le phénomène de sensitivité peut être interprété comme une instabilité : les coques minces à courbure totale positive, encastrées par une partie du bord et libres par le reste "chancellent", à la différence de celles encastrées par tout le bord ("voûtes"), qui sont "*fermes*". Par ailleurs cette instabilité sous-tend des difficultés pratiquement rédhibitoires pour le calcul numérique déjà au niveau de l'implémentation, forcément approchée, des données du problème.

Les notations sont classiques.

On utilise la convention de sommation des indices répétés.

c et C désignent des constantes diverses, indépendantes de ε, pouvant varier d'une formule à une autre.

Les variables soulignées désignent des "vecteurs" à trois composantes (parfois deux) :

$$\underline{u} = (u_1, u_2, u_3)$$

et de même

$$\mathcal{D}(\underline{\Omega}) = (\mathcal{D}(\Omega))^3$$

Parfois on utilisera une flèche pour désigner des "vecteurs à quatre composantes"

$$\vec{\varphi} = (\varphi_1, \varphi_2, \varphi_3, \varphi_4)$$

On considère deux variables d'espace, x_1, x_2 et l'on note

$$\begin{aligned} \partial_\alpha &= \partial/\partial x_\alpha \quad , \quad \alpha = 1,2 \\ \partial_n &= \text{ dérivée normale} \end{aligned}$$

2. Les problemes P^ϵ et P^0

Soit Ω un ouvert borné et connnexe de $\mathbb{R}^2$, de frontière

$$\partial\Omega = \Gamma_0 \cup \Gamma_1, \quad \Gamma_0 \text{ et } \Gamma_1 \text{ de mesure non nulle.} \tag{2.1}$$

Soit

$$\underline{u} = (u_1, u_2, u_3) \tag{2.2}$$

le "vecteur déplacement". On définit les composantes $\gamma_{\alpha\beta}$ du "tenseur déformation" par :

$$\begin{cases} \gamma_{11}(\underline{u}) &= \partial_1 u_1 - u_3 \\ \gamma_{22}(\underline{u}) &= \partial_2 u_2 - u_3 \\ \gamma_{12}(\underline{u}) &= \gamma_{21}(\underline{u}) = \dfrac{1}{2}(\partial_2 u_1 + \partial_1 u_2) \end{cases} \tag{2.3}$$

Remarque 2.1. Les coefficients de u_3 dans les expressions (2.3) ont été pris égaux à $(-1, -1, 0)$ pour simplifier les calculs explicites de la section 4. On peut prendre en général $(-b_{11}, -b_{22}, 0)$ avec b_{11} et b_{22} positifs ; ce sont les coefficients de la deuxième forme fondamentale d'une surface à courbure totale positive ("elliptique") en prenant comme courbes coordonnées les lignes de courbure.

Soit $a^{\alpha\beta\lambda\mu}$ et $b^{\alpha\beta\lambda\mu}$ deux systèmes de coefficients satisfaisant aux propriétés de symétrie et positivité analogues à celles des coefficients d'élasticité :

$$a^{\alpha\beta\lambda\mu} = a^{\beta\alpha\lambda\mu} = a^{\lambda\mu\alpha\beta} \tag{2.4}$$

$$a^{\alpha\beta\lambda\mu}\eta_{\lambda\mu}\eta_{\alpha\beta} \geq C \mid \eta \mid^2 \quad \forall\ \eta_{\alpha\beta} \text{ symétrique} \tag{2.5}$$

(et de même pour $b^{\alpha\beta\lambda\mu}$).

On définit l'espace :

$$V = \{\underline{v} \in H^1(\Omega) \times H^1(\Omega) \times H^2(\Omega) \; ; \; v_1 = v_2 = v_3 = \partial_n v_3 = 0 \text{ sur } \Gamma_0\} \tag{2.6}$$

et les formes bilinéaires et continues sur V :

$$a(\underline{u}, \underline{v}) = \int_\Omega a^{\alpha\beta\lambda\mu} \gamma_{\lambda\mu}(\underline{u}) \gamma_{\alpha\beta}(\underline{v}) dx \tag{2.7}$$

$$b(\underline{u}, \underline{v}) = \int_\Omega b^{\alpha\beta\lambda\mu} (\partial_\lambda \partial_\mu u_3)(\partial_\alpha \partial_\beta v_3) dx \tag{2.8}$$

Lemme 2.2. *Il existe $c > 0$ telle que*

$$a(\underline{v}, \underline{v}) + b(\underline{v}, \underline{v}) \geq c \parallel \underline{v} \parallel_V^2 \quad \forall \underline{v} \in V \tag{2.9}$$

Ce lemme est démontré dans le cas général des coques (avec termes d'ordre inférieur et coefficients variables) dans [2].

Démonstration du Lemme 2.2. Par application réitérée de l'inégalité de Poincaré, compte tenu des conditions aux limites de (2.6), on a :

$$b(\underline{v}, \underline{v}) \geq c \parallel v_3 \parallel_{H^2(\Omega)}^2 . \tag{2.10}$$

En majorant les produits croisés :

$$2 \left| (\partial_1 u_1) u_3 \right| \leq \eta (\partial_1 u_1)^2 + \eta^{-1} (u_3)^2$$

avec $\eta < 1$ mais proche de 1, on a :

$$|\partial_1 u_1 - u_3| + C \, |u_3|^2 \geq c \left[(\partial_1 u_1)^2 + | \, u_3 \, |^2 \right] ,$$

si bien que

$$a(\underline{v}, \underline{v}) + b(\underline{v}, \underline{v}) \geq$$
$$c(\|v_3\|_{H^2}^2 + \|\partial_1 u_1\|_{L^2}^2 + \|\partial_2 u_2\|_{L^2}^2 + \|\partial_2 u_1 + \partial_1 u_2\|_{L^2}^2$$

et (2.9) résulte de l'inégalité de Korn [4] pour (v_1, v_2).

Lemme 2.3. *$a(\underline{v}, \underline{v})^{1/2}$ est une norme sur V.*

Démonstration. Il suffit de démontrer que $\underline{v} \in V$ et

$$\left\{ \begin{array}{lll} \gamma_{11}(\underline{v}) & \equiv \partial_1 v_1 - v_3 & = 0 \\ \gamma_{22}(\underline{v}) & \equiv \partial_2 v_2 - v_3 & = 0 \\ 2\gamma_{12}(\underline{v}) & \equiv \partial_2 v_1 + \partial_1 v_2 & = 0 \end{array} \right. \tag{2.11}$$

entraînent $\underline{v} = 0$. En éliminant v_3 :

$$\begin{cases} \partial_1 v_1 - \partial_2 v_2 = 0 \\ \partial_2 v_1 + \partial_1 v_2 = 0 \end{cases} \tag{2.12}$$

Prolongeons v_1 et v_2 avec valeurs nulles dans un petit domaine D au-delà de Γ_0. A cause des conditions aux limites de (2.6), les fonctions prolongées (dénotées par $\tilde{v}_\alpha$) satisfont à (2.12) dans le domaine prolongé $\tilde{\Omega}$. Mais de (2.12) on déduit que $\Delta\tilde{v}_\alpha = 0$ dans $\tilde{\Omega}$, et $\tilde{v}_\alpha = 0$ sur D, donc $\tilde{v}_\alpha = 0$, $\alpha = 1, 2$; ensuite $v_3 = 0$ découle de (2.11).

Nous construisons alors l'espace de Hilbert V^a, complété de V pour la norme $a(\underline{v}, \underline{v})^{1/2}$.

Pour définir des données de Dirichlet non nulles sur Γ_0, on se donne

$$\vec{\varphi} = (\varphi_1, \varphi_2, \varphi_3, \varphi_4) \in \left\{H^{1/2}(\Gamma_0) \times H^{1/2}(\Gamma_0) \times H^{3/2}(\Gamma_0) \times H^{1/2}(\Gamma_0)\right\} \tag{2.13}$$

et soit

$$\Phi = \{\Phi_1, \Phi_2, \Phi_3\} \in \left\{H^1(\Omega) \times H^1(\Omega) \times H^2(\Omega)\right\} \tag{2.14}$$

un relèvement de $\vec{\varphi}$, c'est-à-dire que

$$\Phi_j|_{\Gamma_0} = \varphi_j, \quad j = 1, 2, 3, \qquad \partial_n \Phi_3|_{\Gamma_0} = \varphi_4. \tag{2.15}$$

On définit alors pour $\varepsilon > 0$:

Problème. P^ε : *Trouver $\underline{u}^\varepsilon \in V$ tel que, $\forall \underline{v} \in V$:*

$$a(\underline{u}^\varepsilon, \underline{v}) + \varepsilon b(\underline{u}^\varepsilon, \underline{v}) = -a(\underline{\Phi}, \underline{v}) - \varepsilon b(\underline{\Phi}, \underline{v}) \tag{2.16}$$

(naturellement $\underline{\Phi}$ n'est pas un élément de V les expressions $a(\underline{\Phi}, \underline{v})$ et $b(\underline{\Phi}, \underline{v})$ désignent les intégrales analogues à (2.7) et (2.8)).

Pour $\varepsilon = 0$ nous définissons le *problème limite :*

Problème. P^0 : *Trouver $\underline{u}^0 \in V^a$ tel que $\forall$ $\underline{v} \in V^a$*

$$a(\underline{u}^0, \underline{v}) = -a(\underline{\Phi}, \underline{v}) \tag{2.17}$$

On démontre aisément (en utilisant en particulier le lemme 2.2) que les premiers membres de (2.16) et (2.17) sont coercifs sur V et V^a respectivement. De même, les seconds membres sont des fonctionnelles continues sur V et V^a respectivement. En vertu du théorème de Lax-Milgram on a :

Proposition 2.4 *La solution. u^ε de P^ε pour $\varepsilon > 0$ fixé, ainsi que $\underline{u}^0$, solution de $\underline{P}^0$ existent et sont uniques.*

Le problème P^ε ainsi que sa limite $\underline{P}^0$ ont été formulés en terme de $\underline{u}^\varepsilon$ et $\underline{u}^0$ de façon à faire intervenir des espaces vectoriels. Les inconnues du "problème physique de coques" sont en fait

$$\begin{cases} \underline{U}^\varepsilon = \underline{\Phi} + \underline{u}^\varepsilon \in H^1 \times H^1 \times H^2 + V \\ \underline{U}^0 = \underline{\Phi} + \underline{u}^0 \in H^1 \times H^1 \times H^2 + V^a. \end{cases} \tag{2.18}$$

Clairement $\underline{U}^\varepsilon$ satisfait à

$$U_j^\varepsilon|_{\Gamma_0} = \varphi_j \quad j = 1,2,3, \quad \partial_n U_3^\varepsilon|_{\Gamma_0} = \varphi_4 \tag{2.19}$$

au sens des traces.

En ce qui concerne $\underline{u}^0$ et $\underline{U}^0$, la situation est plus complexe, car V^a n'est pas un espace classique, comme nous verrons plus tard. Il est néanmoins immédiat de vérifier que V^a contient les éléments de la forme $(0,0,\psi)$ avec $\psi \in L^2(\Omega)$. Cela nous permet de donner une définition *équivalente* du problème limite, faisant intervenir

$$\widehat{\underline{\Phi}} = (\Phi_1, \Phi_2, 0) \in H^1(\Omega) \times H^1(\Omega) \times \{0\} \tag{2.20}$$

au lieu de (2.14). Définissons :

Problème $\hat{P}^0$ - *Trouver* $\widehat{\underline{u}}^0 \in V^a$ *tel que* $\forall \underline{v} \in V^a$,

$$a(\widehat{\underline{u}}^0, \underline{v}) = -a(\widehat{\underline{\Phi}}, \underline{v}) \tag{2.21}$$

qui définit aussi $\widehat{\underline{u}}^0$ *de façon unique. Comme* $(0,0,\Phi_3) \in V^a$, *on a*

$$\underline{U}^0 = \underline{\Phi} + \underline{u}^0 = \widehat{\underline{\Phi}} + \widehat{\underline{u}}^0 \in H^1 \times H^1 \times \{0\} + V^a \tag{2.22}$$

si bien que le "problème physique" pour U^0 *peut se formuler avantageusement sous la forme* $\hat{P}^0$, *qui est indépendante de* Φ_3, *c'est-à-dire des données* φ_3 *et* φ_4.

La "formulation classique" en terme d'équation et conditions aux limites des problèmes P^0 ou $\hat{P}^0$, qui est formelle à cause du caractère non classique de V^a, sera donnée dans la section 6. On y trouvera également une démonstration du fait que les conditions aux limites sur Γ_1 ne satisfont pas à la condition de recouvrement.

Remarque 2.5. Il est clair de (2.17) que la solution $\underline{u}^0$ dépend continûment de $\vec{\varphi}$ dans les topologies de (2.13). Aussi l'application

$$\vec{\varphi} \to \underline{U}^0 = \underline{\Phi} + \underline{u}^0 \tag{2.23}$$

est continue dans les espaces :

$$\begin{aligned} H^{1/2}(\Gamma_0) \times H^{1/2}(\Gamma_0) \times H^{3/2}(\Gamma_0) \times H^{1/2}(\Gamma_0) \to \\ \to H^1(\Omega) \times H^1(\Omega) + H^2(\Omega) + V^a \end{aligned} \tag{2.24}$$

Cette relation peut être simplifiée en vertu de (2.22), puisque $\underline{U}^0$ dépend seulement de φ_1 et φ_2 ; l'application

$$(\varphi_1, \varphi_2) \to \underline{U}^0 = \widehat{\underline{\Phi}} + \widehat{\underline{u}}^0 \tag{2.25}$$

est continue dans les espaces :

$$H^{1/2}(\Gamma_0) \times H^{1/2}(\Gamma_0) \to H^1(\Omega) \times H^1(\Omega) \times \{0\} + V^a \,. \tag{2.26}$$

3. La Perturbation Singulière $\varepsilon \searrow 0$

Théorème 3.1. *Soit $\underline{u}^\varepsilon$ et $\underline{u}^0$ les solutions de $\underline{P}^\varepsilon$ et $\underline{P}^0$ respectivement. Alors,*

$$\underline{u}^\varepsilon \to u^0 \quad \text{dans } V^a \text{ fort } (\varepsilon \searrow 0). \tag{3.1}$$

Ce théorème entre dans le cadre classique des perturbations singulières [6-7]. Nous le démontrons ici pour plus de clarté.

Démonstration du théorème 3.1. En prenant $\underline{v} = \underline{u}^\varepsilon$ dans (2.16) :

$$a(\underline{u}^\varepsilon, \underline{u}^\varepsilon) + \varepsilon b(\underline{u}^\varepsilon, \underline{u}^\varepsilon) \leq C\,(\| \underline{u}^\varepsilon \|_{V^a} + \varepsilon \| \underline{u}^\varepsilon \|_V)\,. \tag{3.2}$$

en minorant le premier membre en vertu de (2.9) on obtient :

$$\varepsilon \| \underline{u}^\varepsilon \|_V^2 + \frac{1}{2} \| \underline{u}^\varepsilon \|_{V^a}^2 \leq C \| \underline{u}^\varepsilon \|_{V^a} + \frac{\varepsilon}{2} \| \underline{u}^\varepsilon \|_V^2 + \varepsilon C^2/2 \tag{3.3}$$

d'où, pour ε suffisamment petit :

$$\| u^\varepsilon \|_{V^a} \leq C \tag{3.4}$$

$$\| u^\varepsilon \|_V \leq C\varepsilon^{-1/2} \tag{3.5}$$

Alors par extraction d'une sous-suite (mais on verra que toutes les sous-suites ont même limite, si bien qu'il s'agit de la suite entière) :

$$\underline{u}^\varepsilon \to \underline{u}^* \quad \text{dans } V^a \text{ faible.} \tag{3.6}$$

Fixons $\underline{v} \in V$ dans (2.16) et faisons tendre ε vers 0. En vertu de (3.5) et (3.6) on a

$$\begin{array}{rcl} a(\underline{u}^\varepsilon, \underline{v}) & \to & a(\underline{u}^*, \underline{v}) \\ | \, \varepsilon b(\underline{u}^\varepsilon, \underline{v}) \, | & \leq & C\varepsilon^{1/2} \to 0 \end{array}$$

si bien que

$$a(\underline{u}^*, \underline{v}) = -a(\underline{\Phi}, \underline{v}). \tag{3.7}$$

Or, V est dense dans V^a, si bien que l'on peut prendre $\underline{v} \in V^a$ dans (3.7), et en comparant avec (2.17) on a $\underline{u}^* = \underline{u}^0$. Par conséquent, (3.6) est la convergence cherchée, mais dans la topologie faible. Pour prouver la convergence forte, en utilisant (2.16) et (2.17) avec $\underline{v} = \underline{u}^\varepsilon$, il vient :

$$\begin{aligned} &\varepsilon b(\underline{u}^\varepsilon, \underline{u}^\varepsilon) + a(\underline{u}^\varepsilon - \underline{u}^0, \underline{u}^\varepsilon - \underline{u}^0) = \\ &= a(\underline{\Phi}, u^\varepsilon) - \varepsilon b(\underline{\Phi}, \underline{u}^\varepsilon) - a(\underline{\Phi}, \underline{u}^0). \end{aligned} \tag{3.8}$$

et en faisant $\varepsilon \searrow 0$, le second membre tend vers zero en vertu de (3.5) et (3.6), donc

$$\| \underline{u}^\varepsilon - \underline{u}^0 \|^2_{V^a} \rightarrow 0.$$

4. Exemples de solutions des problèmes limites $\underline{P}^0$ ou $\widehat{\underline{P}}^0$: Exemples qui né sont pas des distributions

Par un calcul explicite, on peut travailler directement avec $\underline{U}^0$, que nous écrivons dans cette section $\underline{U}$, aucune ambiguité n'étant à craindre. Considérons le problème $\widehat{\underline{P}}^0$. Admettons que pour une donnée "régulière"

$$(\varphi_1, \varphi_2) \in (H^{1/2}(\Gamma_0))^2 \tag{4.1}$$

on puisse trouver $\underline{U} = (U_1, U_2, U_3)$ fonctions "régulières" satisfaisant à

$$U_1|_{\Gamma_0} = \varphi_1, \qquad U_2|_{\Gamma_0} = \varphi_2 \tag{4.2}$$

et

$$\gamma_{\alpha,\beta}(\underline{U}) = 0 \qquad \alpha, \beta = 1, 2. \tag{4.3}$$

Il est clair que $\widehat{\underline{u}}^0 = \underline{U} - \widehat{\underline{\Phi}}$ est la solution de (2.21) et $\underline{U}$ elle-même est la "solution physique" (voir (2.22)).

En plus, si l'on connaît une famille de solutions dans le cadre qui vient d'être décrit, on pourra (dans certains cas) passer à la limite dans les topologies indiquées dans (2.25), (2.26), obtenant ainsi des solutions dans des cas plus généraux.

Considérons comme exemple le cas où

$$\Omega = (0, 1) \times 2\pi - Tore \tag{4.4}$$

où, si l'on veut, sur la bande $(0,1) \times \mathbb{R}$, en imposant des conditions de 2π-périodicité en x_2. Soit

$$\begin{cases} \Gamma_0 = \{0\} \times 2\pi - Tore \\ \Gamma_1 = \{1\} \times 2\pi - Tore \end{cases} \tag{4.5}$$

Ecrivons le système (4.3) :

$$\begin{cases} \partial_1 U_1 - U_3 = 0 \\ \partial_2 U_2 - U_3 = 0 \\ \partial_2 U_1 + \partial_1 U_2 = 0 \end{cases} \tag{4.6}$$

Par élimination on voit que U_1 et U_2 sont des fonctions harmoniques. On en déduit facilement des solutions en variables séparées. En se donnant

$$\begin{cases} \underline{\varphi}^n = (\varphi_1^n, \varphi_2^n) \qquad \text{avec} \\ \varphi_1^n = cos\ nx_2 \ ; \qquad \varphi_2^n = 0 \end{cases} \tag{4.7}$$

on a les solutions : $\underline{U}^n = (U_1^n, U_2^n, U_3^n)$ avec

$$\begin{cases} U_1^n = cos\ nx_2 . ch\ nx_1 \\ U_2^n = sin\ nx_2 . sh\ nx_1 \\ U_3^n = n\ cos\ nx_2 . sh\ nx_1 \end{cases} \tag{4.8}$$

Clairement, pour la donnée (série formelle)

$$\underline{\varphi} = \sum_{n=1}^{\infty} \alpha_n \underline{\varphi}^n \tag{4.9}$$

on a la solution (série formelle) :

$$\underline{U} = \sum_{n=1}^{\infty} \alpha_n \underline{U}^n. \tag{4.10}$$

Naturellement la série converge dans L^2 (resp. H^1) si

$$\sum_{n=1}^{\infty} \alpha_n^2 < \infty \qquad (resp. \sum_{n=1}^{\infty} n^2 \alpha_n^2 < \infty) \tag{4.11}$$

et l'on déduit facilement de la théorie des espaces d'interpolation [8] qu'elle converge dans $H^{1/2}$ si

$$\sum_{n=1}^{\infty} n\ \alpha_n^2 < \infty. \tag{4.12}$$

On déduit alors des considérations au début de cette section le résultat suivant :

Théorème 4.1. *Si les coefficients α_n satisfont à (4.12), la série $\underline{\varphi}$ de (4.9) converge dans $(H^{1/2}(2\pi - Tore))^2$, et $\underline{U}$ définie par (4.10) converge dans* $(H^1(\Omega))^2 \times \{0\} + V^a$ et $\widehat{\underline{u}} = \underline{U} - \widehat{\underline{\Phi}}^0$ *est alors la solution de* $\hat{P}_0$.

Naturellement, on peut aussi dire que $\underline{U}$ est la "solution physique" correspondante. On a alors :

Théorème 4.2. *L'espace des solutions $\underline{U}$ lorsque $\underline{\varphi}$ parcourt $(H^{1/2}(2\pi - Tore))^2$ contient des élements qui ne sont pas des distributions sur Ω.*

Corollaire 4.3. On déduit (voir en particulier (2.25), (2.26)) du théorème 4.2 que V^a n'est pas contenu dans $\mathcal{D}(\Omega)'$.

Démonstration du théorème 4.2. Construisons une fonction $\underline{\theta} \in \underline{\mathcal{D}(\Omega)}$ de la façon suivante :

$$\theta = (\theta_1, \theta_2, \theta_3)\,, \qquad \theta_2 = \theta_3 = 0 \tag{4.13}$$

$$\theta_1(x_1, x_2) = \zeta(x_1) \sum_{n=1}^{\infty} e^{-\gamma n} cos\ n x_2 \tag{4.14}$$

où $\gamma > 0$ est une constante indéterminée pour l'instant, et ζ est une fonction positive de $\mathcal{D}(0,1)$ qui s'annule pour x_1 inférieur à une certaine constante β. On vérifie sans peine en utilisant des expressions analogues à (4.11) que la série de (4.14) représente une fonction de $C^\infty(2\pi - Tore)$.

Pour prouver le théorème par l'absurde, il suffit de montrer qu'avec la constante γ convenablement choisie, l'expression

$$\langle \Sigma\ \alpha_n \underline{U}^n, \theta \rangle_{\mathcal{D}'(\Omega), \mathcal{D}(\Omega)} \tag{4.15}$$

diverge pour certaines suites α_n satisfaisant à (4.12). Calculons :

$$\langle \underline{U}^n, \theta \rangle = \left\langle U_1^n, \zeta(x_1) \sum_1^{\infty}\ e^{-\gamma k} cos\ k x_2 \right\rangle =$$

$$= \pi \int_\alpha^1 ch\ n x_1 . \zeta(x_1) e^{-\gamma n} dx_1 \geq \pi \int_\beta^1 ch\ n\beta\ e^{-\gamma n} \zeta(x_1) dx_1$$

et en choisissant par exemple, $\gamma < \beta/2$, il vient :

$$\langle \underline{U}^n, \theta \rangle \geq \frac{\pi}{2}\ e^{n\beta/2} \int_0^\ell \zeta(x_1) dx_1 = c\ e^{n\beta/2}.$$

Alors,

$$\Sigma\ \alpha_n \langle U^n, \theta\rangle \geq C \sum_1^{\infty} \alpha_n\ e^{n\delta/2}$$

qui diverge manifestement pour certaines suites α_n satisfaisant à (4.12).

Remarque 4.4. On obtient facilement des variantes du théorème 4.2 en changeant la condition (4.12) pour faire intervenir des données plus régulières, même de classe C^∞.

5. Commentaires sur la sensitivité et le calcul numerique

Dans le cadre de la remarque 2.5, en prenant comme données $\underline{\varphi}, = (\varphi_1, \varphi_2)$, la solution $\underline{U}^0$ dépend continûment de $\underline{\varphi}$ dans les topologies des espaces cités dans (2.26). Mais la topologie de V^a étant extrêmement grossière, la "forme" de la solution peut être très modifiée par de petites variations des données. Par exemple, si φ_1, φ_2 subissent des variations $(\delta\varphi_1, \delta\varphi_2) \in H^{1/2}(\Gamma_0)$ aussi petites que l'on veut, on obtient une variaiton de la solution, $\delta\underline{U}^0$ qui n'est pas en général une distribution. On peut donner des exemples analogues à ceux de la démonstration du théorème 4.2 ou de la remarque 4.4. Les erreurs sont de la forme (4.10) avec des coefficients α_n satisfaisant à (4.12), et on voit de (4.8) que les coefficients de Fourier de n'importe quelle section $x_1 = cte. > 0$ croissent exponentiellement avec n.

Le commentaire qui précède prend toute sa force dans le cas du *calcul numérique par éléments finis.* L'espace V^a est obtenu par complétion de V, si bien qu'une discrétisation de V convient parfaitement à V^a. Il s'agit, bien entendu, de discrétisation de $H^1(\Omega) \times H^1(\Omega) \times H^2(\Omega)$ avec les conditions aux limites cinématiques spécifiées dans (2.13). Dans le cadre de la remarque 2.5 on peut simplifier en prenant une discrétisation de $H^1(\Omega) \times H^1(\Omega) \times L^2(\Omega)$ avec les conditions aux limites cinématiques de u_1, u_2 et en ignorant celles de u_3. Alors, en prenant (φ_1, φ_2) dans $(H^{1/2})^2$ les approximations Galerkin convergent dans $H^1 \times H^1 \times \{0\} + V^a$ fort. Néanmoins pour imposer les conditions aux limites

$$U_1 = \varphi_1, \quad U_2 = \varphi_2 \quad \text{sur } \Gamma_0 \tag{5.1}$$

il faudra nécessairement remplacer φ_1, φ_2 par des fonctions approchées. Si pour fixer les idées, φ_1, φ_2 sont de classe C^∞ et l'on utilise une discrétisation P_1, les fonctions effectivement imposées seront affines par morceaux appartenant donc à $H^1(\Gamma_0)$ mais non pas à $H^2(\Gamma_0)$. Cela implique que l'on introduit $\delta\varphi_1, \delta\varphi_2$ de classe H^1 mais non pas H^2. En se référant à la section 4 le développement de $\delta\underline{U}$ sera de la forme (4.10) avec

$$\Sigma\ n^2\alpha_n^2 < \infty \quad \text{mais} \quad \Sigma n^4\alpha_n^2 = \infty, \tag{5.2}$$

ce qui, compte tenu de (4.8), *disloque de façon drastique* la solution cherchée, même si la perturbation est petite. Ce qui précède concerne naturellement la solution exacte avec données modifiées ; on devra y ajouter l'erreur produite par l'approximation numérique.

6. Equations et conditions aux limites du problème limite: La condition de recouvrement n'est pas satisfaite sur Γ_1

Dans cette section nous donnons les équations et conditions aux limites *formelles* satisfaites par la solution $\underline{U} = \underline{\Phi} + \underline{u}^0$ de P^0. "Formelle" signifie ici "obtenues par des intégrations par parties formelles", qui ne sont nullement justifiées étant donnée la nature non classique de l'espace V^a.

Définissons les "tensions" ou contraintes :

$$T^{\alpha\beta}(\underline{u}) = a^{\alpha\beta\lambda\mu}\gamma_{\lambda\mu}(\underline{u}) \quad , \quad \alpha, \beta = 1, 2. \tag{6.1}$$

qui sont symétriques en α, β. On peut alors transformer la forme bilinéaire $a(\underline{u}, \underline{v})$ définie dans (2.3), (2.7) comme suit :

$$\begin{aligned} a(\underline{u}, \underline{v}) &= \int_\Omega T^{\alpha\beta}(\underline{u})\gamma_{\alpha\beta}(\underline{v})dx = \int_\Omega T^{\alpha\beta}\frac{1}{2}(\partial_\alpha v_\beta + \partial_\beta v_\alpha)dx - \\ &- \int_\Omega (T^{11} + T^{22})v_3 dx = \int_\Omega T^{\alpha\beta}\partial_\beta v_\alpha - \int_\Omega (T^{11} + T^{22})v_3 dx = \\ &= \int_{\partial\Omega} T^{\alpha\beta} n_\beta v_\alpha ds - \int_\Omega (\partial_\beta T^{\alpha\beta})v_\alpha dx - \int_\Omega (T^{11} + T^{22})v_3 dx \end{aligned}$$

On en déduit immédiatement :

Proposition 6.1. *Le problème. P^0, écrit en terme de $\underline{U} = \Phi + \underline{u}^0$, est formellement équivalent aux équations*

$$\partial_\beta T^{\alpha\beta}(\underline{U}) = 0, \quad \alpha = 1, 2 \quad \text{sur } \Omega \tag{6.2}$$

$$-T^{11}(\underline{U}) - T^{22}(\underline{U}) = 0 \quad \text{sur } \Omega \tag{6.3}$$

avec les conditions aux limites

$$U_\alpha = \varphi_\alpha, \quad \alpha = 1, 2, \quad \text{sur } \Gamma_0 \tag{6.4}$$

$$T^{\alpha\beta}(\underline{U})n_\beta = 0, \quad \alpha = 1, 2, \quad \text{sur } \Gamma_1 \tag{6.5}$$

Le caractère non classique de ce problème provient de la propriété suivante, que nous démontrons par souci de clarté :

Proposition 6.2. *Les conditions aux limites (6.5) sur Γ_1 ne satisfont pas à la condition de recouvrement pour le système (6.2), (6.3).*

Démonstration. Pour étudier la condition de recouvrement dans un point de Γ_1, on prend des coordonnées locales x_1 normale à Γ_1 vers Ω, x_2 tangente et l'on étudie des solutions dans le demi-plan $x_1 > 0$ du système et des conditions aux limites "gelés" au point considéré et en ne prenant en considération que les termes d'ordre principal. Dans le cas qui nous occupe , d'après les propriétés tensorielles de $a^{\alpha\beta\lambda\mu}$ et vectorielles de (u_1, u_2) le système dans les nouveaux axes a la même forme, avec u_1 et u_2 normal et tangent à Γ_1 (c'est-à-dire à $x_1 = 0$), les coefficients ayant changé mais satisfaisant toujours à (2.4), (2.5). Les termes sont tous principaux, avec les indices d'équations $s_1 = s_2 = 1$, $s_3 = 0$ et les indices d'inconnues $t_1 = t_2 = 1$, $t_3 = 0$.

On cherche alors des solutions du système (6.2), (6.3) dans le demi-plan $x_1 > 0$ de la forme

$$\underline{U}(x_1, x_2) = \underline{v} e^{i(\zeta x_1 + \xi x_2)} \tag{6.6}$$

où $\underline{v} \in \mathbf{C}^3$, $\xi \in \mathbb{R}\backslash\{0\}$, $\zeta \in \mathbf{C}$ avec $Im\zeta > 0$ (c'est-à-dire, des solutions sinusoïdales en x_2 et exponentiellement décroissantes vers l'intérieur du domaine). On impose en plus la condition aux limites (6.5) sur $x_1 = 0$. La condition de recouvrement est satisfaite si et seulement si la seule solution de ce type est $\underline{U} = 0$.

Pour prouver qu'il y a d'autres solutions, on commence par construire des fonctions $\underline{U}$ de la forme (6.6) satisfaisant à

$$\gamma_{\alpha\beta}(\underline{U}) = 0 \qquad \alpha, \beta = 1, 2 \tag{6.7}$$

qui devient :

$$\begin{cases} i\zeta v_1 - v_3 = 0 \\ i\xi v_2 - v_3 = 0 \\ i\xi v_1 + i\zeta v_2 = 0 \end{cases} \tag{6.8}$$

qui a bien des solutions non nulles, par exemple :

$$\underline{v} = (1, i, -1), \quad \xi = 1, \quad \zeta = i. \tag{6.9}$$

Une fois que l'on a des $\underline{U} \neq 0$ satisfaisant à (6.7), en les multipliant par $a^{\alpha\beta\lambda\mu}$ on voit qu'elles satisfont aussi à (6.2), (6.3) et (6.5) puisque tous les $T^{\alpha\beta}$ sont nuls en vertu de (6.1).

Bibliographie

[1] S. Agmon, A. Douglis and L. Nirenberg, "Estimates near the boundary for solutions of elliptic partial differential equations satisfying general boundary conditions.II", *Commun. Pure Appl. Math.*, **17**, p. 35–92 (1964).

[2] M. Bernadou et P.G. Ciarlet, "Sur l'ellipticité du modèle linéaire des coques de W.T. Koiter" in Computing methods in science and engineering, R. Glowinski, J.L. Lions editors, pp. 89-136, *Lecture Notes in Economics and Math. Systems*, **134**, Springer, (1976).

[3] P.G. Ciarlet et E. Sanchez-Palencia, "Un théorème d'existence et d'unicité pour les équations des coques membranaires", *Compt. Rend. Acad. Sci. Paris, série* I, **317**, p. 801–805 (1993).

[4] G. Duvaut et J.L. Lions, "Les inéquations en mécanique et en physique", Dunod, Paris (1972).

[5] A.L. Goldenveizer, V.B. Lidskii et P.E. Tovstik, "Vibrations libres des coques élastiques minces" (en russe), Nauka, Moscou (1979).

[6] D. Huet, "Phénomènes de perturbation singulière dans les problèmes aux limites", *Ann. Inst. Fourier*, **10**, p. 61–150 (1960).

[7] J.L. Lions, "Perturbations singulières dans les problèmes aux limites et en contrôle optimal", *Lectures Notes in Math.*, vol. 323, Springer, Berlin (1973).

[8] J.L. Lions et E. Magenes, "Problèmes aux limites non homogènes et applications", vol. I-III, Dunod, Paris (1968-70).

[9] J.L. Lions et E. Sanchez-Palencia, "Problèmes aux limites sensitifs", *Compt. Rend. Acad. Sci. Paris, série* I, **319**, p. 1021–1026 (1994).

[10] J.L. Lions et E. Sanchez-Palencia, publication systématique en préparation sur la sensitivité.

[11] F. Niordson, "Shell theory", North-Holland, Amsterdam (1985).

[12] E. Sanchez-Palencia, "Asymptotic and spectral properties of a class of singular-stiff problems", *Jour. Math. Pures Appl.*, **71**, p. 379–406 (1992).

J.-L. Lions
Collège de France
3 rue d'Ulm
75005 Paris
and
E. Sanchez-Palencia
Laboratoire de Modélisation en Mécanique
Université Paris 6 et CNRS
4 place Jussieu
75252 Paris

Contrôlabilite exacte frontière de l'équation des ondes en presence de singularités

Mary Teuw Niane

1. Notations et position du problème

Soit Ω un domaine polygonal non vide; on note $(S_i)_{0\leq i\leq N}$ l'ensemble des sommets de Ω auquel on a adjoint un ensemble fini non vide de points C placés sur les côtés. On note Γ_i (resp. Γ_N) le côté compris entre les points S_i et S_{i+1} (resp. entre S_N et S_0); on note τ_i (resp. ν_i) le vecteur unitaire tangent à Γ_i pointé vers S_{i+1} (resp. le vecteur normal unitaire extérieur à Ω sur le côté Γ_i); plus généralement, sauf en un nombre fini de points de la frontière, on note τ (resp. ν) le vecteur unitaire tangent (resp. normal) à Ω au point $x \in \partial\Omega$; enfin ω_i désigne l'angle interne à Ω au sommet S_i mesuré à partir de Γ_i. On subdivise la frontière de Ω en deux parties $\Gamma_D = \bigcup_{j\in J_D} \Gamma_j$ et $\Gamma_N = \bigcup_{j\in J_N} \Gamma_j$ avec $J_D \cup J_N = \{0, —, N\}$ et $J_D \neq \phi$ et $J_N \neq \phi$. On fait l'hypothèse:

$$\bar{\Gamma}_D \cap \bar{\Gamma}_N = \{S_j/j \in J_{DN}\} \neq \phi\,. \tag{H1}$$

Soit T un réel strictement positif, on note:

$$Q =]0,T[\times\Omega, [; \Gamma =]0,T[\times\partial\Omega, \Sigma_N =]0,T[\times\Gamma_N, \Sigma_D =]0,T[\times\Gamma_D\,.$$

Enfin si E est un espace vectoriel normé, on note E^* son dual topologique.

On considère le problème suivant:

Pour tout (y_0, y_1) dans un espace à déterminer, existe-t-il des contrôles u, v définis sur le bord tels que si y est la solution de

$$\begin{aligned} y'' - \Delta u = 0 \ \ &dans\ \mathbb{Q} \\ \gamma y = u \ \ &sur\ \Sigma_D \\ \gamma\frac{\partial y}{\partial\nu} = v \ \ &sur\ \Sigma_N \\ y(0) = y_0\,,\ y'(0) &= y_1 \end{aligned} \tag{E1}$$

alors

$$y(T) = y'(T) = 0\,.$$

L'étude de ce problème sous l'hypothèse (H1) a été faite par Pierre Grisvard [2] qui l'a résolu lorsqu'il existe un point $x_0 \in \mathbb{R}^2$ tel que en tout points $S_j \in \bar{\Gamma}_D \cap \bar{\Gamma}_N$, où le contrôle change de type, vérifie:

i) $\omega_j < \pi$

ou bien

ii) $\omega_j = \pi$ et $\nu_j \cdot (S_j - x_0) = 0$ et $\tau_j \cdot (S_j - x_0) \geq 0$ si $S_j \in S_D$ ou bien $\tau_j \cdot (S_j - x_0) \leq 0$ si $S_j \in S_N$.

Ces deux conditions imposent une géométrie à Ω et le choix de Σ_D et de Σ_N n'est pas libre. Ces conditions excluent qu'un changement de type de contrôle ait lieu en un point non convexe. Lorsque la frontière de Ω est régulière le choix de Σ_D et de Σ_N est fortement limité par la condition (ii).

M. Moussaoui [6] a, dans le même ordre d'idées, obtenu la contrôlabilité frontière avec des conditions géométriques moins restrictives. M.T. Niane–O. Seck [7], [8], ont établi sans condition géométrique la contrôlabilité exacte en ajoutant des contrôles internes à support sur de petits voisinages des points de $\bar{\Gamma}_N \cap \bar{\Gamma}_D$ pour des données initiales dans

$$L^2(\Omega) \times \left\{u \in H^1(\Omega)/\gamma u|_{\Gamma_D} = 0\right\}^* .$$

Dans ce travail, on montre que la méthode H.U.M. de J.L. Lions [5] permet de résoudre le problème lorsque les données initiales sont dans certains espaces avec des conditions géométriques moins restictives, en particulier lorsque Ω est un domaine *régulier*, on établit la contrôlabilité exacte frontière *sans condition géométrique.*

2. Enoncé du résultat principal

Soit $\varepsilon > 0$, tel que la boule euclidienne ouverte $B(S_i, \varepsilon)$ de centre S_i et de rayon ε vérifie pour tout $i \in \{0, \dots, N-1\}$:

$$B(S_i, \varepsilon) \cap B(S_{i+1}, \varepsilon) = \phi \text{ et } B(S_N, \varepsilon) \cap B(S_0, \varepsilon) = \phi .$$

On note S_D les points de $\bar{\Gamma}_D \cap \bar{\Gamma}_N$ en aval de Γ_D et S_N les points de $\bar{\Gamma}_D \cap \bar{\Gamma}_N$ en aval de Γ_N.

Si $S_i \in S_D$ on note S_i^ε le point de Γ_i défini par $\{S_i^\varepsilon\} = \Gamma_i \cap \partial B(S_i, \varepsilon)$; si $S_i \in S_N$, on note S_i^ε le point de Γ_{i-1} défini par $\{S_i^\varepsilon\} = \Gamma_{i-1} \cap \partial B(S_i, \varepsilon)$. On note Ω^ε le domaine polygonal Ω avec comme sommets

$$S \cup \left\{S_i^\varepsilon / S_i \in \bar{\Gamma}_D \cap \bar{\Gamma}_N\right\} ;$$

on définit

$$\bar{\Gamma}^\varepsilon = \bigcup_{S_i \in S_D} [S_i, S_i^\varepsilon] \cup \bigcup_{S_i \in S_N} [S_i^\varepsilon, S_i]$$
$$\Gamma_D^\varepsilon = \Gamma_D \cup \bar{\Gamma}^\varepsilon \text{ et } \Gamma_N^\varepsilon = \Gamma_N \setminus \bar{\Gamma}^\varepsilon .$$

On note

$$V = \{u \in H^1(\Omega)/\gamma\, u|_{\Gamma_D} = 0\}$$
$$V^\varepsilon = \{u \in H^1(\Omega)/\gamma\, u|_{\Gamma_D^\varepsilon} = 0\} .$$

On considère les deux opérateurs A et A^ε définis par

$$D(A) = \left\{u \in V/\exists\, f \in L^2(\Omega) \text{ et } \forall\, v \in V; \int_\Omega \nabla u \cdot \nabla v\, dx = \int_\Omega f v\, dx\right\}$$

avec $\forall\, u \in D(A)\ Au = -\Delta u$

$$D(A^\varepsilon) = \left\{u \in V^\varepsilon/\exists\, f \in L^2(\Omega) \text{ et } \forall\, v \in V^\varepsilon; \int_\Omega \nabla u \cdot \nabla v\, dx = \int_\Omega f v\, dx\right\}$$

avec $\forall\, u \in D(A^\varepsilon)\ \ A^\varepsilon u = -\Delta u$.

Soit $(v_0, v_1) \in V \times L^2(\Omega)$, on note v la solution de l'équation:

$$\begin{aligned} v'' - \Delta v = 0 &\text{ dans } Q \\ \gamma v = 0 &\text{ sur }]0,T[\times \Gamma_D \\ \gamma \frac{\partial v}{\partial \nu} = 0 &\text{ sur }]0,T[\times \Gamma_N \\ v(0) = v_0\ ,\ v'(0) &= v_1\ . \end{aligned} \tag{E2}$$

On fait l'hypothèse suivante:

$$\text{Si } i \in I_D \cup I_N \text{ alors } \omega_i < 2\pi \text{ et si } i \in I_{DN} \text{ alors } \pi \le \omega_i < 3\pi/2\ . \tag{H2}$$

On peut énoncer le résultat principal de ce travail:

Théorème. *Si Ω vérifie les hypothèses* (H1) *et* (H2) *alors il existe $T_0 > 0$ tel que pour tout $(y_0, y_1) \in L^2(\Omega) \times V^{\varepsilon^*}$, il existe $(v_0, v_1) \in V^\varepsilon \times L^2(\Omega)$ tel que si* *v est la solution de* (E2) *correspondante alors la solution y de*

$$\begin{aligned} &y'' - \Delta y = 0 \ \ dans\ \mathbb{Q} \\ &\gamma y = (x - x_0)\cdot \nu \frac{\partial v}{\partial \nu} \ \ sur\ \Sigma_D \\ &\gamma y = (x - x_0)\cdot \tau \frac{\partial v}{\partial \tau} - (1/2)(T-t)v' \ \ sur\ ]0,T[\times \bar{\Gamma}^{\epsilon} \\ &\gamma \frac{\partial y}{\partial \nu} = \frac{d}{dt}[(x-x_0)\cdot \nu \gamma v'] + \frac{\partial}{\partial \tau}\left[(x-x_0)\cdot \nu \frac{\partial v}{\partial \tau}\right] \ \ sur\ ]0,T[\times \Gamma_N^{\epsilon} \\ &y(0) = y_0, y'(0) = y_1 \ \ dans\ \Omega \end{aligned} \tag{E3}$$

vérifie $y(T) = 0$ *et* $y'(T) = 0$.

Remarque. Le contrôle est réparti sur tout Σ_N comme dans le cas *très régulier* (J.-L. Lions [5], Triggiani [10]) par contre sur Σ_D le support du contrôle est plus étendu que dans le cas *très régulier*.

3. Quelques formules d'intégration par parties

Soit $x = (x_1, x_2)$ un point de $\mathbb{R}^2$, on note $\|x\|$, la norme euclidienne de x et θ la mesure de l'angle défini par x et $0x_1$.

Soit $G_R = \{x \in \mathbb{R}^2 / 0 < \|x\| < R,\ 0 < \theta < \omega\}$. On pose

$$G_R^0 = \partial G_R \cap \{x \in \mathbb{R}^2/\theta = 0\};\ G_R^{\omega} = \partial G_R \cap \{x \in \mathbb{R}^2/\theta = \omega\}$$

et

$$C_R = \partial G_R \cap \{x \in \mathbb{R}^2/\|x\| = R\}\ .$$

On a: $\partial G_R = G_R^0 \cup G_R^{\omega} \cup C_R$. Soit B_0 (resp. B_1) l'opérateur non borné défini par $D(B_0) = \{u \in H_0^1(\Omega)/ -\Delta u \in L^2(\Omega)\}$ tel que $\forall\ u \in D(B_0)$, $B_0 u = -\Delta u$ (resp.

$$\Big\{u \in H^1(\Omega, \gamma u|_{G_R^0} = 0 / \ \exists\ f \in L^2(G_R)\ \forall\ v \in H^1(G_R), \gamma v|_{G_R^0} = 0 \Rightarrow$$
$$\int_{G_R} \nabla u \cdot \nabla v\, dx = \int_{G_R} f\, v\, dx\Big\}$$

tel que

$$\forall\ u \in D(B_1),\ B_1 u = -\Delta u\ .$$

Soients

$$u \in C\big(0,T,D(B_0)\big) \cap C^1\big(0,T,H_0^1(G_R)\big) \cap L^2(0,T,L^2(G_R))$$

et

$$v \in C\big(0,T,D(B_1)\big) \cap C^1\big(0,T,\{u \in H^1(G_R)/\gamma u|_{G_R^0} = 0\}\big) \cap L^2(0,T,L^2(G_R))$$

ayant leur support dans $B(0,R_0)$ avec $R_0 < R$ et vérifiant respectivement les équations suivants:

$$\begin{aligned} &\frac{\partial^2 u}{\partial t^2} - \Delta u = f \;\text{ dans }\;]0,T[\times G_R \\ &\gamma u = 0 \;\text{ sur }\;]0,T[\times \partial G_R \\ &u(0) = u_0 \;\text{ dans }\; G_R \\ &u'(0) = u_1 \;\text{ dans }\; G_R \end{aligned} \tag{E4}$$

$$\begin{aligned} &\frac{\partial^2 v}{\partial t^2} - \Delta v = g \\ &\gamma v = 0 \;\;]0,T[\times G_R^0 \\ &\frac{\partial v}{\partial \nu} = 0 \;\;]0,T[\times G_R^\omega \\ &v(0) = v_0 \\ &v(0) = v_1\,. \end{aligned} \tag{E5}$$

On a alors la formule d'intégration par parties préliminaire suivante:

Lemme 1. *Si $\pi \leq \omega < \frac{3\pi}{2}$, pour tout $x_0 \in \mathbb{R}^2$, on a*

$$\begin{aligned} &\int_{G_R} \big[u'(x-x_0)\cdot\nabla v + v'(x-x_0)\cdot\nabla u\big]\,dx\Big|_0^T \\ &+2\int_0^T\int_{G_R} u'v'\,dx\,dt - \int_0^T\int_{G_R^0}(x-x_0)\cdot\nu\frac{\partial u}{\partial\nu}\frac{\partial v}{\partial\nu}d\sigma\,dt \\ &-\int_0^T\int_{G_R^\omega}(x-x_0)\cdot\tau\frac{\partial u}{\partial\nu}\frac{\partial v}{\partial\tau}d\sigma\,dt \\ &=\int_0^T\int_{G_R}\big[f(x-x_0)\cdot\nabla v + g(x-x_0)\cdot\nabla v\big]\,dx\,dt\,. \end{aligned}$$

Preuve. Soit $\varepsilon > 0$ tel que $0 < \varepsilon < R_0$, on pose $G_{R\varepsilon} = G_R \backslash \bar{B}(0,\varepsilon)$,

$$G^0_{R\varepsilon} = \partial G_{R\varepsilon} \cap \{x \in \mathbb{R}^2/\theta = 0\}, G^\omega_{R\varepsilon} = \partial G_{R\varepsilon} \cap \{x \in \mathbb{R}^2/\theta = \omega\}$$

et $\gamma_\varepsilon = G_R \cap \partial\overline{B}(0,\varepsilon)$.

D'après P. Grisvard [ ], on a:

$$\begin{aligned} u &= u_R + c(u)\ \eta\ r^{\frac{\pi}{\omega}} \sin \frac{\pi}{\omega}\theta \\ v &= v_R + c(v)\ \eta\ r^{\frac{\pi}{2\omega}} \sin \frac{\pi}{2\omega}\theta \end{aligned}$$

où η est une fonction de troncature à support dans $B(0, R_0)$, $c(u)$ et $c(v)$ sont des constantes réelles et u_R et v_R appartiennent à $H^2(\Omega)$.

Dans $G_{R\varepsilon}$, u et v sont dans H^2, on peut donc appliquer les formules de Green. On a:

$$\begin{aligned} I_\varepsilon &= \int_0^T \int_{G_{R\varepsilon}} [f(x-x_0)\cdot\nabla v + g(x-x_0)\cdot\nabla u]\,dx\,dt \\ &= \int_{G_{R\varepsilon}} [u'(x-x_0)\cdot\nabla v + v'(x-x_0)\cdot\nabla u]\,dx\Big|_0^T \\ &- \int_0^T \int_{G_{R\varepsilon}} (x-x_0)\cdot\nabla[u'v']\,dx\,dt \\ &+ 2\int_0^T \int_{G_{R\varepsilon}} \nabla u\cdot\nabla v\,dx\,dt + \int_0^T \int_{G_{R\varepsilon}} (x-x_0)\cdot\nabla[\nabla u\cdot\nabla v]\,dx\,dt \\ &- \int_0^T \int_{\partial G_{R\varepsilon}} \left[\frac{\partial u}{\partial\nu}(x-x_0)\cdot\nabla v + \frac{\partial v}{\partial\nu}(x-x_0)\cdot\nabla u\right]\,dx\,dt \end{aligned}$$

d'où en intégrant à nouveau par parties et en regroupant les termes on obtient:

$$\begin{aligned} I_\varepsilon = & \int_{G_{R\varepsilon}} [u'(x-x_0)\cdot\nabla v + v'(x-x_0)\cdot\nabla u]\,dx\Big|_0^T \\ & + 2\int_0^T \int_{G_{R\varepsilon}} u'v'\,dx\,dt - \int_{\partial G_{R\varepsilon}} (x-x_0)\cdot\nu\,\gamma u'\,\gamma v'\,d\sigma\,dt \\ & + \int_0^T \int_{\partial G_{R\varepsilon}} [-\frac{\partial u}{\partial\nu}(x-x_0)\cdot\nabla v \frac{\partial v}{\partial\nu}(x-x_0)\cdot\nabla u \\ & + (x-x_0)\cdot\nu\nabla u\cdot\nabla v]\,d\sigma\,dt\,. \end{aligned}$$

Le passage à limite lorsque ε tend vers 0 dans les intégrales sur $G_{R\varepsilon}$ et sur $]0,T[\times G_{R\varepsilon}$ est trivial. Comme $\partial G_{R\varepsilon} = G^0_{R\varepsilon} \cup G^\omega_{R\varepsilon} \cup \gamma_\varepsilon$, on peut décomposer les intégrales sur le bord en trois parties et étudier chacune séparément:

$$\begin{aligned}J_{0\varepsilon} &= -\int_0^T \int_{G^0_{R\varepsilon}} (x-x_0)\cdot\nu\,\gamma u'\,\gamma v'\,d\sigma\,dt \\ &+ \int_0^T \int_{G^0_{R\varepsilon}} \Big[-\frac{\partial u}{\partial \nu}(x-x_0)\cdot\nabla v - \frac{\partial v}{\partial \nu}(x-x_0)\cdot\nabla u \\ &+ (x-x_0)\cdot\nu\nabla u\cdot\nabla v\Big]d\sigma\,dt \\ &= -\int_0^T \int_{G^0_{R\varepsilon}} (x-x_0)\cdot\nu\frac{\partial u}{\partial \nu}\frac{\partial v}{\partial \nu}d\sigma\,dt\end{aligned}$$

on a:

$$\frac{\partial u}{\partial \nu} = \frac{\partial u_R}{\partial \nu} - \frac{\pi}{\omega}c(u) - \frac{1}{r^{1-\frac{\pi}{\omega}}}\eta$$

où $\frac{\partial u_R}{\partial \nu} \in H^{1/2}(G^0_R)$ et $\frac{1}{r^{1-\frac{\pi}{\omega}}} \in L^p(G^0_R)$ pour $p < \frac{1}{1-\frac{\pi}{\omega}}$ d'où grâce aux injections de Sobolev $\frac{\partial u}{\partial \nu} \in L^p(G^0_R)$ pour tout $1 \le p < \frac{1}{1-\frac{\pi}{\omega}}$; on a aussi:

$$\frac{\partial v}{\partial \nu} = \frac{\partial v_R}{\partial \nu} - \frac{\pi}{2\omega}c(v)\frac{1}{r^{1-\frac{\pi}{2\omega}}}\eta$$

donce $\frac{\partial v}{\partial \nu} \in L^q(G^0_R)$ pour $1 \le q < \frac{1}{1-\frac{\pi}{2\omega}}$ si $\pi \le \omega < 3\pi/2$, il existe q_0 et p_0 tels que $1 < q_0 < \frac{1}{1-\frac{\pi}{2\omega}}$, $2 < p_0 < \frac{1}{1-\frac{\pi}{\omega}}$ et $\frac{1}{p_0} + \frac{1}{q_0} = 1$; en effet, on pose $q_0 = 1+\alpha$, avec $\alpha > 0$, on a $p = \frac{1+\alpha}{\alpha}$ donc $\frac{\omega-\pi}{\pi} < \alpha < \frac{\pi}{2\omega-\pi}$ ce qui est possible si $\pi \le \omega < \frac{3\pi}{2}$. Ainsi $\frac{\partial u}{\partial \nu}\frac{\partial v}{\partial \nu} \in L^1(G^0_R)$ donc

$$\lim_{\varepsilon\to 0} J_{0\varepsilon} = -\int_0^T \int_{G^0_R} (x-x_0)\cdot\nu\frac{\partial u}{\partial \nu}\frac{\partial v}{\partial \nu}d\sigma\,dt\,.$$

Si $J_{\omega\varepsilon}$ est l'intégrale prise sur $G^\omega_{R\varepsilon}$, on a

$$J_{\omega\varepsilon} = -\int_0^T \int_{G^\omega_{R\varepsilon}} (x-x_0)\cdot\tau\frac{\partial u}{\partial \nu}\frac{\partial v}{\partial \tau}d\sigma\,dt\,.$$

Le même raisonnement que précédement conduit à

$$\lim_{\varepsilon\to 0} J_{\omega\varepsilon} = -\int_0^T \int_{G^\omega_R} (x-x_0)\cdot\tau\frac{\partial u}{\partial \nu}\frac{\partial v}{\partial \tau}d\sigma\,dt\,.$$

Maintenant considérons l'intégrale prise sur γ_ε on a:

$$\begin{aligned} J_{\gamma_\varepsilon} = &- \int_0^T \int_{\gamma_\varepsilon} (x - x_0) \cdot \nu \, \gamma u' \, \gamma v' \, d\sigma \, dt \\ &- \int_0^T \int_{\gamma_\varepsilon} \frac{\partial u}{\partial \nu} (x - x_0) \cdot \nabla v \, d\sigma \, dt - \int_0^T \int_{\gamma_\varepsilon} \frac{\partial v}{\partial \nu} (x - x_0) \cdot \nabla u \, d\sigma \, dt \\ &+ \int_0^T \int_{\gamma_\varepsilon} (x - x_0) \cdot \nu \nabla u \cdot \nabla v \, d\sigma \, dt \, . \end{aligned}$$

Notons dans l'ordre $J^i_{\gamma_\varepsilon} i = 1$ à 4 chacune de ces intégrales. Comme $D(B_0) \cap$ vect $\{C^1(\bar{\Omega}); \eta r^{\frac{\pi}{\omega}} \sin \frac{\pi}{\omega} \theta\}$ est dense dans $D(B_0)$, (respectivement $D(B_1) \cap$ vect $\{C^1(\bar{\Omega}); \eta r^{\frac{\pi}{2\omega}} \sin \frac{\pi}{2\omega} \theta\}$ est dense dans $D(B_1)$), il suffit pour obtenir la limite de $J^i_{\gamma_\varepsilon}$ lorsque ε tend vers 0, de considérer l'intégrale prise sur le terme singulier. Ainsi, on a pour $J^1_{\gamma_\varepsilon}$ à considérer

$$\int_0^T \int_0^{2\pi} (x - x_0) \cdot \nu \, c'(u) c'(v) \varepsilon^{\frac{3\pi}{2\omega}} \sin \frac{\pi}{2\omega} \theta \sin \frac{\pi}{\omega} \theta \varepsilon \, d\theta \, dt$$

qui tend évidement vers 0.

Pour $J^2_{\gamma\varepsilon}$, on a:

$$\begin{aligned} \int_0^T \int_0^{2\pi} &\Big[-c(u) \frac{\pi}{\omega} \cos \frac{\pi}{\omega} \theta \varepsilon^{\frac{\pi}{\omega} - 1} (x - x_0) \cdot \tau \, c(v) \varepsilon^{\frac{\pi}{2\omega} - 1} \sin \frac{\pi}{2\omega} \theta \\ &+ c(u) \frac{\pi}{\omega} \cos \frac{\pi}{\omega} \theta \varepsilon^{\frac{\pi}{\omega} - 1} (x - x_0) \cdot \nu \, c(v) \varepsilon^{\frac{\pi}{2\omega} - 1} \frac{\pi}{2\omega} \cos \frac{\pi}{2\omega} \theta \Big] \varepsilon \, d\theta \, dt \, . \end{aligned}$$

Ce term est de l'ordre de $\varepsilon^{\frac{3\pi}{2\omega} - 1}$, comme $\omega < \frac{3\pi}{2}$, il tend vers 0. Il en est de même des autres intégrales $J^i_{\gamma_\varepsilon}$ pour $i = 3$ à 4.

Aussi $\lim_{\varepsilon \to 0} J_{\gamma\varepsilon} = 0$. Ce qui achève la preuve du lemme 1. □

Proposition 1. *Si* Ω *vérifie* (H1) *et* (H2), *si* $x_0 \in \mathbb{R}^2$ *alors pour tous*

$$f \in L^2 \left(0, T, D(A^{\varepsilon 1/2}) \right)$$

et

$$g \in L^2 \left(0, T, D(A^{1/2}) \right) \, ,$$

si u et v sont les solutions respectives de

$$\begin{aligned} &u'' - \Delta u = f \ \textit{dans } \mathbb{Q} \\ &\gamma u = 0 \ \textit{sur } \Gamma_D^\varepsilon \\ &\frac{\partial u}{\partial \nu} = 0 \ \textit{sur } \Gamma_N^\varepsilon \\ &u(0) - u_0 \, , \ u'(0) = u_1 \ \textit{dans } \Omega \, , \end{aligned} \tag{E6}$$

où $u_0 \in D(A^\varepsilon)$ et $u_1 \in V^\varepsilon$ et de

$$\begin{aligned} &v'' - \Delta v = f \ \textit{dans } \mathbb{Q} \\ &\gamma v = 0 \ \textit{sur } \Gamma_D \\ &\frac{\gamma v}{\gamma \nu} = 0 \ \textit{sur } \Gamma_N \\ &v(0) = v_0 \, , v(0) = v_1 \ \textit{dans } \Omega \, , \end{aligned} \tag{E7}$$

où $v_0 \in D(A)$ et $v_1 \in V$ alors

$$\begin{aligned} &\int_\Omega [u'(x - x_0) \cdot \nabla v + v'(x - x_0) \cdot \nabla u] dx \Big|_0^T \\ &\quad + 2 \int_0^T \int_\Omega u'v' - \int_0^T \int_{\Gamma_D} (x - x_0) \cdot \nu \frac{\partial u}{\partial \nu} \frac{\partial v}{\partial \nu} \, d\sigma \, dt \\ &\quad + \int_0^T \int_{\Gamma_N^\varepsilon} \left[-(x - x_0) \cdot \nu \, \gamma u' \, \gamma \nu' + (x - x_0) \cdot \nu \frac{\partial u}{\partial \tau} \frac{\partial v}{\partial \tau} \right] d\sigma \, dt \\ &\quad - \int_0^T \int_{\bar{\Gamma}^\varepsilon} (x - x_0) \cdot \tau \frac{\partial u}{\partial \nu} \frac{\partial v}{\partial \tau} d\sigma \, dt \\ &\quad = \int_0^T \int_\Omega [f(x - x_0) \cdot \nabla v + g(x - x_0) \cdot \nabla u] dx \, dt \, . \end{aligned}$$

Preuve. Il suffit de localiser la formule, dans ce cas, deux situations se présentent au voisinage du sommet considéré:

1°) cas u et v ne changent pas de conditions, alors les termes de bord sont bien définis et le calcul suit P. Grisvard [2].

2°) cas u ou bien v changent de conditions au bord alors on applique le lemme 1. Ce qui achève la preuve de la proposition 1. □

Pour u et v définis dans la proposition 1, on note

$$E(u, v, t) = \int_\Omega [u'v' + \nabla u \cdot \nabla v] dx \, .$$

On a la

Proposition 2. *Si Ω vérifie* (H1) *et* (H2), *on a:*

$$\begin{aligned}
&\int_\Omega \left[u'(x-x_0)\cdot\nabla v + v'(x-x_0)\cdot\nabla u\right]dx\Big|_0^T \\
&\quad + \frac{1}{2}T\,E(u,v,0) + \frac{1}{2}\int_0^T\int_0^t\int_\Omega (fv'+gu')dx\,ds\,dt \\
&\quad + \frac{1}{2}\int_0^T\int_{\bar\Gamma_\varepsilon}(T-t)\frac{\partial u}{\partial\nu}v'\,d\sigma\,dt \\
&\quad + \frac{1}{2}\int_\Omega v'u\,dx\Big|_0^T - \frac{1}{2}\int_0^T\int_\Omega gu\,dx\,dt \\
&\quad - \int_0^T\int_{\Gamma_N^\varepsilon}(x-x_0)\cdot\nu u'v'd\sigma\,dt - \int_0^T\int_{\Gamma_D}(x-x_0)\cdot\nu\frac{\partial u}{\partial\nu}\frac{\partial v}{\partial\nu}d\sigma\,dt \\
&\quad - \int_0^T\int_{\bar\Gamma^\varepsilon}(x-x_0)\cdot\tau\frac{\partial u}{\partial\nu}\frac{\partial v}{\partial\tau}d\sigma\,dt + \int_0^T\int_{\Gamma_N^\varepsilon}(x-x_0)\cdot\nu\frac{\partial u}{\partial\tau}\frac{\partial v}{\partial\tau}d\sigma\,dt \\
&= \int_0^T\int_\Omega\left[f(x-x_0)\cdot\nabla v + g(x-x_0)\cdot\nabla u\right]dx\,dt\,.
\end{aligned}$$

Preuve. On établit d'abord deux formules. On a par des intégrations par parties du même type que précédement le

Lemme 2.

$$E(u,v,t) = E(u,v,0) + \int_\Omega^T\int_\Omega(fv+gu')dx\,ds + \int_0^t\int_{\bar\Gamma^\varepsilon}\frac{\partial u}{\partial\nu}v'd\sigma\,ds\,.$$

De ce lemme, on déduit le

Lemme 3.

$$\begin{aligned}
\int_0^T E(u,v,t)\,dt &= TE(u,v,o) + \int_0^T\int_0^t\int_\Omega(fv'+gu')dx\,ds\,dt \\
&\quad + \int_0^T\int_{\bar\Gamma^\varepsilon}(T-t)\frac{\partial u}{\partial\nu}v'd\sigma\,dt\,.
\end{aligned}$$

□

Ensuite on remarque

$$\begin{aligned}
\int_0^T\int_\Omega u'v'dx\,dt &= \frac{1}{2}\int_0^T E(u,v,t)\,dt \\
&\quad + \frac{1}{2}\int_0^T\int_\Omega(u'v'-\nabla u\cdot\nabla v)dx\,dt
\end{aligned}$$

or

$$\int_0^T \int_\Omega (u'v' - \nabla u \cdot \nabla v) dx\, dt = \int_\Omega v'u\, dx\Big|_0^T - \int_0^T \int_\Omega gu\, dx\, dt \,.$$

On termine la preuve de la proposition 2 en reportant les formules précédentes dans celle de la proposition 1. □

4. Formulation faible du problème non homogène

Pour $(u_0, u_1, f) \in D(A^\varepsilon) \times V^\varepsilon \times L^2(0,T,D(A^{\varepsilon 1/2}))$ et $(v_0, v_1, g) \in D(A) \times V \times L^2(0,T,D(A^{1/2}))$ avec $g = 0$ en conservant les notations de la proposition 2, on pose:

$$\begin{aligned} a((u_0,u_1,f),(v_0,v_1)) = &\int_0^T \int_{\Gamma_N^\varepsilon} (x - x_0) \cdot \nu u'v' d\sigma\, dt \\ &+ \int_0^T \int_{\Gamma_D} (x - x_0) \cdot \nu \frac{\partial u}{\partial \nu} \frac{\partial v}{\partial \nu} d\sigma\, dt \\ &+ \int_0^T \int_{\bar{\Gamma}^\varepsilon} (x - x_0) \cdot \tau \frac{\partial u}{\partial \nu} \frac{\partial v}{\partial \tau} d\sigma\, dt \\ &- \int_0^T \int_{\Gamma_N^\varepsilon} (x - x_0) \cdot \nu \frac{\partial u}{\partial \tau} \frac{\partial v}{\partial \tau} d\sigma\, dt \\ &- (1/2) \int_0^T \int_{\bar{\Gamma}^\varepsilon} (T - t) \frac{\partial u}{\partial \nu} v' d\sigma\, dt \,. \end{aligned}$$

Sous les hypothèses précédentes a est bien définie et en plus on a:

$$\begin{aligned} \big|a\big((u_0,u_1,f),(v_0,v_1)\big)\big| \le K_T \big[E_0(u_0,u_1) + \|f\|^2_{L^1(0,T,L^2(\Omega))}\big]^{1/2} \\ \times \big[E_0(v_0,v_1)\big]^{1/2} \end{aligned}$$

où

$$E_0(h,k) = \frac{1}{2} \int_\Omega \left[k^2 + \|\nabla h\|^2\right] dx \,.$$

a est donc bilinéaire de

$$\left[D(A^\varepsilon) \times D(A^\varepsilon)^{1/2} \times L^1(0,T,L^2(\Omega))\right] \times \left[D(A) \times D(A^{1/2})\right]$$

et continue pour la norme produit des normes

$$\left[\|\ \|^2_{V^\varepsilon} + \|\ \|^2_{L^2(\Omega)} + \|\ \|^2_{L^1(0,T,L^2(\Omega))}\right]^{1/2} \text{ et } \left[\|\ \|^2_V + \|\ \|^2_{L^2(\Omega)}\right]^{1/2} .$$

Aussi a admet donc une unique extension bilinéaire continue sur

$$\left[V^\varepsilon \times L^2(\Omega) \times L^1(0,T,L^2(\Omega))\right] \times \left[V \times L^2(\Omega)\right] .$$

En appliquant la méthode de transposition de J.L. Lions [5], on obtient: Si $(v_0, v_1) \in V \times L^2(\Omega)$, il existe un unique triplet

$$(\psi_0, \psi_1, y) \in L^2(\Omega) \times V^{\varepsilon^*} \times L^\infty(0,T,L^2(\Omega))$$

vérifiant: pour tout $(u_0, u_1, f) \in V^\varepsilon \times L^2(\Omega) \times L^1(0,T,L^2(\Omega))$

$$\begin{aligned} &\langle \psi_0, u_1 \rangle_{L^2(\Omega),L^2(\Omega)} - \langle \psi_1, u_0 \rangle_{V^{\varepsilon^*},V^\varepsilon} + \int_0^T \int_\Omega y f \, dx \, dt = \\ &a\big((u_0, u_1, f), (v_0, v_1)\big) . \end{aligned}$$

On peut interpréter ce problème sous la forme

$$\begin{aligned} &y'' - \Delta y = 0 \text{ dans } Q \\ &\gamma y = (x - x_0) \cdot \nu \frac{\partial v}{\partial \nu} \text{ sur } \Sigma_D \\ &\gamma y = (x - x_0) \cdot \tau \frac{\partial v}{\partial \tau} - (1/2)(T - t) v' \text{ sur }]0,T[\times \bar{\Gamma}^\varepsilon \qquad \text{(E8)} \\ \gamma \frac{\partial y}{\partial \nu} &= \frac{d}{dt}\left[(x - x_0) \cdot \nu \gamma v'\right] + \frac{\partial}{\partial \tau}\left[(x - x_0) \cdot \nu \frac{\partial v}{\partial \tau}\right] \text{ sur }]0,T[\times \Gamma_N^\varepsilon \\ &y(0) = \psi_0,\ y'(0) = \psi_1,\ y(T) = 0 \text{ et } y'(T) = 0 \text{ dans } \Omega . \end{aligned}$$

En plus la norme de y dans $L^\infty(0,T,L^2(\Omega))$ est continue par rapport à la norme de (v_0, v_1) dans $V \times L^2(\Omega)$. □

Preuve du théorème

Soient $(u_0, u_1), (v_0, v_1) \in V^\varepsilon \times L^2(\Omega)$, on pose

$$b\big((u_0, u_1), (v_0, v_1)\big) = a\big((u_0, u_1, 0), (v_0, v_1)\big) .$$

Proposition 3. *b est une forme bilinéaire continue et coercitive sur $V^\varepsilon \times L^2(\Omega)$.*

Preuve. En effet b est bilinéaire continue d'après les propriétés de a. En reportant dans la proposition 2, le fait que $f = g = 0$ et en appliquant l'inégalité de Poincaré, on a:

$$b\big((v_0, v_1), (v_0, v_1)\big) \geq (T - T_0) E_0(v_0, v_1)$$

où T_0 est une constante indépendante de (v_0, v_1) et $\{E_0(v_0, v_1)\}^{1/2}$ est une norme équivalente à celle de $V^\varepsilon \times L^2(\Omega)$. □

Soit $(y_0, y_1) \in L^2(\Omega) \times V^{\varepsilon^*}$, on considère le problème variationnel suivant: Trouver $(v_0, v_1) \in V^\varepsilon \times L^2(\Omega)$ tel que pour tout $(u_0, u_1) \in V^\varepsilon \times L^2(\Omega)$, on ait

$$b\big((u_0, u_1), (v_0, v_1)\big) - \langle u_1, y_0\rangle_{L^2(\Omega), L^2(\Omega)} + \langle u_0, y_1\rangle_{V^\varepsilon, V^{\varepsilon*}} = 0\,.$$

Soit

$$l\big(u_0, u_1\big) = -\langle u_1, y_0\rangle_{L^2(\Omega), L^2(\Omega)} + \langle u_0, y_1\rangle_{V^\varepsilon, V^{\varepsilon*}}$$

est bien définie et continue sur $V^\varepsilon \times L^2(\Omega)$.

Comme b est coercitive ce problème admet une unique solution

$$\big(v_0, v_1\big) \in V^\varepsilon \times L^2(\Omega)\,.$$

Or pour $\big(v_0, v_1\big)$ appartenant à $V^\varepsilon \times L^2(\Omega)$ qui est contenu dans $V \times L^2(\Omega)$, la solution $\big(\psi_0, \psi_1, y\big)$ de l'équation des ondes non homogène (E8) vérifie:

$$\begin{aligned}&\langle u_1, \psi_0\rangle_{L^2(\Omega), L^2(\Omega)} - \langle u_0, \psi_1\rangle_{V^\varepsilon V^{\varepsilon*}} + \int_0^T \int_\Omega y f\, dx\, dt = \\ &\quad a\big((u_0, u_1, f), (v_0, v_1)\big)\end{aligned}$$

pour tout $(u_0, u_1, f) \in V^\varepsilon \times L^2(\Omega) \times L^1(0, T, L^2(\Omega))$.

En tenant compte du problème variationnel défini par b, on a: pour tout $(u_0, u_1) \in V^\varepsilon \times L^2(\Omega)$

$$\begin{aligned}&\langle u_1, \psi_0\rangle_{L^2(\Omega), L^2(\Omega)} - \langle u_0, \psi_1\rangle_{V^\varepsilon V^{\varepsilon*}} = \\ &\quad \langle u_1, y_0\rangle_{L^2(\Omega), L^2(\Omega)} + \langle u_0, y_1\rangle_{V^\varepsilon, V^{\varepsilon*}}\end{aligned}$$

d'où $\psi_0 = y_0$ et $\psi_1 = y_1$.

Il en résulte donc que pour toute donnée $(y_0, y_1) \in L^2(\Omega) \times V^{\varepsilon^*}$, il y a contrôlabilité exacte frontière, ce qui achève la preuve du théorème. □

Remarques.

1) Lorsque Ω est un ouvert régulier il y a donc contrôlabilité exacte frontière avec un contrôle de type Dirichlet sur Σ_D et un contrôle de type Neumann sur Σ_N.

2) On peut aussi s'affranchir de la condition $\omega < 3\pi/2$ du théorème en modifiant légèrement le choix de Γ_N et de Γ_D dans le cas polygonal.

Bibliographie

[1] Bardos, Lebeau, Rauch, Contrôle et stabilisation dans les problèmes hyperboliques, in [5], 493–557.

[2] P. Grisvard, Contrôlabilité exacte de l'équation des ondes en présence de singularités, *J. Math. Pures et Appl.* **68** (1989), 215–259.

[3] P. Grisvard, *Elliptic Problems in Nonsmooth Domains*, Pitman, 1985.

[4] I. Lasiecka, J.-L. Lions, R. Triggiani, Non homogeneous boundary value problems for second order hyperbolic operators, *J. Math. Pures et Appl.* **65** (1986), 149–192.

[5] J.-L. Lions, *Contrôlabilité Exacte, Perturbations et Stabilisation des Systèmes Distribués*, tome 1, Masson Paris, 1988.

[6] M. Moussaoui, Contrôlabilité exacte des solutions de l'équation des ondes dans un domaine plan fissuré, préprint ENS Lyon, UMPA no. 98, 1993.

[7] M.T. Niane, O. Seck, Exact controllability of the wave equation in a polygonal domain with cracks, in *Control of Partial Differential Equations*, Da Prato-Tubaro, ed., Marcel Dekker, 1994.

[8] M.T. Nianne, O. Seck, Contrôlabilité exacte de l'équation des ondes avec des conditions mêlées, *C.R. Acad. Sciences Paris*, série I, t318 (1994), 945–948.

[9] S. Nicaise, Exact controllability of a pluridimensional coupled problem, *Publ. IRMA*, Lille, vol 20, IV (1990).

[10] R. Triggiani, Exact controllability for wave equation with Neumann boundary control, in *Boundary Control and Boundary Variations*, J.P. Zolézio, ed., Springer Verlag, 1988.

Université de Saint-Louis, BP 234 Saint-Louis, Sénégal

Interpolation and Extrapolation Spaces in Evolution Equations

E. Sinestrari

Summary

In the first part of the paper we show how some methods of functional analysis (i.e. semigroup theory and abstract interpolation) have been used to obtain known and new results in the field of parabolic differential equations of many different types.

In the second part we indicate a new method based on abstract extrapolation theory to study non homogeneous evolution equations and its applications to hyperbolic partial differential equations: in particular to linear (and nonlinear) Volterra integrodifferential equations. These results will be applied to a wave equation with memory effects.

0. Introduction

Thirty years ago P. Grisvard introduced the "complex method" to study the equation

$$Ax + Bx = y \tag{0.1}$$

where $A : D(A) \subset X \to X$ and $B : D(B) \subset X \to X$ are linear operators in a complex Banach space X. The solution of (0.1) was represented by a Dunford integral containing the resolvents $z \to (z - A)^{-1}$ and $z \to (z + B)^{-1}$ and its regularity was studied using the interpolation spaces between $D(A)$ (or $D(B)$) and X: the main result of this method was the so called maximal regularity i.e. when Ax and Bx belong to the same space where y is prescribed.

In a series of papers in collaboration with G. Da Prato the theory was extended and used to study the abstract parabolic equation

$$u'(t) = \Lambda u(t) + f(t), \qquad t \in [0, T] \tag{0.2}$$

where $\Lambda : D(\Lambda) \subset E \to E$ is the generator of an analytic semigroup in a Banach space E (by defining $X = L^p(]0, T[; E)$ or $C([0, T]; E)$, $(Au)(t) = \Lambda u(t)$ and $Bu = -u'$) and applied also to nonautonomous and nonlinear problems. To this purpose new classes of Banach spaces were introduced: the so-called continuous interpolation spaces and the extrapolation spaces. The importance of these papers lies in the fact that for the first time maximal regularity results were obtained in a general Banach space (and not only in particular ones as the Hilbert spaces or L^p or $C^{\alpha,\alpha/2}$ spaces) and for the complete symmetry between space and time regularity suggested by the formulation of (0.2), as a particular case of (0.1). This influenced the subsequent research even when (0.2) was studied by different methods.

In the first part of this paper we want to give a survey of later developments of the theory of the abstract parabolic evolution equations and their application to PDEs with special emphasis on the importance (and even necessity) of interpolation spaces.

In the second part we use extrapolation theory to study abstract hyperbolic equations using a method of homogenization and perturbation; then we apply the abstract results to the study of Volterra integrodifferential equations.

1. Interpolation spaces

$(E, \|\cdot\|)$ will denote a complex Banach space and $\mathcal{L}(E)$ the Banach algebra of linear bounded operators of E into itself.

We will need the usual spaces of functions $u : [a,b] \to E$ such as $C^n([a,b];E)$, $n \in \mathbb{N}$, $C^\alpha([a,b];E)$, $\alpha \in]0,1[$ or the spaces of (classes) of functions $L^p(]a,b[;E)$, $1 \le p \le \infty$ and the Sobolev spaces $W^{r,p}(]a,b[;E)$, $r \in]0,+\infty[$. In addition we will denote by $B([a,b];E)$ the space of bounded functions from $[a,b]$ to E and by $h^\alpha([a,b];E)$ the space of little-Hölder continuous functions i.e. the closure of $C^1([a,b];E)$ in $C^\alpha([a,b];E)$ (for its characterization see [34]).

In the sequel we will need two generalizations of the definition of the generator of a strongly continuous semigroup of linear bounded operators in E.

1.1 Definition. A linear operator $\Lambda : D(\Lambda) \subset E \to E$ such that there exists $\omega \in \mathbb{R}$ verifying

$$(\omega, +\infty) \subset \rho(\Lambda), \tag{1.1}$$

where $\rho(\Lambda)$ denotes the resolvent set of Λ, and

$$M = \sup\{\| (\lambda-\omega)^n(\lambda-\Lambda)^{-n} \|_{\mathcal{L}(E)}\ ,\ \lambda > \omega,\ n \in \mathbb{N}\} < \infty \tag{1.2}$$

is called a *Hille-Yosida operator*; when the condition $\overline{D(\Lambda)} = E$ also holds then Λ is the generator of a strongly continuous semigroup $e^{t\Lambda}$ (see [14]).

1.2 Definition. A linear operator $\Lambda : D(\Lambda) \subset E \to E$ such that there exists $\omega \in \mathbb{R}$ and $\theta \in]\frac{\pi}{2}, \pi[$ verifying

$$\{z \in C; |arg(z-\omega)| < \theta\} \subset \rho(\Lambda) \tag{1.3}$$

$$M = \sup\{\| (z-\omega)(z-\Lambda)^{-1} \|_{\mathcal{L}(E)} \ , \ |arg(z-\omega)| < \theta\} < \infty \tag{1.4}$$

is called a *sectorial operator*; in this case Λ is the generator of a semigroup which is analytic from $]0, +\infty[$ to $\mathcal{L}(E)$ and such that $\lim_{t\to 0} e^{t\Lambda}x = x$ if $x \in \overline{D(\Lambda)}$; hence $e^{t\Lambda}$ is strongly continuous for $t \geq 0$ if $\overline{D(\Lambda)} = E$ (see [35]). In the sequel we will assume (only for simplicity) that (1.1)-(1.2) or (1.3)-(1.4) hold for some $\omega < 0$.

Let us recall some characterizations of the *real interpolation spaces* $D_\Lambda(\theta, p)$ for $0 < \theta < 1$, $1 \leq p \leq \infty$ (these correspond to the spaces $S(p, 1-\theta, D_\Lambda \ ; \ p, -\theta, X)$ according to [25]):

(i) If $(0, +\infty) \subset \rho(\Lambda)$ and $\sup\{\| \lambda(\lambda-\Lambda)^{-1} \|_{\mathcal{L}(E)} \ , \ \lambda > 0\} < \infty$ then

$$D_\Lambda(\theta, p) = \{x \in E \ ; \ \int_0^{+\infty} \| t^\theta \Lambda(t-\Lambda)^{-1} x \|^p \frac{dt}{t} \ < \infty\}$$

(see [19]);

(ii) if Λ is the generator of a bounded and strongly continuous semigroup $e^{t\Lambda}$, then

$$D_\Lambda(\theta, p) = \{x \in E \ ; \ \int_0^{+\infty} \| \frac{e^{t\Lambda}x - x}{t^\theta} \|^p \frac{dt}{t} \ < \infty\}$$

(see [23]);

(iii) if Λ is a sectorial operator then

$$D_\Lambda(\theta, p) = \{x \in E \ ; \ \int_0^{+\infty} \| t^{1-\theta} \Lambda e^{t\Lambda} x \|^p \frac{dt}{t} \ < \infty\}$$

(see [4], [35]).

In [10] the closure $D_\Lambda(\theta)$ of $D(\Lambda)$ in $D_\Lambda(\theta, \infty)$ $(0 < \theta < 1)$ has been characterized (and called a *continuous interpolation space*): under the assumptions of (ii) or (iii) we have respectively

$$D_\Lambda(\theta) = \{x \in E \ ; \ \lim_{t\to 0} \frac{e^{t\Lambda}x - x}{t^\theta} = 0\}$$

$$D_\Lambda(\theta) = \{x \in E \ ; \ \lim_{t \to 0} t^{1-\theta} \Lambda e^{t\Lambda} x = 0\}.$$

When E is a Banach space as $L^p(\Omega)$ or $C(\overline{\Omega})$ and Λ is a differential operator (with suitable domain, taking into account the boundary conditions), the interpolation and continuous interpolation spaces can be completely described by means of Sobolev spaces of fractional order, Hölder or little Hölder spaces (see [20], [27], [2]).

2. Linear abstract parabolic equation

Let $\Lambda : D(\Lambda) \subset E \to E$ be a sectorial operator and $u_0 \in E$: we will consider in this section the abstract Cauchy problem

$$(ACP) \qquad \begin{cases} u'(t) &= \Lambda u(t) + f(t) \ , \ 0 < t \leq T \\ u(0) &= u_0 \end{cases}$$

under the assumption that $f : [0,T] \to E$ is (at least) continuous. A *classical solution* of (ACP) is a function $u \in C^0([0,T]; E) \cap C^1(]0,T]; E)$ satisfying (ACP): if $u \in C^1([0,T]; E)$ and $(ACP)_1$ holds also for $t = 0$ the solution is called *strict.*

Let us collect the main regularity results for (ACP) under additional assumptions on u_0 and f.

2.1 Theorem.
(i) *If* $u_0 \in \overline{D(\Lambda)}$ *and* $f \in C^\alpha([0,T]; X)$*; then there exists a unique classical solution* u *of* (ACP) *and* $u', \Lambda u \in C^\alpha([\epsilon,T]; X)$, $\forall \epsilon \in]0,T[$
(ii) *the solution is strict and* $u', \Lambda u \in C^\alpha([0,T]; X)$ *if and only if*

$$u_0 \in D(\Lambda) \ \ \textit{and} \ \ \Lambda u_0 + f(0) \in D_\Lambda(\alpha, \infty). \tag{2.1}$$

In this case we have also

$$u' \in B([0,T]; D_\Lambda(\alpha, \infty)). \tag{2.2}$$

(iii) *If* $u_0 \in \overline{D(\Lambda)}$ *and* $f \in C([0,T]; X) \cap B([0,T]; D_\Lambda(\alpha,\infty))$ *then there exists a unique classical solution* u *of* (ACP) *and* $u', \Lambda u \in C(]0,T]; X) \cap B(]\epsilon,T]; D_\Lambda(\alpha,\infty))$ $\forall \epsilon \in]0,T[$.
(iv) *The solution is strict and* $u', \Lambda u \in C([0,T]; X) \cap B([0,T]; D_\Lambda(\alpha,\infty))$ *if and only if*

$$u_0 \in D(\Lambda) \ \ \textit{and} \ \ \Lambda u_0 \in D_\Lambda(\alpha, \infty) \tag{2.3}$$

In this case we have also

$$\Lambda u \in C^\alpha([0,T]; X). \tag{2.4}$$

For a proof see [35], when $\overline{D(\Lambda)}$ the case (i) was proved in advance in [21] while the case (ii) with $u_0 = f(0) = 0$ is a corollary of the sum of operator's method (see [9]).

The parts (ii) and (iv) are called also maximal (or optimal) regularity results because u' and Λu belong to the same space as f; note that to this end the conditions (2.1)(or(2.3)) must be verified so that a maximal regularity result (of this type) needs the use of the spaces $D_\Lambda(\theta, \infty)$ (see in contrast e.g. the statement in [7] page 150).

One can ask whether condition $f \in C([0,T]; X)$ is sufficient (supposing for simplicity that $u_0 = 0$) to get a strict solution of (ACP); for general X the answer is negative by virtue of a result in [3] showing that if (ACP) has a strict solution for each $f \in C([0,T]; X)$ (and $\Lambda \notin \mathcal{L}(X)$) then X must contain a subspace isomorphic to c_0: in particular X must be nonreflexive. On the other hand the following theorem has been proved in [10] by means of the continuous interpolation spaces.

2.2 Theorem. *If $u_0 \in D_\Lambda$, $\Lambda u_0 \in D_\Lambda(\theta)$ and $f \in C([0,T]; D_\Lambda(\theta))$ then (ACP) has a unique solution such that $u', \Lambda u \in C([0,T]; D_\Lambda(\theta))$.*

There are many other results obtained by interpolation theory concerning the mild solution of (ACP) for which we refer to ([35]).

We did not consider problem (ACP) when f is not continuous in $[0,T]$: for the case when $f \in L^p(]0,T[; X)$ with X as a Hilbert space we refer to [15] and [24]; finally for the case when X is a general Banach space, see [9],[17] and [18].

3. Parabolic partial differential equations

The abstract regularity results of theorem 2.1 can be applied to initial-boundary value problems of parabolic type as

$$\begin{cases} u_t(t,x) = \mathcal{A}u(t,x) + f(t,x) , & (t,x) \in [0,T] \times \overline{\Omega} \\ \mathcal{B}u(t,x) = 0 , & (t,x) \in [0,T] \times \partial\Omega \\ u(0,\dot{x}) = u_0(x) , & x \in \Omega \end{cases} \tag{3.1}$$

where $\Omega \subset \mathbb{R}^n$ is bounded, $\mathcal{A}$ is an elliptic operator and $\mathcal{B}$ is an operator acting on the boundary of Ω. By choosing

$$\begin{cases} E = C(\overline{\Omega}) , & \Lambda u = \mathcal{A}u \\ D_\Lambda = \{u \in C(\overline{\Omega}) , & \mathcal{B}u = 0 \ on \ \partial\Omega \ ; \ \mathcal{B}u \in C(\overline{\Omega})\} \end{cases} \tag{3.2}$$

one can apply (under suitable regularity assumptions on the coefficients and on Ω) Theorem because Λ is a sectorial operator by virtue of

[39] and $D_\Lambda(\theta,\infty)$ can be characterized as a subset of Hölder continuous functions (see [27]); in this way it is possible to give another proof of the results obtained by potential theoretical methods (see [22]) when f is Hölder continuous in both variables (see [30]) and also to prove for the first time the existence of the strict solution when f is Hölder continuous only with respect to t (Th. 6.1 of [35]) or to x (see [38] and [28]).

The first use of the interpolation spaces in nonlinear problems seems to have been made in [37] were a semilinear problem was studied in the space $E = L^p(\Omega)$ hence (in contrast with the choice of $E = C(\overline{\Omega})$) one has to limit a priori the growth of the nonlinearities and the classical solution verifies the equation only a.e. in Ω.

On the other hand when $E = C(\overline{\Omega})$ and the boundary condition is of Dirichlet type, the domain of Λ is not dense and one is forced to use the theory of sectorial operators (unless one considers only the case $\Omega = \mathbb{R}^n$; see e.g. [8]).

The benefits of the regularity results are apparent in other types of equations. For instance the time dependent linear equation

$$u'(t) = \Lambda(t)u(t) + f(t) \ , \ \ t \in [0,T] \tag{3.3}$$

where for each t, $\Lambda(t)$ is a sectorial operator with domain independent on t can be solved by a perturbation argument thanks to the maximal regularity under the same assumptions as the classical theory of Sobolevski and Tanabe but avoiding the construction of the fundamental solution (see [13]). More recently the fundamental solution has been obtained without using the integral equation theory (as done by the classical theories) but with the aid of the time regularity and the crossed regularity (i.e.(2.2)); see [29]. For a survey on the nonautonomous equation (3.3) we refer to [1].

There are many other applications of the abstract regularity results e.g. to:

(i) fully nonlinear parabolic equations

$$u'(t) = f(t, u(t)) \ , \ \ t \in [0,T]$$

where $f_u(t,0) = \Lambda$ is a sectorial operator and $f(t,\cdot)$ is defined in $D(\Lambda)$

(ii) Volterra integrodifferential equations

$$u'(t) = f(t, u(t)) + \int_0^t g(t,s,u(s))\, ds \ , \ \ t \in [0,T]$$

(iii) Delay equations

$$u'(t) = Au(t) + Au(t-r) + \int_{-r}^0 a(\theta)A(t+\theta)u(\theta)\, d\theta + f(t) \ , \ \ t \in [0,T]$$

For details see the references quoted in [36] and [29].

4. Extrapolation spaces

In 1982 G. Da Prato and P. Grisvard introduced the extrapolation spaces in connection with the study of a nonautonomous linear Cauchy problem where the operators have a time dependent domain (see [11] and [12]). Another definition was given by R .Nagel (see [31]) and subsequently other authors used these two definitions; for references and a comparison between them we refer to [33]. We will use an adaptation of Nagel's definition to Hille-Yosida operators.

In this section we will assume that $A_0 : D(A_0) \subset X_0 \to X_0$ is the generator of a strongly continuous semigroup $T_0(t), t \geq 0$ in the Banach space $(X_0, \|\cdot\|)$ (with growth bound $\omega < 0$: see Definition 1.1).

4.1 Definition. The *extrapolation space* of X_0 (associated with the operator A_0) is the completion of $Y_0 := (X_0, \|\cdot\|_{-1})$ where $\| x \|_{-1} := \| A_0^{-1} x \|$, $x \in X_0$, denoted by X_{-1}. The *extrapolated semigroup* $T_{-1}(t)$, $t \geq 0$ in X_{-1} is the (unique) continuous extension of $T_0(t) : Y_0 \subset X_{-1} \to X_{-1}$.

The following properties can be proved (see [32]).

4.2 Proposition. *The semigroup* T_{-1} *is strongly continuous and* $\| T_{-1}(t) \|_{\mathcal{L}(X_{-1})} = \| T_0(t) \|_{\mathcal{L}(X)}$. *If* $A_{-1} : D(A_{-1}) \subset X_{-1} \to X_{-1}$ *is the generator of* T_{-1} *then*

(i) $D(A_{-1}) = X_0$

(ii) A_{-1} *is an isometry from* X_0 *to* X_{-1}

(iii) $(A_{-1})^{-1} \in \mathcal{L}(X_{-1})$

Let now $A : D(A) \subset X \to X$ *be a Hille-Yosida operator in a Banach space* $(X, \|\cdot\|)$ *with* $\omega < 0$. *(see Section 1).*

Setting

$$\begin{cases} X_0 & = (\overline{D(A)}, \|\cdot\|) \\ D(A_0) & = \{x \in D(A)\ ,\ Ax \in \overline{D(A)}\} \\ A_0 x & = Ax\ ,\ x \in D(A_0) \end{cases} \tag{4.1}$$

we deduce that A_0 is the generator of a strongly continuous semigroup T_0 on X_0. The extrapolation space of X_0 will be also called the extrapolation space of X (associated with the operator A). It turns out that X lies between X_0 and X_{-1}; more precisely X_{-1} is also the completion of $(X, \|\cdot\|_{-1})$. In addition A_{-1} is an extension of A hence

$$(A_{-1})^{-1}(X) = D(A). \tag{4.2}$$

5. Homogenization of linear equations

In this section we use a technique (which goes back to R.K. Miller) to reduce a nonhomogeneous problem in a Banach space E

$$\begin{cases} u'(t) &= \Lambda u(t) + f(t), \quad t \geq 0 \\ u(0) &= u_0 \end{cases} \tag{5.1}$$

to a homogeneous one

$$\begin{cases} U'(t) &= \mathcal{A}U(t), \quad t \geq 0 \\ U(0) &= U_0 \end{cases} \tag{5.2}$$

in the Banach space $Z = E \oplus L^1(\mathbb{R}_+; E)$ where $\mathcal{A}$ generates a semigroup in Z so that the existence of the solution of problem (5.1) is deduced from that of problem (5.2). But we can also use restriction and perturbation theorems for the generator $\mathcal{A}$ to solve (5.2) in different spaces and with different $\mathcal{A}$: this yields several types of regularity results for (5.1) and also the possibility of substituting f with suitable operators (e.g. a Volterra integrodifferential operator as we will show).

We will assume in this section that $\Lambda : D(\Lambda) \subset E \to E$ is the generator of a semigroup $e^{\Lambda t}$ in the Banach space $(E, \|\cdot\|)$.

We will consider in the space $L^1(\mathbb{R}_+, E)$ the translation semigroup:

$$(S(t)f)(s) := f(t+s) \quad , \quad s \geq 0$$

whose generator is the a.e. derivation with domain $W^{1,1}(\mathbb{R}_+, E)$. From this we deduce the following theorem

5.1 Theorem. *Setting for* $f \in L^1(\mathbb{R}, E)$ *and* $t \geq 0$

$$(e^{\Lambda} * f)(t) := \int_0^t e^{\Lambda(t-s)} f(s)\, ds,$$

let us define in the Banach space $Z = E \oplus L^1(\mathbb{R}_+, E)$ *the semigroup*

$$G(t)\begin{pmatrix} x \\ f \end{pmatrix} = \begin{pmatrix} e^{\Lambda t}x + (e^{\Lambda} * f)(t) \\ S(t)f \end{pmatrix}. \tag{5.3}$$

Then its generator $\mathcal{A}:D(\mathcal{A})\subset Z \to Z$ *is given by*

$$\left\{\begin{array}{lll} D(\mathcal{A}) & = & D(\Lambda)\oplus W^{1,1}(\mathbb{R}_+,E) \\ \mathcal{A}\begin{pmatrix} x \\ f \end{pmatrix} & = & \begin{pmatrix} \Lambda x + f(0) \\ f' \end{pmatrix} \end{array}\right. \tag{5.4}$$

Hence if $u_0 \in D(\Lambda)$ *and* $f \in W^{1,1}(\mathbb{R}_+,E)$, *problem (5.1) has a unique strict solution* u *where* $U(t) = \begin{pmatrix} u(t) \\ v(t) \end{pmatrix} = G(t)\begin{pmatrix} u_0 \\ f \end{pmatrix}$ *is the solution of (5.2).*

5.2 Theorem. *Let* F *be a Banach space such that* $F \hookrightarrow E$ *and there exists* $c > 0$ *verifying*

$$\| \frac{1}{t}\int_0^t e^{\Lambda s}x\, ds \|_{D(\Lambda)} \le C \| x \|_F \tag{5.5}$$

for each $x \in F$ *and* $t > 0$. *Given* $u_0 \in D(\Lambda)$ *and* $f \in W^{1,1}(\mathbb{R}_+;F)$ *such that*

$$\Lambda u_0 + f(0) \in D(\Lambda) \tag{5.6}$$

there exist a unique $u \in C^1(\mathbb{R}_+;D(\Lambda))$ *solution of (5.1).*

Proof. Estimate (5.5) implies the existence of $C' > 0$ such that for each $f \in L^1(\mathbb{R}_+;F)$ and $t > 0$ we have

$$\left\{\begin{array}{l} \| (e^{\Lambda t} * f)(t) \|_{D(\Lambda)} \le C' \| f \|_{L^1(]0,t[;F)} \\ \lim_{t\to 0} \| (e^{\Lambda t} * f)(t) \|_{D(\Lambda)} = 0 \end{array}\right. \tag{5.7}$$

This can be shown first for step functions with compact support then for each $f \in L^1(\mathbb{R}_+;F)$ by density.

By using (5.7) one deduces that the space

$$Z_F := D(\Lambda)\oplus L^1(\mathbb{R}_+;F)$$

is invariant for the semigroup G (defined in (5.3)) and its restriction G_F is a strongly continuous semigroup in Z_F with a generator $\mathcal{A}_F : D(\mathcal{A}_F) \subset Z_F \to Z_F$ given by

$$\left\{\begin{array}{lll} D(\mathcal{A}_F) & = & \left\{\begin{pmatrix} x \\ f \end{pmatrix} \in D(\Lambda)\oplus W^{1,1}(\mathbb{R}_+;F)\ ;\ \Lambda x + f(0) \in D(\Lambda)\right\} \\ \mathcal{A}_F\begin{pmatrix} x \\ f \end{pmatrix} & = & \begin{pmatrix} \Lambda x + f(0) \\ f' \end{pmatrix} \end{array}\right. \tag{5.8}$$

Hence if $U_0 = \begin{pmatrix} u_0 \\ f \end{pmatrix}$ satisfies (5.6), problem

$$\begin{cases} U'(t) &= \mathcal{A}_F U(t) \quad , \quad t \geq 0 \\ U(0) &= U_0 \end{cases}$$

has a unique strict solution $U(t) := \begin{pmatrix} u(t) \\ v(t) \end{pmatrix}$ and so u is a strict solution of (5.1).

Let us recall now the definition of the Favard class associated to a bounded semigroup $e^{\Lambda t}$ (see [6]).

5.3 Definition. Let T_0 be a semigroup in the Banach space $(X_0, \| \cdot \|)$ with $\omega < 0$. Its *Favard class* is the Banach space

$$Fav(T_0) := \{x \in X_0 \; ; \; \sup_{t>0} \frac{1}{t} \| T_0(t)x - x \| < \infty\}$$

with norm

$$\| x \|_F := \sup_{t>0} \frac{1}{t} \| T_0(t)x - x \|$$

Hence $Fav(T_0) \hookrightarrow X_0$.

Note that $Fav(T_0) = D_{A_0}(1, \infty)$ if A_0 is the generator of T_0. We have $D(A_0) \hookrightarrow F$ (with equality if X_0 is reflexive; see [6]). More generally we have the following result

5.4 Proposition. *Let F a Banach space such that $F \subset X_0$. Then $F \hookrightarrow Fav(T_0)$ if and only if there exists $c > 0$ verifying*

$$\| \frac{1}{t} \int_0^t T_0(s)x \, ds \|_{D(A_0)} \leq C \| x \|_F \tag{5.9}$$

for each $x \in F$ and $t > 0$.

Proof. Let $\| T_0(t) \|_{L(X_0)} \leq M_0$, $t \geq 0$; if $x \in X_0$ and $t > 0$ we have

$$\begin{aligned} \| \frac{1}{t} \int_0^t T_0(s)x \, ds \|_{D(A_0)} &= \| \frac{1}{t} \int_0^t T_0(s)x \, ds \| \\ &+ \frac{1}{t} \| T_0(t)x - x \| \leq M_0 \| x \| + \frac{1}{t} \| T_0(t)x - x \| \end{aligned}$$

and from this the conclusion follows.

5.5 Theorem. *Let $A : D(A) \subset X \to X$ be a Hille-Yosida operator in a Banach space $(X, \| \cdot \|)$ and denote by T_0 the strongly continuous* *semigroup generated by the part* A_0 *of* A *in* $X_0 = \overline{D(A)}$ *(see (4.1)). If* X_{-1} *is the extrapolation space of* X_0 *and* T_{-1} *the extrapolated semigroup of* T_0 *then*

(i) $A_{-1}(Fav(T_0)) = Fav(T_{-1})$

(ii) $\| A_{-1}x \|_{Fav(T_{-1})} = \| x \|_{Fav(T_0)}$ *for* $x \in Fav(T_0)$

(iii) $D(A) \subseteq Fav(T_0) \hookrightarrow X_0 \subseteq Fav(T_{-1})$

Proof. We know that A_{-1} is an isometry from X_0 to X_{-1} (see (ii) of Proposition): hence for $t > 0$ and $x \in X_0$ we have

$$\| T_0(t)x - x \| = \| A_{-1}(T_0(t)x - x) \|_{-1} = \| T_{-1}(t)A_{-1}x - A_{-1}x \|_{-1}$$

which implies (i) and (ii). Let us prove the first inclusion of (iii): if $x \in D(A)$ we have $\lim_{\lambda \to \infty}(\lambda - \omega)(\lambda - A_0)^{-1}x = x$; and so from (1.2) we deduce for $t > 0$ and $\lambda > \omega$

$$\begin{aligned}
&\| (T_0(t) - I)(\lambda - \omega)(\lambda - A_0)^{-1}x \| \\
&= \| A_0 \int_0^t T_0(s)(\lambda - \omega)(\lambda - A_0)^{-1}x \, ds \| \\
&= \| \int_0^t T_0(s)(\lambda - \omega)A(\lambda - A)^{-1}x \, ds \| \\
&\leq tM_0M \| Ax \|,
\end{aligned}$$

hence for $\lambda \to +\infty$

$$\| T_0(t)x - x \| \leq tM_0M \| Ax \|$$

i.e. $x \in Fav(T_0)$.

As A_{-1} is an extension of A from $D(A) \subseteq Fav(T_0)$ and (i) we deduce $X \subseteq Fav(T_{-1})$.

As an application of the preceding results let us give another proof of the following generalization of Phillips' theorem (see [14]).

5.6 Theorem. *Let* $A : D(A) \subset X \to X$ *be a Hille-Yosida operator. Given* $u_0 \in D(A)$ *and* $f \in W^{1,1}(\mathbb{R}_+; X)$ *verifying*

$$Au_0 + f(0) \in \overline{D(A)} \tag{5.10}$$

there exists a unique $u \in C^1(\mathbb{R}_+; X)$ *solution of*

$$\begin{cases} u'(t) &= Au(t) + f(t) \quad , \quad t \in [0, T] \\ u(0) &= u_0 \end{cases} \tag{5.11}$$

Proof. By virtue of Proposition and (iii) of Theorem we see that we can apply Theorem to $E = X_{-1}$, $\Lambda = A_{-1}$, $D(\Lambda) = \overline{D(A)}$ and $F = X$: hence (5.11) has a solution $u \in C^1([0,T];X)$ for each $u_0 \in \overline{D(A)}$ and $f \in W^{1,1}([0,T];X)$ such that $A_{-1}u_0 + f(0) \in \overline{D(A)}$; but $f(0) \in X$ hence $A_{-1}u_0 \in X$. From (4.2) we get $u_0 \in D(A)$.

5.7 Remark. Let $A : D(A) \subset X \to X$ be a Hille-Yosida operator and T_{-1} the extrapolated semigroup associated to A (see (4.1)).

From the proof of Theorem we deduce that in the space

$$Z_X := \overline{D(A)} \oplus L^1(\mathbb{R}_+, X) \tag{5.12}$$

the semigroup

$$G_X(t)\begin{pmatrix} x \\ f \end{pmatrix} = \begin{pmatrix} T_{-1}(t)x + (T_{-1} * f)(t) \\ S(t)f \end{pmatrix} \tag{5.13}$$

has a generator $\mathcal{A}_X : D(\mathcal{A}_X) \subset Z_X \to Z_X$ given by

$$\begin{cases} D(\mathcal{A}_X) &= \left\{\begin{pmatrix} x \\ f \end{pmatrix} \in D(A) \oplus W^{1,1}(\mathbb{R}_+;X) \;;\; Ax + f(0) \in \overline{D(A)}\right\} \\ \mathcal{A}_X &= \begin{pmatrix} Ax + f(0) \\ f' \end{pmatrix} \end{cases} \tag{5.14}$$

If in the proof of theorem we substitute X with $Fav(T_{-1})$ we deduce the following

5.8 Theorem. *Let $A : D(A) \subset X \to X$ be a Hille-Yosida operator. For each $u_0 \in Fav(T_0)$ and $f \in W^{1,1}(\mathbb{R}_+; Fav(T_{-1}))$ such that*

$$A_{-1}u_0 + f(0) \in \overline{D(A)}$$

problem

$$\begin{cases} u'(t) &= A_{-1}u(t) + f(t) \quad , \quad t \geq 0 \\ u(0) &= u_0 \end{cases}$$

has a unique solution $u \in C^1(\mathbb{R}_+;X) \cap C(\mathbb{R}_+;F)$.

6. Volterra integrodifferential equations

In this section we will use the matrix-setting of the previous section to deduce by a perturbation theorem of semigroup generators the existence and uniqueness of strict solutions of the Volterra integrodifferential equation

$$\begin{cases} u'(t) &= Au(t) + \int_0^t a(t-s)Au(s)\,ds + f(t) \;,\; t \geq 0 \\ u(0) &= u_0. \end{cases} \tag{6.1}$$

We will assume that

$$A : D(A) \subset X \to X \quad \text{is a Hille} - \text{Yosida operator} \tag{6.2}$$

$$a \in L^1(\mathbb{R}_+) \tag{6.3}$$

$$\begin{aligned} a &\in BV(\mathbb{R}_+) \; i.e. \; for\ each\ T > 0, \\ V_0^T(a) &:= \left\{ \sum_{k=1}^{n} |a(t_k) - a(t_{k-1})| \; ; \; 0 = t_0 < \ldots < t_n = T \right\} < \infty \end{aligned} \tag{6.4}$$

and

$$V_0^\infty(a) = \lim_{T \to +\infty} V_0^T(a) < \infty$$

From this we deduce (see Lemma A.1 of [5]) for $t \geq 0$:

$$\begin{cases} |a(t)| &\leq |a(0)| + V_0^\infty(a) \\ \int_0^{+\infty} |a(t+s) - a(s)|\,ds &\leq t \cdot V_0^\infty(a). \end{cases} \tag{6.5}$$

6.1 Theorem. *Let $A : D(A) \subset X \to X$ be a Hille-Yosida operator and Z_X , $\mathcal{A}_X$ and let G_X be as in Remark . Under the assumptions (6.3)-(6.4) let us define $C : D(A) \to L^1(\mathbb{R}_+; X)$ as*

$$Cx = a(\cdot)Ax \quad , \quad x \in D(A)$$

and $B : D(\mathcal{A}_X) \to Z_X$ as

$$B\begin{pmatrix} x \\ f \end{pmatrix} = \begin{pmatrix} 0 \\ Cx \end{pmatrix}$$

for $\begin{pmatrix} x \\ f \end{pmatrix} \in D(A) \oplus W^{1,1}(\mathbb{R}_+; X)$ such that $Ax + f(0) \in \overline{D(A)}$. Then B is continuous from $D(\mathcal{A}_X)$ to $Fav(G_X)$ and $\mathcal{A}_X + B : D(\mathcal{A}_X) \subset Z_X \to Z_X$ is the generator of a strongly continuous semigroup in Z_X.

Proof. Given $\begin{pmatrix} x \\ f \end{pmatrix} \in D(\mathcal{A})$ and $t > 0$ we deduce from (5.7) (with $\Lambda = A_{-1}$ and $F = X$) and (6.5)

$$\begin{aligned} \frac{1}{t} \| (G_X(t) - I)B \begin{pmatrix} x \\ f \end{pmatrix} \|_{Z_X} = & \| \frac{1}{t} \int_0^t T_{-1}(t-s)a(s)Ax \, ds \| \\ & + \frac{1}{t} \int_0^{+\infty} |a(t+s) - a(s)| \, \| Ax \| \, ds \\ & \leq C' \| Ax \| (|a(0)| + V_0^\infty(a)) + V_0^\infty(a) \| Ax \| \end{aligned}$$

As

$$\| Ax \| \leq \| Ax + f(0) \| + \| f' \|_{L^1(\mathbb{R}_+;X)} = \| \mathcal{A}_X \begin{pmatrix} x \\ f \end{pmatrix} \|_{Z_X} \tag{6.6}$$

we deduce the first part of the theorem: the second one is a consequence of example 3 of [16].

We are now in position to prove the mentioned results on problem (6.1).

6.2 Theorem. *Let (6.2)-(6.4) hold. Given* $u_0 \in D(A)$ *and* $f \in W^{1,1}(\mathbb{R}_+, X)$ *such that* $Au_0 + f(0) \in \overline{D(A)}$*, problem (6.1) has a unique solution* $u \in C^1(\mathbb{R}_+, X)$*.*

Proof. From Theorem we have that $U(t) := \begin{pmatrix} x(t) \\ f(t) \end{pmatrix} := e^{t(\mathcal{A}_X + B)} \begin{pmatrix} u_0 \\ f \end{pmatrix}$ is the unique solution of $U'(t) = (\mathcal{A}_X + B)U(t)$, $U(0) = \begin{pmatrix} u_0 \\ f \end{pmatrix}$ and $U \in C^1(\mathbb{R}_+; Z_X) \oplus C(\mathbb{R}_+; D(\mathcal{A}_X))$; hence we have $x \in C^1(\mathbb{R}_+; X) \cap C(\mathbb{R}_+; D(A))$ and

$$\begin{cases} x'(t) = Ax(t) + F(t)(0) \ , \ t \geq 0 \\ x(0) = u_0. \end{cases} \tag{6.7}$$

As

$$e^{t(\mathcal{A}_X + B)} \begin{pmatrix} u_0 \\ f \end{pmatrix} = e^{t\mathcal{A}_X} \begin{pmatrix} u_0 \\ f \end{pmatrix} + \int_0^t e^{(t-s)\mathcal{A}_X} B e^{s(\mathcal{A}_X + B)} \begin{pmatrix} u_0 \\ f \end{pmatrix} ds$$

we get

$$F(t) = S(t)f + \int_0^t S(t-s)a(\cdot)Ax(s) \, ds$$

and so

$$F(t)(0) = f(t) + \int_0^t a(t-s)Ax(s) \, ds$$

which can be inserted in (6.7) and the conclusion follows.

7. Wave equation

In this section we want to apply the preceding abstract theory to the partial integrodifferential equation

$$\begin{cases} w_{tt}(t,x) &= w_{xx}(t,x) + \int_0^t a(t-s)w_{xx}(s,x)\,ds \ + f(t,x), \\ & \quad (t,x) \in [0,T] \times [0,\ell] \\ w(0,x) &= u_0(x) \\ w_t(0,x) &= u_1(x) \\ w(t,o) &= u(t,\ell) = 0 \end{cases} \tag{7.1}$$

We seek the classical solutions of this problem (and not only in the sense of L^2), and so we must consider in one space dimension (see [26]).

7.1 Theorem. *Suppose that* $a \in L^1(\mathbb{R}_+) \cap BV(\mathbb{R}_+)$, $f \in W^{1,1}(\mathbb{R}_+; C(0,\ell))$ *and let*

$$u_0 \in C^2(0,\ell) \ , \ \ u_1 \in C'(0,\ell) \tag{7.2}$$

be such that

$$u_0(0) = u_0(\ell) = u_1(0) = u_1(\ell) = u_0''(0) + f(0,0) = u_0''(\ell) + f(0,\ell) = 0. \tag{7.3}$$

Then problem (7.1) has a unique solution $w \in C^2(\mathbb{R}_+ \times [0,\ell])$.

Proof. Setting

$$u = \frac{1}{2}(w_t + w_x) \ , \ \ v = \frac{1}{2}(w_t - w_x) \tag{7.4}$$

we reduce (7.1) to the system

$$\begin{cases} u_t(t,x) &= u_x(t,x) + \frac{1}{2}\int_0^t a(t-s)[u_x(s,x) - v_x(s,x)]\,ds + \frac{1}{2}f(t,x) \\ v_t(t,x) &= -v_x(t,x) + \frac{1}{2}\int_0^t a(t-s)[u_x(s,x) - v_x(s,x)]\,ds + \frac{1}{2}f(t,x) \\ u(0,x) &= \frac{1}{2}[u_1(x) + u_0'(x)] \\ v(0,x) &= \frac{1}{2}[u_1(x) - u_0'(x)] \\ u(t,0) + v(t,0) &= 0 \\ u(t,\ell) + v(t,\ell) &= 0 \end{cases} \tag{7.5}$$

Let us consider now in the Banach space

$$X = C(0,\ell) \oplus C(0,\ell)$$

the operator $A : D(A) \subset X \to X$ defined as

$$\begin{cases} D(A) = \left\{ \begin{pmatrix} u \\ v \end{pmatrix} \in C^1(0,\ell) \oplus C^1(0,\ell) \ ; \right. \\ \qquad\qquad\qquad\qquad \left. (u+v)(0) = (u+v)(\ell) = 0 \right\} \\ A \begin{pmatrix} u \\ v \end{pmatrix} = \begin{pmatrix} u' \\ -v' \end{pmatrix} \end{cases} \tag{7.6}$$

Given $\lambda > 0$ and $V \in X$, the resolvent equation $\lambda U - AU = V$ can be explicity solved and we find $\| U \|_X \leq \frac{1}{\lambda} \| V \|_X$: hence A is a Hille-Yosida operator.

If we define $\tilde{A} : D(A) \subset X \to X$ as

$$\tilde{A} \begin{pmatrix} u \\ v \end{pmatrix} = \begin{pmatrix} u' - v' \\ u' - v' \end{pmatrix} \tag{7.7}$$

and

$$F = \frac{1}{2} \begin{pmatrix} f \\ f \end{pmatrix} \quad , \quad U_0 = \frac{1}{2} \begin{pmatrix} u_1 + u_0' \\ u_1 - u_0' \end{pmatrix} \tag{7.8}$$

system (7.5) can be written as

$$\begin{cases} U'(t) = AU(t) + \frac{1}{2} \int_0^t a(t-s) \tilde{A} u(s) \, ds + F(t) \quad , t \geq 0 \\ U(0) = U_0 \end{cases} \tag{7.9}$$

We can proceed now as in the proof of Theorem by substituting the operators C and B by $\tilde{C} : D(A) \to L^1(\mathbb{R}_+; X)$, $\tilde{C}z := a(\cdot)\tilde{A}z$ and $\tilde{B} : D(\mathcal{A}_X) \to Z_X$, $\tilde{B} \begin{pmatrix} z \\ \varphi \end{pmatrix} = \begin{pmatrix} 0 \\ \tilde{C}z \end{pmatrix}$; as for each $\begin{pmatrix} u \\ v \end{pmatrix} \in D(A)$ we have

$$\| \tilde{A} \begin{pmatrix} u \\ v \end{pmatrix} \| \leq 2 \| A \begin{pmatrix} u \\ v \end{pmatrix} \| \tag{7.10}$$

we can conclude that $\mathcal{A}_X + \tilde{B} : D(\mathcal{A}_X) \subset Z_X \to Z_X$ is the generator of a strongly continuous semigroup. Now (7.2)-(7.3) imply

$$U_0 \in D(A) \quad , \quad AU_0 + F(0) \in \overline{D(A)} \tag{7.11}$$

hence (as in Theorem 5.1) we deduce that problem (7.9) has a unique solution $U = \begin{pmatrix} u \\ v \end{pmatrix} \in C^1(\mathbb{R}_+; X)$. From this we obtain a solution $w \in C^2(\mathbb{R}_+ \times [0,\ell])$ of (7.1) by setting

$$w(t,x) = \int_0^t [u(s,x) + v(s,x)] \, ds + u_0(x)$$

7.2 Remark. Let us observe that conditions (7.2)-(7.3) are necessary for the existence of a solution $w \in C^2([0,+\infty[\times[0,\ell])$ of problem (7.1).

References

[1] P. Acquistapace, Abstract linear non autonomous parabolic equations: a survey, *Dekker Lecture Notes in Pure and Appl. Math.* **148** (1993), 1–19.

[2] P. Acquistapace, B. Terreni, Hölder classes with boundary conditions as interpolation spaces, *Math. Zeit.* **195** (1987) 451–471.

[3] J.B. Baillon, Charactère borné de certains générateurs de semigroupes linéaires dans les espaces de Banach, *C.R. Acad. Sci. Paris* **290** (1980), 757–760.

[4] H. Berens, P.L. Butzer, Approximation theorems for semigroup operators in intermediate spaces, *Bull. Amer. Math .Soc.* **70** (1964), 689–692.

[5] H. Brezis, *Opérateurs maximaux monotones et semi-groupes de contractions dans les espaces de Hilbert,* North-Holland, 1972.

[6] P.L. Butzer, H. Berens, *Semigroups of Operators and Approximation*, Springer, 1967.

[7] P. Clement, H.J.A.M. Heijmans, S. Angenent, C.J. Van Duijn, and B. De Pagter, *One-Parameter Semigroups,* North-Holland, 1987.

[8] D. Daners, P. Koch Medina, *Abstract Evolution Equations, Periodic Problems and Applications*, Longman, 1992.

[9] G. Da Prato, P. Grisvard, Sommes d'opérateurs linéaires et équations différentielles opérationnelles, *J. Math. Pures et Appl.* **54** (1975), 305–387.

[10] G. Da Prato, P. Grisvard, Equations d'évolution abstraites non linéaires de type parabolique, *Annali Mat. Pura e Appl.* **120** (1979), 329–396.

[11] G. Da Prato, P. Grisvard, On extrapolation spaces, *Rend. Accad. Naz. Lincei* **72** (1982), 330–332.

[12] G. Da Prato, P. Grisvard, Maximal regularity for evolution equations by interpolation and extrapolation, *J. Funct. Anal.* **58** (1984), 107–124.

[13] G. Da Prato, E. Sinestrari, Hölder regularity for non autonomous abstract parabolic equations, *Israel J. Math.* **42** (1982) 1–19.

[14] G. Da Prato, E. Sinestrari, Differential operators with non dense domain, *Annali Sc. Norm. Sup. Pisa* **14** (1987), 285–344.

[15] L. De Simon, Un'applicazione della teoria degli integrali singolari allo studio delle equazioni differenziali lineari astratte del primo ordine, *Rend. Sem. Mat. Univ. Padova* **34** (1964), 205–232.

[16] W. Desch, W. Schappacher, On relatively bounded perturbations of linear c_0-semigroups, *Annali Sc. Norm. Sup. Pisa* **11**, (1984) 327–341.

[17] G. Di Blasio, Linear parabolic evolution equations in L^p-spaces. *Annali Mat. Pura Appl.* **138** (1984), 55–104.

[18] G. Dore, A. Venni, On the closedness of the sum of two closed operators, *Math. Zeit.* **196** (1987), 189–201.

[19] P. Grisvard, Commutativité de deux foncteurs d'interpolation et applications, *J. Math. Pures et Appl.* **45** (1966), 143–206.

[20] P. Grisvard, Equations différentielles abstraites, *Ann. Scient. Ec. Norm. Sup.* **2** (1969), 311–395.

[21] T. Kato, *Perturbations Theory for Linear Operators*,. Springer, 1966.

[22] O. Ladyzhenskaya,V. Solonnikov, N. Uralceva, Linear and quasilinear equations of parabolic type, *Am. Math. Soc.* **1968**.

[23] J. L. Lions, Théorèmes de trace et d'interpolation, *Ann. Sc. Norm. Sup. Pisa Z* **13** (1959), 389–403.

[24] J. L. Lions, E. Magenes, *Problèmes aux limites non homogènes et applications*, Dunod, **1968**.

[25] J. L. Lions, J. Peetre, *Sur une classe d'espaces d'interpolation*, IHES, **1964**.

[26] W. Littman, The wave operator and L_p norms, *J. Math. Mech.* **12** (1963), 55–68.

[27] A. Lunardi, Characterization of interpolation spaces between domains of elliptic operators and spaces of continuous functions, *Math. Nachr.* **121** (1985), 295–318.

[28] A. Lunardi, Maximal space regularity in non homogeneous initial boundary value parabolic problem, *Numer. Funct. Anal. Optim.* **10** (1989), 323–339.

[29] A. Lunardi, *Analytic semigroups and optimal regularity in parabolic problems,* preprint 1994.

[30] A. Lunardi, E. Sinestrari, W. Von Wahl, A semigroup approach to the time-dependent parabolic initial boundary value problem, *Diff. Int. Eq.* **5** (1992), 1275–1306.

[31] R. Nagel, Sobolev spaces and semigroups, *Semesterbericht Funktionalanalysis Tübingen*, (Sommersemester 1983) 1–19.

[32] R. Nagel, E. Sinestrari, Inhomogeneous volterra integrodifferential equations for Hille-Yosida operators, *Dekker Lecture Notes Pure Appl. Math.* no. 150 (1994), 51–70.

[33] J. Neerven, The adjoint of a semigroup of linear operators, *Lecture Notes in Mathematics* no. 1529, Springer, (1992).

[34] E. Sinestrari, Continuous interpolation spaces and spatial regularity in non linear Volterra integrodifferential equations, *J. Int. Eq.* **5** (1983), 287–308.

[35] E. Sinestrari, On the abstract Cauchy problem of parabolic type in spaces of continuous functions, *J. Math. Anal. Appl.* **107** (1985), 16–66.

[36] E. Sinestrari, *Optimal Regularity for Parabolic Equations and its Applications,* in: Semigroups, theory and applications. Pitman Research Notes in Mathematics, H. Brezis,M. G. Crandall, F. Kappel, eds: **152** (1986), 217–233.

[37] E. Sinestrari, P. Vernole, Semilinear evolution equations in interpolation spaces, *Nonlin. Anal.* **1** (1977), 249–261.

[38] E. Sinestrari, W. Von Wahl, On the solution of the first boundary value problem for linear parabolic equations, *Proc. Roy. Soc. Edinburgh* **108A** (1988), 339–355.

[39] B. Stewart, Generation of analytic semigroups by strongly elliptic operators under general boundary conditions, *Trans. Amer. Math. Soc.* **259** (1980), 299–310.

Dipartimento di Matematica
Università di Roma "La Sapienza"
P. le Aldo Moro 2
00185 Roma, Italy

Localisation des singularités sur la frontière et partitions de l'unité

Martin Zerner

Abstract

Un trait caractéristique, qui fait une partie de l'intérêt de l'œuvre scientifique de Pierre Grisvard, est sa capacité, à propos d'un problème particulier, à mettre en œuvre des mathématiques très diverses et parfois inattendues dans ce contexte. Je me suis posé le problème dont je vais parler à partir d'un passage de son livre (Grisvard 1992). On ne s'étonnera pas s'il m'a amené un peu loin de ce point de départ.

1. Le problème

Grisvard considère le problème suivant où Ω est un ouvert polygonal dont les Γ_j $(j = 1, \dots, k)$ sont les côtés, $\mathcal{D}$ et $\mathcal{N}$ forment une partition de $\{1, \dots, k\}$, ∂_ν et ∂_τ désignent les dérivées normale et tangentielle respectivement, enfin α_j et β_j sont des réels:

$$\begin{cases} \Delta u = f & \text{sur } \Omega \\ u = 0 & \text{sur } \Gamma_j \text{ si } j \in \mathcal{D} \\ \partial_\nu u = \alpha_j \partial_\tau u + \beta_j u & \text{sur } \Gamma_j \text{ si } j \in \mathcal{N}. \end{cases}$$

Les problèmes correspondants dans des secteurs angulaires ont été traités, il s'agit maintenant de passer à Ω. Pour cela, Grisvard utilise une partition de l'unité $\{\eta_i\}$ par des fonctions $\mathcal{C}^\infty$ dont le support ou bien ne coupe qu'un des Γ_j au plus, ou bien contient un sommet et ne coupe que les deux côtés adjacents. Mais il faut pour que $\eta_i u$ vérifie les conditions à la frontière que:

$$\partial_\nu \eta_i = \alpha_j \partial_\tau \eta_i \quad \text{sur } \Gamma_j \text{ si } j \in \mathcal{N}.$$

Dans cette situation particulière, il est assez facile de s'assurer de l'existence d'une partition de l'unité ayant ces propriétés. Cependant je me suis posé la question de la généralité de ce résultat d'existence, un problème qui ne semble pas avoir été abordé par les géomètres. Les résultats que je présenterai sont partiels, ils suffisent à mettre en évidence une situation très contrastée.

Nous supposerons toujours donnés un ouvert Ω de $\mathbb{R}^n$, dans cet ouvert un certain nombre d'hypersurfaces $\mathcal{C}^\infty$ notées S_i et, sur chacune de ces hypersurfaces, un champ $\mathcal{C}^\infty$ de vecteurs v_i à valeurs dans $\mathbb{R}^n$ transversal à l'hypersurface. Nous identifierons les champs de vecteurs et les opérateurs différentiels qui leur sont associés; en d'autres termes, le champ de vecteurs a ayant pour composantes a_j, nous poserons $af = \Sigma a_j \partial_j f$. Nous pouvons alors définir les conditions suivantes portant sur la fonction f :

(C$_i$) $\forall x \in S_i \quad v_i f = 0.$

Notre problème se formule ainsi:

(P) Pour tout recouvrement ouvert $\{U_\alpha\}$ de Ω existe-t-il une partition de l'unité localement finie $\{\eta_\beta\}$ subordonnée à ce recouvrement où les fonctions η_β vérifient les conditions (C$_i$)?

Signalons une fois pour toute que tout ce qui va être dit en C^∞ sera aussi vrai, et même un peu plus simple, en C^m.

2. Banalités et cas de la dimension deux

Lemme 1. *Si deux fonctions f et g vérifient les conditions* (C$_i$), *il en est de même de leur produit et de f/g là où g ne s'annule pas.*

Démonstration. Simple calcul.

Lemme 2 (de localisation). *Pour que la réponse au problème* (P) *soit positive il faut (évidemment!) et il suffit que tout point x de Ω vérifie la condition:*

(L$_x$) *Pour tout voisinage de x il existe une fonction C^∞ à support contenu dans ce voisinage, non nulle en x et vérifiant les conditions* (C$_i$) .

Démonstration. Quitte à la remplacer par son carré, on peut supposer que la fonction de la propriété (L$_x$) est positive. On procède alors exactement comme dans la démonstration habituelle d'existence d'une partition de l'unité pour construire un ensemble $\{\phi_\beta\}$ de fonctions C^∞ positives à support compact telles que:

- le support de chacune d'elles est contenu dans un ouvert du recouvrement donné;
- tout compact ne rencontre que le support d'un nombre fini d'entre elles;
- en tout point, l'une au moins d'entre elles est non nulle;
- elles vérifient les conditions (C$_i$).

En posant $\eta_\beta = \phi_\beta / \Sigma\phi_\gamma$ on obtient la partition de l'unité cherchée.

Corollaire. *Soit Ω^* l'ouvert Ω privé d'un sous-ensemble discret. Si la réponse au problème* (P) *est positive pour Ω^*, elle l'est aussi pour Ω.*

Démonstration. Soit x appartenant à Ω mais pas à Ω^* et U un voisinage ouvert de ce point. En restreignant U au besoin, on peut supposer que tous ses autres points appartiennent à Ω^*. Soient r, r' deux nombres strictement positifs tels que $r < r'$ et que la boule fermée de centre x et de rayon r' soit contenue dans U. Recouvrons $U - \{x\}$ par la boule ouverte de centre x et de rayon r' privée de x lui même et le complémentaire de la boule fermée de de rayon r. Soit $\{\eta_\beta\}$ une partition de l'unité subordonnée à ce recouvrement, où les η_β vérifient les conditions (C_i). Posons $f = 1$ sur la boule de centre x et de rayon r, et ailleurs $f = \Sigma'\eta_\beta$ où la somme est étendue aux β tels que le support de η_β rencontre la couronne $\{y; r \leq \|x - y\| \leq r'\}$. Cette fonction a les propriétés voulues pour vérifier la condition (L_x).

Lemme 3. *Tout point appartenant à une S_i au plus vérifie la condition* (L_x) .

Démonstration. Si x n'appartient à aucune S_i, c'est évident. Sinon, on peut faire un changement de coordonnées de façon que x soit l'origine, S_i l'hyperplan $x_1 = 0$ et que la condition (C_i) s'écrive:

$$\partial_1 f = \sum_{k=2}^{n} a_k \partial_k f.$$

Prenons alors une fonction indéfiniment dérivable ϕ d'une variable égale à un sur un voisinage de l'origine et à support assez petit et une fonction ψ indéfiniment différentiable de $n-1$ variables, non nulle à l'origine et elle aussi de support assez petit. La fonction:

$$f(x) = \phi(x_1)\left[\psi(x_2, \ldots, x_n) + x_1 \sum_{k=2}^{n} a_k \partial_k \psi\right]$$

aura les propriétés voulues.

Ouverts dont le cas est ainsi réglé. Ce sont les polygones curvilignes entendus comme ouverts dont la frontière consiste en un nombre fini d'arcs C^∞ d'un seul côté desquels ils se trouvent et en sommets qui sont des points isolés de la frontière. Ces sommets peuvent être des points de rebroussement, même d'ordre infini. Le cas d'une fissure est exclu.

En dimensions supérieures, on ne peut appliquer les résultats obtenus jusqu'ici qu'aux ouverts dont les singularités de la frontière sont isolées.

3. L'intersection de deux hypersurfaces et la dimension trois

Lemme 4. *Soient données quatre fonctions indéfiniment différentiables, f_1^j ($j = 0, 1$) sur l'hyperplan S_1 d'équation $x_1 = 0$ et f_2^j ($j = 0, 1$) sur l'hyperplan S_2 d'équation $x_2 = 0$. Supposons vérifiées les conditions de compatibilité:*

$$f_1^0(0, x_3, \ldots, x_n) = f_2^0(0, x_3, \ldots, x_n) \tag{1}$$

$$\partial_2 f_1^0(0, x_3, \ldots, x_n) = f_2^1(0, x_3, \ldots, x_n) \tag{2}$$

$$f_1^1(0, x_3, \ldots, x_n) = \partial_1 f_2^0(0, x_3, \ldots, x_n) \tag{3}$$

$$\partial_2 f_1^1(0, x_3, \ldots, x_n) = \partial_1 f_2^1(0, x_3, \ldots, x_n). \tag{4}$$

Il existe alors une fonction f indéfiniment dérivable sur $\mathbb{R}^n$ vérifiant:

$$f_{|S_i} = f_i^0, \quad \partial_i f_{|S_i} = f_i^1 \quad (i = 1, 2).$$

De plus, le support de f peut être pris dans un voisinage arbitraire de la réunion des supports des quatre fonctions données.

Démonstration. On construit d'abord, comme dans la démonstration du lemme 3, une fonction g vérifiant $g_{|S_2} = f_2^0$ et $\partial_2 g_{|S_2} = f_2^1$. On pose sur S_1:

$$h_0 = f_1^0 - g(0, x_2, \ldots, x_n)$$
$$h_1 = f_1^1 - \partial_1 g(0, x_2, \ldots, x_n)$$

et partout: $h = h_0 + x_1 h_1$. Un calcul ennuyeux mais facile montre que h et sa dérivée par rapport à x_2 s'annulent sur S_2. La fonction $g + h$ vérifie donc les relations annoncées. Il suffit de la multiplier par une fonction égale à un sur un voisinage de la réunion des supports pour obtenir la fonction f de l'énoncé.

Proposition 1. *Soit x appartenant à S_i, S_j $(i \neq j)$ et aucune autre des S_l. Notons S l'intersection de ces deux hypersurfaces et supposons qu'en x le sous-espace vectoriel engendré par v_i et v_j ne contient pas de vecteur (non nul) tangent à cette intersection. Supposons de plus que soit vérifiée l'une ou l'autre des deux hypothèses suivantes:*

(H1) *v_i n'est pas tangent à S_j ou v_j n'est pas tangent à S_i.*

(H2) *Sur un voisinage de x dans S, le champ de vecteurs v_i est tangent à S_j et v_j à S_i et le crochet $[v_i, v_j]$ est combinaison linéaire de v_i et v_j.*

Alors $(\mathrm{L_x})$ est vérifiée.

Commentaire. La première hypothèse concernant les champs de vecteurs est indispensable. Si elle n'était pas vérifiée, une fonction vérifiant les conditions $(\mathrm{C_i})$ et $(\mathrm{C_j})$ serait constante sur les trajectoires d'un champ de vecteurs tangent à S et ne pourrait pas s'annuler en dehors d'un voisinage arbitraire de x sans s'annuler aussi en x. Les hypothèses supplémentaires peuvent être affaiblies mais elles ne peuvent pas être supprimées. En effet, si le champ de vecteurs v_i est tangent à S_j et v_j à S_i, le crochet est déterminé par leurs restrictions respectives à S_i et S_j. De plus, d'après la première hypothèse, on a $[v_i, v_j] = \lambda v_i + \mu v_j + w$ où w est tangent à S. Si f est annulée par v_i sur S_i et v_j sur S_j, on a donc $wf = 0$ sur S.

Démonstration. On choisit un système de coordonnées où x est l'origine et S_i et S_j les hyperplans $x_1 = 0$ et $x_2 = 0$. La démonstration consiste à construire quatre fonctions vérifiant les hypothèses du lemme 4 dont nous conservons les notations. Les conditions $(\mathrm{C_i})$ et $(\mathrm{C_j})$ deviennent:

$$f_1^1 = a_{1,2}\partial_2 f_1^0 + \sum_{j=3}^{n} a_{1,j}\partial_j f_1^0, \tag{5}$$

$$f_2^1 = a_{2,1}\partial_1 f_2^0 + \sum_{j=3}^{n} a_{2,j}\partial_j f_2^0. \tag{6}$$

Nous ne disposons donc que du choix de f_1^0 et f_2^0. Pour assurer la condition (1) nous choisissons une fonction $\phi \in \mathcal{D}(S)$ de support assez petit et nous poserons $f_{1|S}^1 = f_{2|S}^0 = \phi$.

Les conditions (2) et (3) deviennent, compte tenu de (5) et (6):

$$\partial_2 f_1^0 = a_{2,1}\partial_1 f_2^0 + \sum_{j=3}^{n} a_{2,j}\partial_j f_2^0 \quad \text{sur } S,$$

$$\partial_1 f_2^0 = a_{1,2}\partial_2 f_1^0 + \sum_{j=3}^{n} a_{1,j}\partial_j f_1^0 \quad \text{sur } S.$$

d'où l'on tire (toujours sur S):

$$\partial_2 f_1^0 = \sum_{j=3}^{n} b_{1,j}\partial_j \phi, \tag{7}$$

$$\partial_1 f_2^0 = \sum_{j=3}^{n} b_{2,j}\partial_j \phi \tag{8}$$

(le déterminant $1 - a_{1,2}a_{2,1}$ ne s'annule pas, c'est la traduction analytique du fait que v_i et v_j n'engendrent pas de vecteur tangent à S).

Au point où nous en sommes, nous savons que les conditions (2) et (3) équivalent à (7) et (8), donc à prendre pour les dérivées normales sur S de f_1^0 et f_2^0 certaines valeurs déterminées par ϕ. Pour vérifier (4), on dérive (5) et (6) par rapport à x_2 et x_1 respectivement et on tient compte de (7) et (8) dans les équations obtenues. Cela donne en fin de compte une relation de la forme:

$$a_{1,2}\partial_2^2 f_1^0 - a_{2,1}\partial_1^2 f_2^0 = \sum_{j=3}^{n}\sum_{k=3}^{n} c_{j,k}\partial_j\partial_k\phi + \sum_{j=3}^{n} c'_j\phi. \tag{9}$$

Le fait que v_i soit tangent à S_j se traduit par $a_{1,2} = 0$. Sous l'hypothèse (H2) , la relation (9) se réduit après calcul à:

$$w\phi = 0,$$

où le vecteur w est celui qui a été introduit dans le commentaire et qui ici est nul par hypothèse. Il suffit alors d'étendre f_1^0 et f_2^0 à S_1 et S_2 respectivement de façon à vérifier les relations (7) et (8), ainsi que nous l'avons déjà fait deux fois (démonstrations des lemmes 3 et 4).

Sous l'hypothèse (H1), la relation (9) détermine une des dérivées normales secondes de f_1^0 et f_2^0 sur S (la situation est légèrement différente selon que l'un des coefficients $a_{1,2}$ et $a_{2,1}$ s'annule ou pas). Supposons que celle de f_1^0 soit f_1^2. On posera:

$$f_1^0(x_2,\ldots,x_n) = \psi(x_2)\Big[f_1^0(0,x_3,\ldots,x_n) + x_2 f_1^1(0,x_3,\ldots,x_n) + \frac{x_2^2}{2} f_1^2(x_3,\ldots,x_n)\Big].$$

Les ouverts auxquels s'applique cette proposition sont ceux dont la frontière ne comporte comme singularités que des arêtes sans rebroussement et des points isolés, en particulier en dimension trois tout ouvert difféomorphe à un polyèdre. Voici quelques autres exemples:

(i) la demi boule: $\{x \in \mathbb{R}^n; \Sigma x_j^2 < R^2 \text{ et } x_n > 0\}$.

(ii) On se donne un ouvert polygonal Ω_0 et une fonction ϕ indéfiniment dérivable strictement positive sur $]0,1]$ vérifiant $\lim_{t\to 0}\phi(t) = 0$ et on pose:

$$\Omega = \{x \in \mathbb{R}^3; x_3 \in]0,1[\text{ et } (x_1,x_2) \in \phi(x_3)\Omega_0\}.$$

(iii) $\Omega = \{x \in \mathbb{R}^n; x_1 \in]-1,1[,\ x_2 \in]0,1[\text{ et } 0 < x_3 < x_2^2 + x_1^2 x_2\}$ (les hypersurfaces frontières ne sont pas en position générale à l'origine, mais c'est un point isolé).

4. Les difficultés des dimensions supérieures

A ce point, la situation se complique. Nous laisserons de cêté le cas de l'intersection de trois hypersurfaces. Il pose un problème spécifique et n'aurait à mon avis d'intérêt que dans le cadre d'un résultat plus général qui m'échappe.

Considérons quatre hypersurfaces dans $\mathbb{R}^5$. Grâce à l'hypothèse de position générale, on se ramène à la situation suivante. Les hypersurfaces S_i sont les hyperplans $x_i = 0$ où i va de 1 à 4, nous noterons encore S leur intersection. Quitte à multiplier chaque champ de vecteurs par un facteur scalaire, les conditions ($\mathrm{C_i}$) prennent la forme:

($\mathrm{C_i'}$) $\partial_i f_{|S_i} = \sum_{j\neq i} a_{i,j}\partial_j f_{|S_i} \quad i = 1\ldots 4$

(la somme va de 1 à 5).

Ce système de quatre équations détermine, dans le cas général où son déterminant ne s'annule pas, les restrictions à S des dérivées premières de f par rapport aux x_i pour i au plus égal à quatre sous la forme:

$$\partial_i f_{|S} = c_i \phi' \tag{10}$$

où ϕ désigne la restriction de f à S et les c_i sont des fractions rationnelles des $a_{i,j}$.

Dérivons maintenant (C_i') par rapport à x_j, où j est compris entre un et quatre et différent de i, et restreignons à S. Nous obtenons, compte tenu de (10):

$$(\mathrm{D_{i,j}})\quad \partial_i\partial_j f_{|S} - \textstyle\sum_{k\neq i} a_{i,k}\partial_j\partial_k f_{|S} =$$

$$\left(\textstyle\sum_{k\neq i}\partial_j a_{i,k}c_k + \partial_j a_{i,5} + a_{i,5}\partial_5 c_j\right)f' + a_{i,5}c_j f''$$

où les sommes sont prises pour k allant de un à quatre.

Nous numéroterons ces douze équations par ordre lexicographique des (i,j). Elles doivent être vérifiées par les dix fonctions $\partial_i\partial_j f_{|S}$. Il s'agit d'examiner leurs conditions de compatibilité qui sont de la forme $A\phi''+B\phi'=0$ où A est un polynôme en les $a_{i,j}$ et B une forme linéaire par rapport à leurs dérivées, à coefficients polynômes.

Considérons d'abord le déterminant D_1 du système formé par les dix premières équations. Son écriture demanderait une ou plusieurs dizaines de pages. Cependant le système de calcul formel maple (dans la version dont je dispose) est capable de le calculer et aussi (ça lui est plus pénible) de vérifier qu'il est non nul (en tant que polynôme en les $a_{i,j}$). Il s'agit alors de voir si les déterminants des matrices 11×11 obtenues à partir de la précédente en ajoutant une ligne formée des coefficients de la onzième ou de la douzième équation et une colonne formée des seconds membres correspondants s'annulent. Il suffit d'ailleurs d'en examiner un: l'autre s'en déduit par une simple transposition d'indices. Par linéarité, il suffit aussi de mettre dans la dernière colonne d'abord $a_{i,5}c_j$, puis le coefficient de ϕ'. Le premier de ces déterminants est nul (toujours selon maple), j'en ignore la raison profonde. Montrons que le second D_2 ne l'est pas. Il suffit pour cela de considérer un cas particulier. Supposons donc que tous les $a_{i,j}$ sont constants sauf un des $a_{4,k}$, avec k au plus égal à trois, qui ne dépend que de x_2. On a alors simplement:

$$D_2 = \partial_2 a_{4,k} c_k D_1.$$

Il reste à s'assurer que le facteur c_k n'est pas nul. Or c'est une forme linéaire par rapport aux $a_{i,5}$ à coefficients fractions rationnelles en les $a_{i,j}$ ($j < 5$) non identiquement nulle. De la symétrie par permutation des indices au plus égaux à quatre, il résulte que si un de ces coefficients, en particulier celui de $a_{4,5}$, est nul, alors tous sont nuls. Impossible.

La conclusion de ces considérations élémentaires mais quelque peu complexes est donc que pour la compatibilité du système des ($D_{i,j}$) il faut que s'annule ou bien D_1 ou bien $B\phi'$, où B est un polynôme non nul en les $a_{i,j}$ et leurs dérivées. Mais ϕ' ne peut pas être nulle a cause des conditions de support et de non nullité à l'origine. Revenant aux coordonnées initiales, on peut résumer de la façon suivante.

Proposition 2. *Si parmi les S_i il y en a quatre qui se coupent en position générale, il faut pour que le problème* (P) *ait une réponse positive que les coefficients des champs de vecteurs et leurs dérivées annulent au moins un polynôme à coefficients C^∞.*

On voit donc que la solution du problème posé est extrêmement sensible à la dimension, par l'intermédiaire du nombre de parties régulières de la frontière qui peuvent se couper en position générale. En dimension deux, nous avons un résultat très général. En dimension trois des conditions nécessaires sur les champs de vecteurs apparaissent, mais elles n'éliminent qu'une sous-variété (dans un sens qu'il serait facile mais pas très utile de préciser) des champs de vecteurs. A partir de la dimension cinq, ce n'est au contraire que si les champs de vecteurs appartiennent à une telle sous-variété que le problème est possible.

5. Remarques diverses

Si le problème est à peu près réglé en dimension trois, il reste des cas particuliers où la technique des partitions de l'unité ne s'applique pas. Il faut alors passer par la méthode plus ardue de l'étude des données au bord inhomogènes et des théorèmes de trace. D'ailleurs il semble bien qu'il en soit ainsi pour la plupart des problèmes d'ordre supérieur à deux.

Un type de singularité à la frontière qui n'a pas été envisagé dans ce qui précède est celui où on a deux conditions aux dérivées obliques différentes de part et d'autre d'une sous-variété régulière S de la frontière Γ, elle aussi régulière. Soient donc deux champs de vecteurs v et w définis sur Γ et transversaux et x un point de S. Si nous voulons trouver une fonction f non nulle en x, de support arbitrairement petit et vérifiant $vf = 0$ d'un côté de S et $wf = 0$ de l'autre, il faut supposer que v et w n'engendrent pas de vecteur (non nul) tangent à S. C'est aussi suffisant. En effet, ramenons nous une fois de plus à la situation où l'équation de Γ est $x_1 = 0$, celles de S, $x_1 = x_2 = 0$ et écrivons les conditions sur f sous la forme $\partial_1 f = af$ pour $x_2 < 0$ et $\partial_1 f = bf$, où a et b sont des champs de vecteurs tangents à Γ; $a - b$ ne doit pas être tangent à S. On peut construire une fonction ϕ indéfiniment différentiable sur Γ, de support aussi petit qu'on veut et vérifiant $(a-b)\phi = 0$ pour $-\varepsilon < x_2 < \varepsilon$ (se donner ϕ sur S, la prolonger en la prenant constante sur les trajectoires de $a - b$, puis multiplier par une fonction de x_2 seul). Soit enfin θ une fonction C^∞ nulle lorsque la variable est plus petite que $-\varepsilon'$ et égale à un lorsqu'elle est plus grande que ε' (où $\varepsilon' < \varepsilon$). On prendra f égale à ϕ sur Γ et y vérifiant $\partial_1 f = [1-\theta)a + \theta b]f$.

Terminons par quelques considérations simples sur les ouverts "vraiment" lipschitziens, en nous limitant à un cas particulier. Soient Ω un ouvert lipschitzien du plan et u une fonction C^∞ sur $\bar{\Omega}$ dont la dérivée normale est nulle. En un point de la frontière Γ qui est limite essentielle de points de discontinuités de la dérivée normale, les deux dérivées partielles de u s'annulent. Si les discontinuités de la normale sont denses sur la frontière, u devra y être constante. Prenons en particulier la solution du problème de Neumann:

$$\begin{cases} \Delta u - f & \text{sur } \Omega \\ \partial_\nu u = 0 & \text{sur } \Gamma \\ \int u(s)ds = 0 \\ \Gamma. \end{cases}$$

Toute solution C^1 sera aussi solution du problème de Dirichlet. Ce résultat était sans doute déjà connu, mais pas de moi.

Référence

[1] Grisvard, P., *Singularities in Boundary Value Problems*, Masson, Paris et Springer-Verlag, Berlin, 1992.

Université de Nice-Sophia Antipolis
U.F.R. Sciences Exactes Mathématiques (U.A. 168 du CNRS),
Parc Valrose, B.P. 71 06108 Nice Cedex 2
tél. 92 07 62 24 p.9324
fax: 93 51 79 74
email: zerner math.unice.fr

Progress in Nonlinear Differential Equations and Their Applications

Editor

Haim Brezis
Département de Mathématiques
Université P. et M. Curie
4, Place Jussieu
75252 Paris Cedex 05
France
and
Department of Mathematics
Rutgers University
New Brunswick, NJ 08903
U.S.A.

Progress in Nonlinear Differential Equations and Their Applications is a book series that lies at the interface of pure and applied mathematics. Many differential equations are motivated by problems arising in such diversified fields as Mechanics, Physics, Differential Geometry, Engineering, Control Theory, Biology, and Economics. This series is open to both the theoretical and applied aspects, hopefully stimulating a fruitful interaction between the two sides. It will publish monographs, polished notes arising from lectures and seminars, graduate level texts, and proceedings of focused and refereed conferences.

We encourage preparation of manuscripts in some form of TeX for delivery in camera-ready copy, which leads to rapid publication, or in electronic form for interfacing with laser printers or typesetters.

Proposals should be sent directly to the editor or to: Birkhäuser Boston, 675 Massachusetts Avenue, Cambridge, MA 02139

PNLDE 5 Composite Media and Homogenization Theory
Gianni Dal Maso and Gian Fausto Dell'Antonio, editors

PNLDE 6 Infinite Dimensional Morse Theory and Multiple Solution Problems
Kung-ching Chang

PNLDE 7 Nonlinear Differential Equations and their Equilibrium States, 3
N.G. Lloyd, W.M. Ni, L.A. Peletier, J. Serrin, editors

PNLDE 8 Introduction to Γ–Convergence
Gianni Dal Maso

PNLDE 9 Differential Inclusions in Nonsmooth Mechanical Problems: Shocks and Dry Frictions
Manuel D. P. Monteiro Marques

PNLDE 10 Periodic Solutions of Singular Lagrangian Systems
Antonio Ambrosetti and Vittorio Coti Zelati

PNLDE 11 Nonlinear Waves and Weak Turbulence with Applications in Oceanography and Condensed Matter Physics
N. Fitzmaurice, D. Gurarie, F. McCaugham, and W. A. Woyczynski, editors

PNLDE 12 Semester on Dynamical Systems: Euler International Math Institute, St. Petersburg, 1991
Kuksin, Lazutkin, and Poeschel, editors

PNLDE 13 Ginzburg-Landau Vortices
F. Bethuel, H. Brezis, and F. Hélein

PNLDE 14 Variational Methods for Image Segmentation
Jean-Michel Morel and Sergio Solomini

PNLDE 15 Topological Nonlinear Analysis: Degree, Singularity, and Variations
Michele Matzeu and Alfonso Vignoli, editors

PNLDE 16 Analytic Semigroups and Optimal Regularity in Parabolic Problems
A. Lunardi

PNLDE 17 Blowup for Nonlinear Hyperbolic Equations
Serge Alinhac

PNLDE 18 The Heat Kernel Lefschetz Fixed Point Formula for the Spin-c Dirac Operator
J. J. Duistermaat

PNLDE 19 Chaos and Nonlinear Dynamical Systems
I. Hoveijn

PNLDE 20 Topics in Geometry: In Memory of Joseph D'Atri
Simon Gindikin, editor

PNLDE 21 Partial Differential Equations and Mathematical Physics: The Danish-Swedish Analysis Seminar, 1995
Lars Hörmander and Anders Melin, editors

PNLDE 22 Partial Differential Equations and Functional Analysis: In Memory of Pierre Grisvard
J. Cea, D. Chenais, G. Geymonat, and J.-L. Lions, editors

PNLDE 23 Geometry of Harmonic Maps
Yuanlong Xin